S0-DUV-714

DATA PROCESSING
IN
BIOLOGY AND GEOLOGY

THE SYSTEMATICS ASSOCIATION PUBLICATIONS

1. BIBLIOGRAPHY OF KEY WORKS FOR THE IDENTIFICATION OF THE BRITISH FAUNA AND FLORA
3rd edition (1967)
Edited by G. J. Kerrich, R. D. Meikle *and* Norman Tebble

2. THE SPECIES CONCEPT IN PALAEONTOLOGY (1965)
Edited by P. C. Sylvester-Bradley, B.Sc., F.G.S.

3. FUNCTION AND TAXONOMIC IMPORTANCE (1959)
Edited by A. J. Cain, M.A., D.Phil., F.L.S.

4. TAXONOMY AND GEOGRAPHY (1962)
Edited by David Nichols, M.A., D.Phil.

5. SPECIATION IN THE SEA (1963)
Edited by J. P. Harding *and* Norman Tebble

6. PHENETIC AND PHYLOGENETIC CLASSIFICATION (1964)
Edited by V. H. Heywood, Ph.D., D.Sc. *and* J. McNeill, B.Sc., Ph.D.

7. ASPECTS OF TETHYAN BIOGEOGRAPHY (1967)
Edited by C. G. Adams *and* D. V. Ager

8. THE SOIL ECOSYSTEM (1969)
Edited by J. G. Sheals

LONDON. Published by the Association

Price per volume: No. 1, 30s (*cloth*), *24s* (*paper*)*; Nos. 2–6, 20s; No. 7, 36s; No. 8, 50s.* (*U.S.A. No. 1, $4.25* (*cloth*), *$3.50* (*paper*)*; Nos. 2–6, $3; No. 7, $5; No. 8, $6.60. Free of charges*)

Available from E. W. Classey Ltd., 353 Hanworth Road, Hampton, Middlesex

SYSTEMATICS ASSOCIATION SPECIAL VOLUMES

1. THE NEW SYSTEMATICS (1940)
Edited by Julian Huxley
(Reprint in preparation)

2. CHEMOTAXONOMY AND SEROTAXONOMY (1968)*
Edited by J. G. Hawkes

* Published by Academic Press for the Systematics Association

THE SYSTEMATICS ASSOCIATION
SPECIAL VOLUME No. 3

DATA PROCESSING IN BIOLOGY AND GEOLOGY

Proceedings of a Symposium held at the Department of Geology, University of Cambridge 24–26 September, 1969

Edited by
J. L. CUTBILL
Department of Geology, University of Cambridge, England

1971

Published for the
SYSTEMATICS ASSOCIATION
by
ACADEMIC PRESS • LONDON • NEW YORK

ACADEMIC PRESS INC. (LONDON) LTD
Berkeley Square House
Berkeley Square
London, W1X 6BA

U.S. Edition published by
ACADEMIC PRESS INC.
111 Fifth Avenue
New York, New York 10003

Library of Congress Catalog Card Number: 70–117119

SBN: 12 199750 2

Made and printed in Great Britain by
William Clowes and Sons, Limited, London, Beccles and Colchester

Contributors and Symposium Members

Contributors to this volume who attended the Symposium are marked by an asterisk* and contributors who were unable to attend the Symposium are marked by a double asterisk**.

* P. D. ALEXANDER-MARRACK, *Scott Polar Research Institute, Cambridge, England.*

A. F. AMSDEN, *Department of Zoology, National Museum of Wales, Cardiff, Wales.*

I. ARMSTRONG, *British Petroleum Co. Ltd., London, England.*

P. B. ATTEWELL, *Engineering Geology Laboratories, University of Durham, England.*

J. W. AUCOTT, *Geochemistry Division, Institute of Geological Sciences, London, England.*

S. A. BALDWIN, *IBM United Kingdom Ltd., London, England.*

D. R. BARRACLOUGH, *Institute of Geological Sciences, Herstmonceux, Hailsham, Sussex, England.*

M. G. BASSETT, *Department of Geology, The National Museum of Wales, Cardiff, Wales.*

R. S. BAXTER, *30, Causewayside, Cambridge, England.*

** J. H. BEAMAN, *Department of Botany and Plant Pathology, Michigan State University, East Lansing, Michigan, U.S.A.*

R. E. K. BENJAMIN, *Warren Spring Laboratory, Ministry of Technology, Stevenage, Herts, England.*

J. D. BLUNDELL, *Blackburn Museum and Art Gallery, Blackburn, Lancs, England.*

* G. BONHAM-CARTER, *Department of Geology, University of Rochester, Rochester, New York, U.S.A.*

* D. A. BONYUN, *Department of Computer Sciences, Acadia University, Wolfville, Nova Scotia, Canada.*

H. A. BREMNER, *R.T.Z. Services Ltd., London, England.*

H. BRUNTON, *Department of Palaeontology, British Museum (Natural History). London, England.*

C. F. BURK, Jr., *Secretariat for Geoscience Data, Ottawa, Ontario, Canada.*

D. M. BURN, *Department of Zoology, The University, Leeds, England.*

R. CAMBON, *c/o S.M.M. Pennarroya, 93 O'Connell Street, Limerick, Ireland.*

J. F. M. CANNON, *Department of Botany, British Museum (Natural History), London, England.*

B. CAPITANT, *Centre d'informatique géologique, Ecole des Mines, Paris, France.*

V. CAZABAT, *Société Nationale des Pétroles D'Aquitaine, Sce. Mathématiques Appliquées, Pau, France.*

R. F. CHEENEY, *Grant Institute of Geology, Edinburgh, Scotland.*

K. M. CLAYTON, *School of Environmental Sciences, University of East Anglia, Norwich, England.*

H. J. COLIN, *Geological Survey of Northrhine–Westphalia, Krefeld, West Germany.*

B. W. COLLINS, *Commonwealth Geological Liaison Office, London, England.*

G. B. CORBET, *British Museum (Natural History), London, England.*

F. M. COURTNEY, *Soil Survey of England and Wales, Bristol, England.*

G. Y. CRAIG, *Department of Geology, University of Edinburgh, Edinburgh, Scotland.*

** J. J. CROCKETT, *Information Systems Division, Smithsonian Institution, Washington D.C., U.S.A.*

** R. A. CREIGHTON, *Information Systems Division, Smithsonian Institution, Washington D.C., U.S.A.*

* J. L. CUTBILL, *Department of Geology, Sedgwick Museum, University of Cambridge, England.*

* J. C. DAVIS, *Kansas Geological Survey, The University of Kansas, Lawrence, Kansas, U.S.A.*

J. DEACON, *85, Mawson Road, Cambridge, England.*

G. DICKINSON, *Department of Geography, University of Glasgow, Glasgow, Scotland.*

* C. DIXON, *Present Address: Department of Geology, Royal School of Mines, Imperial College of Science and Technology, London, England.*

J. W. C. DOPPERT, *P.T. Caltex Pacific (Indonesia) Ltd., c/o Mr. W. L. Mitcham, 29/30, Old Burlington Street, London, England.*

G. DRAPEAU, *Atlantic Oceanographic Laboratory, Bedford Institute, Dartmouth, Nova Scotia, Canada.*

P. A. ESSELAAR, *Economic Geology Research Unit, University of the Witwatersrand, Johannesburg, South Africa.*

* I. S. EVANS, *Present Address: Department of Geography, University of Durham, England.*

** C. L. Forbes, *Department of Geology, Sedgwick Museum, University of Cambridge, England.*

J. R. Flenley, *Geography Department, The University, Hull, England.*

* P. F. Friend, *Scott Polar Institute, Cambridge, England.*

I. C. J. Galbraith, *British Museum (Natural History), London, England.*

P. A. Garth, *School of Environmental Sciences, University of East Anglia, Norwich, England.*

M. J. A. Garth, *School of Environmental Sciences, University of East Anglia, Norwich, England.*

E. Gill, *Atlas Computing Laboratory, Chilton, Didcot, Berks, England.*

D. J. Gobbett, *Department of Geology, Sedgwick Museum, University of Cambridge, England.*

J. C. Gower, *Department of Statistics, Rothamsted Experimental Station, Harpenden, Herts, England.*

Grandclaude, *CRCG-CNRS Case officielle No. 1, 54 Vandœuvre, France.*

D. M. Greene, *British Antarctic Survey, Botanical Section, Research Gardens Winterbourne, University of Birmingham, England.*

S. W. Greene, *British Antarctic Survey, Botanical Section, Research Gardens Winterbourne, University of Birmingham, England.*

E. F. Greenwood, *City of Liverpool Museum, Liverpool, England.*

A. J. Grindrod, *Department of Geology, University of Newcastle, Newcastle upon Tyne, England.*

T. B. H. Hall, *British Petroleum Co. Ltd., London, England.*

* A. Hallan, *Imperial War Museum, London, England.*

** M. E. Hale, *Department of Botany, Smithsonian Institution, Washington D.C., U.S.A.*

P. J. Hammond, *British Petroleum Co. Ltd., London, England.*

** J. W. Harbaugh, *Stanford University, California, U.S.A.*

* W. B. Harland, *Department of Geology, Sedgwick Museum, University of Cambridge, England.*

P. K. Harvey, *Department of Geology, University of Nottingham, England.*

J. G. Hawkes, *Department of Botany, University of Birmingham, England.*

D. M. Hawkins, *Department of Operational Research, University of Sussex, Brighton, England.*

D. G. Hewett, *The Nature Conservancy, Furzebrook Research Station, Wareham, Dorset, England.*

C. High, *c/o Department of Geography, University of Bristol, England.*

I. L. Hodbod, *Documentation Processing Centre, Quay House, Manchester, England.*

J. G. Holland, *Department of Geology, University of Durham, England.*

A. HOLLOWELL, *Assistant Curator, Natural History, The City Museum, Bristol, England.*

R. J. HOWARTH, *Applied Geochemistry Research Group, Imperial College of Science and Technology, University of London, England.*

C. P. HUGHES, *Department of Geology, Sedgwick Museum, University of Cambridge, England.*

* N. F. HUGHES, *Department of Geology, Sedgwick Museum, University of Cambridge, England.*

D. R. HUNT, *Royal Botanic Gardens, Kew, Surrey, England.*

J. F. HUNTINGTON, *Department of Geology, Royal School of Mines, Imperial College of Science and Technology, London, England.*

H. IVIMEY-COOK, *Institute of Geological Sciences, London, England.*

J. G. T. JONES, *Hoskyns Systems Research Ltd., London, England.*

M. D. JONES, *Department of Geology, City of Leicester Museums, Leicester, England*

** K. A. JOYSEY, *University Museum of Zoology, University of Cambridge, England.*

M. E. KAUFFMAN, *Department of Geology, Franklin and Marshall College, Lancaster, Pennsylvania, U.S.A.*

P. S. KENBER, *IBM United Kingdom Ltd., Research and Education Branch, London, England.*

J. L. KNIGHT, *Commonwealth Geological Liaison Office, London, England.*

G. LEA, *Lea Associates Ltd., London, England.*

F. H. LEROY, *ELFRE, Paris, France.*

* G. LEWIS, *Sheffield City Museums, Sheffield, England.*

* T. V. LOUDON, *Institute of Geological Sciences, London, England.*

D. LOWE, *Department of Zoology, University of Leeds, England.*

M. J. MCCULLAGH, *c/o Geography Department, University of Nottingham, England.*

K. G. MCKENZIE, *British Museum (Natural History), London, England.*

J. MCNEILL, *Plant Research Institute, Canadian Dept. of Agriculture Research Branch, Central Experimental Farm, Ottawa, Ontario, Canada.*

A. E. MARSHALL, *Esso, Begles, France.*

P. M. MATHER, *c/o Geography Department, University of Nottingham, England.*

A. MEDD, *Institute of Geological Sciences, London, England.*

D. F. MERRIAM, *Kansas Geological Survey, The University of Kansas, Lawrence, Kansas, U.S.A.*

M. MITCHELL, *Institute of Geological Sciences, Leeds, England.*

H. M. MOORES, *Edinburgh Regional Computing Centre, Edinburgh, Scotland.*

** L. E. MORSE, *Department of Botany and Plant Pathology, Michigan State University, East Lansing, Michigan, U.S.A.*

M. G. Mortimer, *Department of Geology, University College, University of London, England.*

J. Nicholson, *Scott Polar Research Institute, Cambridge, England.*

B. S. Norford, *Institute of Sedimentary and Petroleum Geology, Geological Survey of Canada, Calgary, Alberta, Canada.*

* R. J. Pankhurst, *Cambridge University Computer Laboratory, Cambridge, England.*

M. E. B. Perrin, *National Fruit Trials, Ministry of Agriculture, Fisheries and Food, Brogdale, Faversham, Kent, England.*

* F. H. Perring, *Biological Records Centre, Monks Wood Experimental Station, Abbots Ripton, Hunts, England.*

* D. J. W. Piper, *Department of Geology, University of Cambridge, Sedgwick Museum, Cambridge, England.*

M. A. Prince, *British Petroleum Co. Ltd., London, England.*

J. C. Pusey, *Office for Scientific and Technical Information, Department of Education and Science, London, England.*

R. E. Randall, *Department of Geography, University of Cambridge, England.*

F. Ramsbottom, *Institute of Geological Sciences, Leeds, England.*

W. A. Read, *Institute of Geological Sciences, Edinburgh, Scotland.*

R. J. Rebbeck, *Hunting Geology and Geophysics Ltd., Boreham Wood, Herts, England.*

P. Richards, *c/o Department of Geography, University of Bristol, Bristol, England.*

R. B. Rickards, *Curator, Sedgwick Museum, Cambridge.*

S. C. Robinson, *Geological Survey of Canada, Ottawa, Ontario, Canada.*

W. D. I. Rolfe, *Hunterian Museum, University of Glasgow, Scotland.*

G. J. Ross, *Department of Statistics, Rothamsted Experimental Station, Harpenden, Herts, England.*

R. Ross, *British Museum (Natural History), London, England.*

O. Saxl, *Grant Institute of Geology, Edinburgh, Scotland.*

L. Schultz, *Biosciences Information Service, Philadelphia, Pennsylvania, U.S.A.*

B. Scott-Barrett, *International Seismological Centre, Edinburgh, Scotland.*

J. G. Sheals, *Department of Zoology, British Museum (Natural History), London, England.*

* S. G. Shetler, *Department of Botany, Smithsonian Institution, Washington, D.C., U.S.A.*

J. Smart, *Department of Zoology, University of Cambridge, England.*

P. H. A. Sneath, *MRC Microbial Systematics Unit, University of Leicester, England.*

J. H. Soper, *Chief Botanist, National Museum of Natural Sciences, National Herbarium of Canada, Ottawa, Ontario, Canada.*

C. Sourisse, *Centre de Recherches S.N.P.A., Pau, France.*

D. A. E. Spalding, *Head Curator of Natural History, Provincial Museum and Archives of Alberta, Edmonton, Alberta, Canada.*

* D. F. Squires, *Marine Sciences Research Center, State University of New York, Stony Brook, New York, U.S.A.*

* G. R. Stevens, *Department of Geology, Acadia University, Wolfville, Nova Scotia, Canada.*

E. A. Strevens, *Arthritis and Rheumatism Council, Field Unit for Epidemiological Investigation, Clinical Sciences Building, Manchester, England.*

A. Sugimura, *Geological Institute, Faculty of Sciences, University of Tokyo, Hongo, Tokyo, Japan.*

* N. Suszynski, *Smithsonian Institution, Washington, D.C., U.S.A.*

J. B. Taylor, *Royal School of Mines, Imperial College of Science and Technology, London, England.*

A. J. Wadge, *Institute of Geological Sciences, Leeds, England.*

E. R. Wallace, *Guest House Hotel, Hills Rd., Cambridge, England.*

** S. M. Walters, *Herbarium, Botany School, University of Cambridge, England.*

T. E. Walters, *British Petroleum Co. Ltd., London, England.*

M. H. Waring, *Department of Geology, University of Cambridge, Sedgwick Museum, Cambridge, England.*

H. R. Warman, *British Petroleum Co. Ltd., London, England.*

A. Warren, *Department of Geography, University College, London, England.*

R. Webster, *Soil Survey of England and Wales, Wycliffe Hall, Oxford, England.*

T. Webster, *N.I.A.B., Cambridge, England.*

* P. Wilkinson, *Department of Geology, University of Sheffield, England.*

W. R. Willcox, *Computer Trials Dept., Central Public Health Laboratory, London, England.*

* D. B. Williams, *Department of Zoology, British Museum (Natural History), London, England.*

J. van der Wolk, *Otterstraat 51, Utrecht, The Netherlands.*

* A. K. Yeats, *Scott Polar Research Institute, Cambridge, England.*

Contents

Introduction

The promise of a computerized information Utopia has been with us some years. So far the impact on geologists and biologists who use systematic data has been slight. Though much of the work on automation is being done in other disciplines, everybody should be aware of what is going on. It is even more important that they should be thinking about their own information problems and demanding solutions. For this reason Franklyn Perring, Gordon Sheals and I were very pleased to be asked by the Systematics Association to organize a symposium at which the achievements and problems of data automation could be discussed. The need for such a meeting was confirmed by the attendance of 135 people from 10 countries.

The choice of contributors reflects a personal view of the importance of particular developments in data automation. Once data have been published they have been lost. Abstract journals, current awareness services and automated libraries all set out to rediscover these lost data and thus treat the symptoms and not the disease. Publication is no longer an adequate method for the distribution of scientific data. Automatic data processing and communication is a new technology which may provide us with alternatives, but it will not solve our problems if we simply use it to speed up what we already do. New methods must come from a critical examination of how and why we collect and use data. This will take some time. Fast computers do not give instant solutions.

Before automated data archives can exist, it is necessary for scientists to use automated filing systems as a normal aid to research. For this to be practical data must be put into machine processable form easily, cheaply and if possible, at the time it is first recorded. Techniques for field use are discussed here by Peter Alexander-Marrack, David Piper and their colleagues (Chapters 1 and 2). More conventional methods for converting text are reviewed by Nicholas Suszynski (Chapter 4). Direct characterization of objects by machine will become more important and John Davis describes a use of lasers for this purpose (Chapter 5). Automation in geochemistry is covered by Peter Wilkinson (Chapter 15) and he includes a discussion of the automation of the mechanical aspects of specimen preparation, a point also taken up by Ken Joysey in his paper on the sectioning of fossils (Chapter 6). Recording of measurements and digitization of shapes are discussed by David Piper (Chapter 7) and also by Ian Evans (Chapter 3) who deal with the problems involved in processing high quality cartographic information. Colin Forbes and Brian Harland (Chapter 19) describe the benefits obtained from provision of good recording forms for routine data.

The next problem that has to be solved is how to represent all this data in the computer. Programming languages for directing the machine have received much attention but far less has been done to develop languages to be used just for saying things in machine processable form. Those now in use by geologists and biologists have been taken up as part of particular statistical or analytical techniques, of programming languages like Fortran, or else inherited from the days of punch cards and are far from suitable. Possibly some of the opposition to automation and to numerical taxonomy comes from an intuitive recognition of this difficulty. Colin Dixon (Chapter 10), Victor Loudon (Chapter 11), David Piper and Brian Harland (Chapter 2), and Andrew Hallan and Geoffrey Lewis (Chapter 17) all take up this problem which was an important theme at the symposium. Graeme Bonham-Carter and John Harbaugh's paper on stratigraphic modelling also looks at this from a rather different point of view (Chapter 12).

Methods for display of machine processable data are discussed by several contributors—Peter Alexander-Marrack, Peter Friend and Tony Yeats (Chapter 1) for graphical plots of stratigraphic sections, David Bonyun and George Stevens (Chapter 13) for stereographic data, and Richard Pankhurst and Max Walter (Chapter 14) for the production of diagnostic keys. However, the techniques of computer displays seem not to have been very fully exploited in geology and biology and much remains to be done.

John Cutbill and David Williams (Chapter 8) describe an attempt to solve some of the problems of producing computer programs to perform all these new processes. In the discussions at Cambridge several people mentioned the difficulties of programming and the incompatibilities of computer systems as serious obstacles to the wide use of automation. The problem of getting a new technique from the experimental stage into routine use is a long way from being solved.

The final theme of the symposium was the management of the vast amounts of data already in existence and the far larger amounts that

n areas which are very expensive and difficult to visit. Because
xpense, and the desire to sample widely, time in any one locality
ted. Moreover, repeated visits to one locality are not only
able because of the time needed, but may be impossible because
weather or ice conditions.
sic step in the detailed recording of sedimentary sequences
s the dividing of the sequence into units for the purposes of
tion. For our work we usually divide into "sets" (or cosets).
ee and Weir (1953) defined a set as "a group of essentially
able strata or cross-strata, separated from other sedimentary
y surfaces of erosion, non-deposition or abrupt change of
er". In actual practice the field definition of these sets is often a
arbitrary matter.
e numbers of varied features must be noted for each set, to
a reasonably complete record of it. These features range from
ize distribution within the set, to the thickness and shape of the
a whole. Our method records, as a matter of routine, thirteen
f characteristics per set, or up to 51 "bits" of information.
use of variations in weathering, mineral content and exposure,
gnition of these features can be time-consuming and requires an
nced observer. Our method does nothing to make this experience
portant.
re exposures are large, as they often are in Arctic areas, enormous
s of sediment are available for study. For instance, just con-
the vertical dimension, cliffs in East Greenland often extend
00 m, and we have found it practical to record only about 100 m
one day. So the planning of a sampling procedure, and the
of an efficient recording technique, is extremely important.

AMPLING TECHNIQUES, THE LOGGING METHOD

importance of making observations systematically, when faced
e enormous bulk of material and the wide range of observations,
e clear, at least to the members of this Association.
r work, localities are chosen because of the existence of exposures
formation being studied, and because of their accessibility. We
hope of adopting a technique aimed at random positioning of
s, but we try to achieve a scatter of localities throughout our

ach locality we sample the exposed rock strata by measuring
cording a continuous "vertical" section. This logging method,
h time-consuming, provides an amount of information which
optimum. It provides more information than the recording of

will shortly be produced. Donald Squires (Chapter 16) examines museums and points out that libraries preserve books and museums preserve objects but nobody preserves data. In this context the description by Stan Shetler and his colleagues (Chapter 18) of Flora North America is most significant. This ambitious project sets out to build a complete data system which will not produce any conventional publications such as floras. Franklyn Perring's description (Chapter 9) of the British Biological Recording Network is important for the same reason and also because it shows that it is the approach that matters rather than the complexity of equipment.

The last contribution, by Norman Hughes (Chapter 20), looks at the role of nomenclature in palaeontological data and has implications outside its immediate context. He shows that the unit of information which we have to handle is quite small—individuals or samples rather than species—and very much smaller than the systematic publications that now form the basic unit of communication. Any successful system must be able to move these micro-documents smoothly from data collection, through a research stage, to a final archive without the trauma of publication. No doubt we are a long way from this happy state but the technology is developing whether we use it or not. I hope that this volume will encourage more people to challenge their own scientific practices and to demand a more efficient service for and from systematic geologists and biologists.

Department of Geology,
University of Cambridge.
October, 1970.

John Cutbill

1. Mark Sensing for Recording
Sedimentological I

P. D. ALEXANDER-MARRACK, P. F. FRI

Scott Polar Research Institute and Departme
Cambridge, Englan

ABSTRACT

The systematic collection of the very lar
required in the study of an ancient sedimenta
tackled by using a logging method, and often, in
been used. We have used a printed mark-sense

The design of the mark-sense form is descri
our experience of the lengths of time, and cost
mark-sense process. The processing costs are v
costs of carrying out the field work, in our case

Our method provides data rapidly availa
analysis of facies profiles, grain-size variation, se
palaeocurrents, and the relations of these featu

COLLECTING SEDIMENTOL

The sediment which has filled a sedim
the investigation of the history of that bas
sediment indicate the former environments
way in which the shape of the basin has ev
an attempt to overcome some of the difficu
tion of sedimentary data.

P. F. Friend has, for some years, been
Sandstone (Devonian) of Spitsbergen. In 1
East Greenland, where we began a program
similar, Old Red Sandstone, type. All along
collecting of sedimentary data from a ra

Systematics Association Special Volume No. 3.
Geology, edited by J. L. Cutbill, 1970, pp. 1–16.

types
of the
is lim
undesi
of bad

A b
involv
descri

McK
confor
units
charac
rather

Larg
provid
grain-s
set as
types

Beca
the rec
experie
less im

Whe
amoun
siderin
over 1
in any
design

S

The
with t
must b

In o
of the
see no
localit
areas.

At e
and re
althou
we fin

observations at isolated points regularly arranged in or through the succession of strata ("point-counting"). "Point-counting" tends to ignore thin sets, and the nature of the transitions from one set to another.

In practice, the logging section is not vertical, or perpendicular to the stratification. Because of the detailed nature of the observations, the observer must stand within 1 m or so of the rocks being recorded. Sets are recorded one by one up a route whose precise direction is determined by the accessibility of parts of the rock face to the observer. We feel that this route is not determined by the primary sedimentary features which we are observing, and should not therefore bias our sampling.

PRINTED FORMS

We have been surprised by the way in which a printed form helps the observer to persist in making routine observations for longer periods of time. Not only does the presence of printed headings relieve the memory, but there is a satisfaction about completing sheet after sheet of standard record form, which does not come from entering records continuously in a notebook.

Before we developed the form which is the main subject of this paper, a printed form was used on which much of the information was plotted graphically in the field. Although this could not be used for machine reading purposes, it has the advantage, not shared by the mark-sense form, of being immediately readable by the observer in the field. This may have introduced slight bias on the part of the observer but particular strata could be relocated, and a general idea of the nature of the stratal record was immediately available.

GENERAL DESIGN AND READING OF THE MARK-SENSE FORM

The idea of developing a mark-sense form to allow machine-reading of sedimentary field records was given to us originally by Dr J. L. Cutbill, Department of Geology, University of Cambridge. We would like to thank him for this initiative, and for continual and very time-consuming help at all stages in this project.

After some consideration, it was decided to use a format designed for the IBM 1232 optical mark page reader. The dimensions of the sheet, as read by the machine, are 11 in (27·9 cm) by $8\frac{1}{2}$ in (21·5 cm), but the sheets were produced with a tear-off "stub" down one of the long sides, and this gave total dimensions of 11 in (27·9 cm) by $10\frac{1}{2}$ in

(26·6 cm) (Fig. 1). This stub gave us the flexibility of a field notebook, in that extra observations and comments could be entered on the stub, if they were not covered by the characteristics printed on the form.

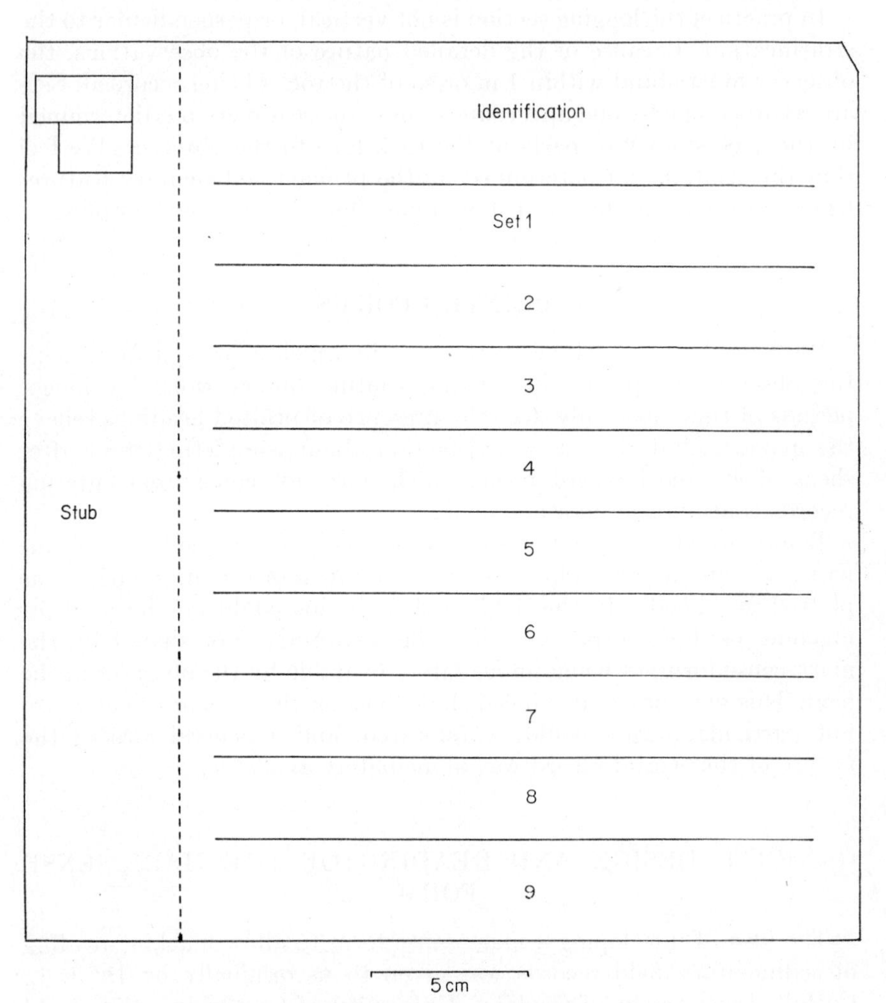

Fig. 1. General layout of mark-sense form.

The reading of the mark-sense form is achieved by the machine when it scans the 1000 critical positions (50 rows of 20 positions) of the form, and records whether each one has a black pencil mark in it or not. The marks must be made by dark pencil (2B is advised), and are thus distinguished from colour printing on the form which shows the positions, and labels them. A column of black bars, printed on the right-hand edge of the form, locates the rows for the machine scanner.

A number of torn, incorrectly or poorly marked and soiled sheets were run, to test the discernment of the reader. The reader accepted most of the damaged forms and was rarely misled by patches of mud or coffee on the sheets. Pencil marks had to be very faint or strongly oblique to the mark space for the reader to ignore them. Accuracy of reading normal marks was very high. However good marks made with a ball-point pen are often not read, and the error rate is higher for HB than for 2B pencils. Best results were obtained if the 1232 is set to accept borderline marks ("multiple response").

One machine design condition is that it cannot be asked to read more than three marks in each group of five positions in any row (though this restriction is removed if the reader is attached directly to a computer).

In general, we designed the use of the positions to achieve simplicity of layout rather than to minimize the number of positions. The forms could therefore be completed by a relatively inexperienced helper listening to the statements of the geologist who was making the observations.

We were very conscious of the need to avoid loss of flexibility when designing the printed form. This was successfully achieved by three means:

(1) providing positions for the recording of a wide range of characteristics;
(2) providing the stub, on which additional notes could be made;
(3) providing the restart position (see below) so that information on one set could be compounded.

We feel that we have not lost flexibility in our recording. However we would point out that our form was designed for a particular type of sedimentary association, dominantly clastic, and mainly of sand and silt grade.

DESIGN OF THE MARK-SENSE FORM, DETAILS OF LAYOUT

Identification box: five rows of twenty positions, only four rows used.

Station number: a system of unique letters (one for each observer) and numbers, for cataloging localities.

Sheet Number: identifies the particular mark-sense sheets used at that station (locality).

Set box: nine per sheet, each box consists of five rows of twenty positions.

STATION
SHEET

IBM 866-0/S 41

STATION NUMBER — SHEET NUMBER

A B C D E F G H J K | 00 10 20 30 40 50 60 70 80 90
000 100 200 300 400 500 600 700 800 900 | 0 1 2 3 4 5 6 7 8 9
00 10 20 30 40 50 60 70 80 90 | WRITE STATION AND SHEET NUMBERS ON STUB
0 1 2 3 4 5 6 7 8 9

0 1 2 3 4 THICK 5 6 7 8 9 | NO. CEM $CaCO_3$ ROCK DOL | FLAT SY.P AS.R PL.X TR.X
·0 ·1 ·2 ·3 ·4 (M) ·5 ·6 ·7 ·8 ·9 | LIN. DEF. CONCR.- CO_3 OR STUB | FOSSILS NO. VERT OTHR TR.
·00 ·05 EXPOSED COVERED RED P.RD GN GY OTHER | NO DIR 2. DIR 1. DIR 000 100 200 300
F.SL M.SL C.SL VF.SS F.SS M.SS C.SS VC.SS CON PEBBLY | 00 10 20 30 40 50 60 70 80 90
GRAD SHP SMTH SCR. TOOL MKS. TR.FOS. MD.CR RIPPLES | STUB ← SPEC RESTART | 1 2 SPARE 4 CANC'L

0 1 2 3 4 THICK 5 6 7 8 9 | NO. CEM $CaCO_3$ ROCK DOL | FLAT SY.P AS.R PL.X TR.X
·0 ·1 ·2 ·3 ·4 (M) ·5 ·6 ·7 ·8 ·9 | LIN. DEF. CONCR.- CO_3 OR STUB | FOSSILS NO. VERT OTHR TR.
·00 ·05 EXPOSED COVERED RED P.RD GN GY OTHER | NO DIR 2. DIR 1. DIR 000 100 200 300
F.SL M.SL C.SL VF.SS F.SS M.SS C.SS VC.SS CON PEBBLY | 00 10 20 30 40 50 60 70 80 90
GRAD SHP SMTH SCR. TOOL MKS. TR.FOS. MD.CR RIPPLES | STUB ← SPEC RESTART | 1 2 SPARE 4 CANC'L

0 1 2 3 4 THICK 5 6 7 8 9 | NO. CEM $CaCO_3$ ROCK DOL | FLAT SY.P AS.R PL.X TR.X
·0 ·1 ·2 ·3 ·4 (M) ·5 ·6 ·7 ·8 ·9 | LIN. DEF. CONCR.- CO_3 OR STUB | FOSSILS NO. VERT OTHR TR.
·00 ·05 EXPOSED COVERED RED P.RD GN GY OTHER | NO DIR 2. DIR 1. DIR 000 100 200 300
F.SL M.SL C.SL VF.SS F.SS M.SS C.SS VC.SS CON PEBBLY | 00 10 20 30 40 50 60 70 80 90
GRAD SHP SMTH SCR. TOOL MKS. TR.FOS. MD.CR RIPPLES | STUB ← SPEC RESTART | 1 2 SPARE 4 CANC'L

0 1 2 3 4 THICK 5 6 7 8 9 | NO. CEM $CaCO_3$ ROCK DOL | FLAT SY.P AS.R PL.X TR.X
·0 ·1 ·2 ·3 ·4 (M) ·5 ·6 ·7 ·8 ·9 | LIN. DEF. CONCR.- CO_3 OR STUB | FOSSILS NO. VERT OTHR TR.
·00 ·05 EXPOSED COVERED RED P.RD GN GY OTHER | NO DIR 2. DIR 1. DIR 000 100 200 300
F.SL M.SL C.SL VF.SS F.SS M.SS C.SS VC.SS CON PEBBLY | 00 10 20 30 40 50 60 70 80 90
GRAD SHP SMTH SCR. TOOL MKS. TR.FOS. MD.CR RIPPLES | STUB ← SPEC RESTART | 1 2 SPARE 4 CANC'L

0 1 2 3 4 THICK 5 6 7 8 9 | NO. CEM $CaCO_3$ ROCK DOL | FLAT SY.P AS.R PL.X TR.X
·0 ·1 ·2 ·3 ·4 (M) ·5 ·6 ·7 ·8 ·9 | LIN. DEF. CONCR.- CO_3 OR STUB | FOSSILS NO. VERT OTHR TR.
·00 ·05 EXPOSED COVERED RED P.RD GN GY OTHER | NO DIR 2. DIR 1. DIR 000 100 200 300
F.SL M.SL C.SL VF.SS F.SS M.SS C.SS VC.SS CON PEBBLY | 00 10 20 30 40 50 60 70 80 90
GRAD SHP SMTH SCR. TOOL MKS. TR.FOS. MD.CR RIPPLES | STUB ← SPEC RESTART | 1 2 SPARE 4 CANC'L

0 1 2 3 4 THICK 5 6 7 8 9 | NO. CEM $CaCO_3$ ROCK DOL | FLAT SY.P AS.R PL.X TR.X
·0 ·1 ·2 ·3 ·4 (M) ·5 ·6 ·7 ·8 ·9 | LIN. DEF. CONCR.- CO_3 OR STUB | FOSSILS NO. VERT OTHR TR.
·00 ·05 EXPOSED COVERED RED P.RD GN GY OTHER | NO DIR 2. DIR 1. DIR 000 100 200 300
F.SL M.SL C.SL VF.SS F.SS M.SS C.SS VC.SS CON PEBBLY | 00 10 20 30 40 50 60 70 80 90
GRAD SHP SMTH SCR. TOOL MKS. TR.FOS. MD.CR RIPPLES | STUB ← SPEC RESTART | 1 2 SPARE 4 CANC'L

0 1 2 3 4 THICK 5 6 7 8 9 | NO. CEM $CaCO_3$ ROCK DOL | FLAT SY.P AS.R PL.X TR.X
·0 ·1 ·2 ·3 ·4 (M) ·5 ·6 ·7 ·8 ·9 | LIN. DEF. CONCR.- CO_3 OR STUB | FOSSILS NO. VERT OTHR TR.
·00 ·05 EXPOSED COVERED RED P.RD GN GY OTHER | NO DIR 2. DIR 1. DIR 000 100 200 300
F.SL M.SL C.SL VF.SS F.SS M.SS C.SS VC.SS CON PEBBLY | 00 10 20 30 40 50 60 70 80 90
GRAD SHP SMTH SCR. TOOL MKS. TR.FOS. MD.CR RIPPLES | STUB ← SPEC RESTART | 1 2 SPARE 4 CANC'L

0 1 2 3 4 THICK 5 6 7 8 9 | NO. CEM $CaCO_3$ ROCK DOL | FLAT SY.P AS.R PL.X TR.X
·0 ·1 ·2 ·3 ·4 (M) ·5 ·6 ·7 ·8 ·9 | LIN. DEF. CONCR.- CO_3 OR STUB | FOSSILS NO. VERT OTHR TR.
·00 ·05 EXPOSED COVERED RED P.RD GN GY OTHER | NO DIR 2. DIR 1. DIR 000 100 200 300
F.SL M.SL C.SL VF.SS F.SS M.SS C.SS VC.SS CON PEBBLY | 00 10 20 30 40 50 60 70 80 90
GRAD SHP SMTH SCR. TOOL MKS. TR.FOS. MD.CR RIPPLES | STUB ← SPEC RESTART | 1 2 SPARE 4 CANC'L

0 1 2 3 4 THICK 5 6 7 8 9 | NO. CEM $CaCO_3$ ROCK DOL | FLAT SY.P AS.R PL.X TR.X
·0 ·1 ·2 ·3 ·4 (M) ·5 ·6 ·7 ·8 ·9 | LIN. DEF. CONCR.- CO_3 OR STUB | FOSSILS NO. VERT OTHR TR.
·00 ·05 EXPOSED COVERED RED P.RD GN GY OTHER | NO DIR 2. DIR 1. DIR 000 100 200 300
F.SL M.SL C.SL VF.SS F.SS M.SS C.SS VC.SS CON PEBBLY | 00 10 20 30 40 50 60 70 80 90
GRAD SHP SMTH SCR. TOOL MKS. TR.FOS. MD.CR RIPPLES | STUB ← SPEC RESTART | 1 2 SPARE 4 CANC'L

Fig. 2. Mark-sense form.

Thickness: 0·05–9·95 m, recordable in 5-cm increments. Thickness greater than 9·95 m noted by using two, or more, set boxes.

Exposed or covered: if part of the section was covered by scree or vegetation, its thickness was noted, and the covered mark made. If a set had a covered base or top, this was shown by completing the entries for the set as usual, but making the covered mark. The difference between a covered base and top was shown by the entry or otherwise of marks indicating basal surface features.

Colour: red, pale red, green, grey, other colour (noted on stub).

Grain-size: fine–coarse siltstone, fine–very coarse sandstone, conglomerate, pebbles (in conjunction with one of preceding).

Basal Surface: gradational, sharp.

Basal Surface: smooth, scour marks, tool marks, trace fossils, mud cracks, ripples.

Carbonate in rock: none, as cement, whole rock (limestone), dolomitic.

Internal structures; flat stratification, symmetrical ripples, asymmetrical ripples, planar cross-stratification, trough cross-stratification.

Lineation: parting (primary current).

Deformation: soft sediment folding.

Concretions: if not carbonate, entered on stub.

Fossils: none, vertebrate, other, trace fossils.

Directions of primary structures: none, ambiguous two ended directions, unidirectional indicator, azimuth of direction.

Stub: marked where entry made in stub.

Specimen: marked where specimen collected, and number of specimen entered on stub.

Restart: marked where next set box was required to contain further information about the same set. If a set contained gradational features (e.g. in grain size), or a number of different features (e.g. in internal sedimentary structures), a restart in the next and subsequent boxes, and the entry of the various features, in stratigraphical order in the first, second, etc. boxes would record this situation.

Spare: 1, 2, 3, 4, used by local agreement to indicate features (e.g. the presence of a concordant igneous body, the presence of a particularly characteristic brown pigmentation).

Cancelled: marked when a mistake could not be easily rectified by erasing marks in the field.

SUMMARY OF EXPERIENCE OF THE MARK-SENSE PROCESS

Below we summarize our experience of the various stages of the mark-sense process with times and costs. We do not include in this summary the considerable time spent in designing and "debugging" the various stages.

(a) Printing forms

The cost for 1000 sheets was £24 (the minimum order being 5000 sheets). Costs are much less with orders for tens of thousands of sheets, reducing to around 70 shillings per thousand.

(b) Field use

The cost of field work, in which 1000 sheets were completed, was £2500. The costs of field work in Greenland are, of course, much higher than in more accessible areas. In our first season we completed 1000 sheets, recording 4000 m of section (a rate of 100 m/6 h mark-sense day per party), 240 party hours of recording. The cost is estimated assuming half of the expedition's work was mark-sensing.

(c) Reading forms

One thousand sheets took 4 h to run. There was no charge in our case. These were read by IBM 1232 optical mark page reader. Data was punched onto cards at a rate of 5 sheets/min. There were three cards per sheet.

(d) Checking

The cost was estimated at £40 for 30 min of computer time. The data (3000 punch cards) is copied onto magnetic tape. From this it is checked and converted into text form, which is stored on disc where it is available for further processing. A print of one page of this document is shown in Fig. 3. During conversion a number of checks are made to ensure that:

(1) the cards are in order;
(2) no five position fields on the sheet have more than three positions marked;
(3) a station and sheet number are entered on the sheet;
(4) either exposed or covered has been marked for each set;
(5) directions of 370° or greater have not been recorded.

Because of the arbitrary nature of the binary pattern produced on the magnetic tape it was necessary to write this checking and conversion program in assembly language, and this was done for us by J. L. Cutbill. Apart from this all other stages were done using pre-existing programs from the Cambridge Geological Data System package written by J. L. Cutbill and D. B. Williams, and the job was run on the ICL 4–50 computer of the University of Wales at Cardiff.

(e) Transfer of data to Cambridge University Titan Computer

Rather than go through the difficulties of translation from $\frac{1}{2}$ in 9-track to 1 in magnetic tapes via 7-track tapes, the documents were punched onto 7-track paper tape at Cardiff, and re-input to Titan and stored on magnetic tape

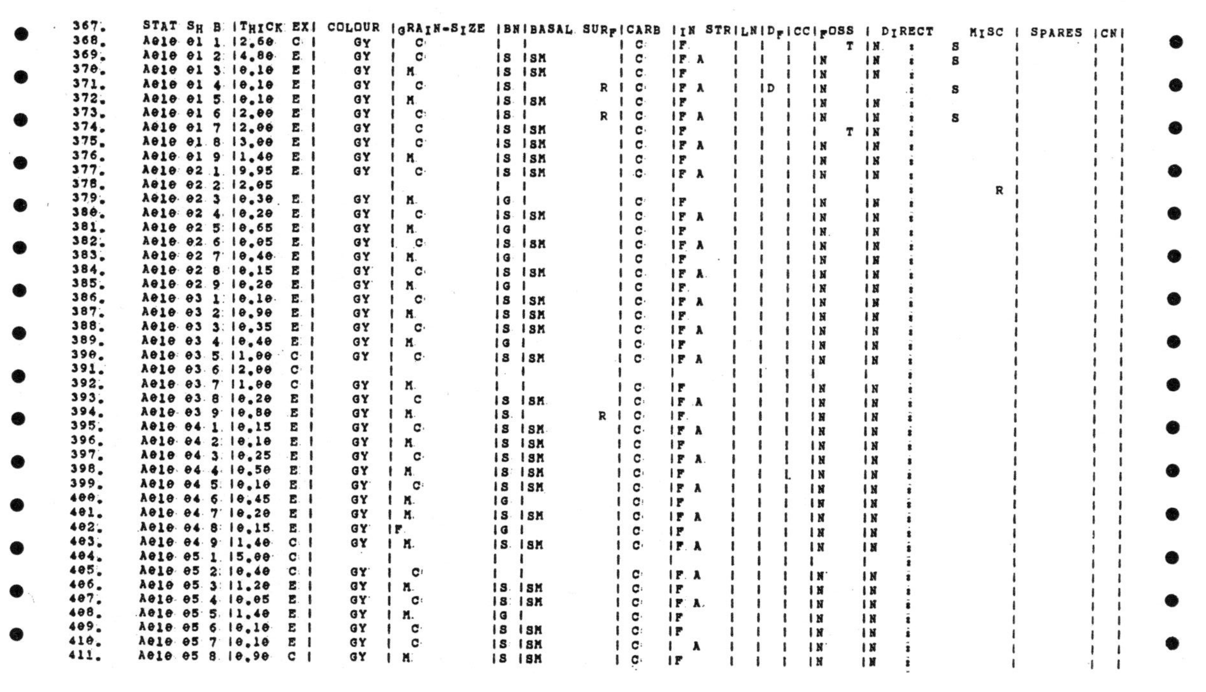

367.	STAT	SH	B	THICK	EX	COLOUR	GRAIN-SIZE	BN	BASAL	SURF	CARB	IN STR	LN	DF	CC	FOSS	DIRECT	MISC	SPARES	CN
368.	A010	01	1	2.60	C	GY	C				C	F				T	N :	S		
369.	A010	01	2	4.80	E	GY	C	S	SM		C	F A				N	N :	S		
370.	A010	01	3	0.10	E	GY	M	S	SM		C	F				N	N :			
371.	A010	01	4	0.10	E	GY	C	S		R	C	F A		D		N	:	S		
372.	A010	01	5	0.10	E	GY	M	S	SM		C	F				N	N :			
373.	A010	01	6	2.00	E	GY	C	S		R	C	F A				N	N :	S		
374.	A010	01	7	2.00	E	GY	C	S	SM		C	F				T	N :			
375.	A010	01	8	3.00	E	GY	C	S	SM		C	F A				N	N :			
376.	A010	01	9	1.40	E	GY	M	S	SM		C	F				N	N :			
377.	A010	02	1	9.95	E	GY	C	S	SM		C	F A				N	N :			
378.	A010	02	2	2.05													:	R		
379.	A010	02	3	0.30	E	GY	M	G			C	F				N	N :			
380.	A010	02	4	0.20	E	GY	C	S	SM		C	F A				N	N :			
381.	A010	02	5	0.65	E	GY	M	G			C	F				N	N :			
382.	A010	02	6	0.05	E	GY	C	S	SM		C	F A				N	N :			
383.	A010	02	7	0.40	E	GY	M	G			C	F				N	N :			
384.	A010	02	8	0.15	E	GY	C	S	SM		C	F A				N	N :			
385.	A010	02	9	0.20	E	GY	M	G			C	F				N	N :			
386.	A010	03	1	0.10	E	GY	C	S	SM		C	F A				N	N :			
387.	A010	03	2	0.90	E	GY	M	S	SM		C	F				N	N :			
388.	A010	03	3	0.35	E	GY	C	S	SM		C	F A				N	N :			
389.	A010	03	4	0.40	E	GY	M	G			C	F				N	N :			
390.	A010	03	5	1.00	C	GY	C	S	SM		C	F A				N	N :			
391.	A010	03	6	2.00	C												:			
392.	A010	03	7	1.00	C	GY	M				C	F				N	N :			
393.	A010	03	8	0.20	E	GY	C	S	SM		C	F A				N	N :			
394.	A010	03	9	0.80	E	GY	M	S		R	C	F				N	N :			
395.	A010	04	1	0.15	E	GY	C	S	SM		C	F A				N	N :			
396.	A010	04	2	0.10	E	GY	M	S	SM		C	F				N	N :			
397.	A010	04	3	0.25	E	GY	C	S	SM		C	F A				N	N :			
398.	A010	04	4	0.50	E	GY	M	S	SM		C	F			L	N	N :			
399.	A010	04	5	0.10	E	GY	C	S	SM		C	F A				N	N :			
400.	A010	04	6	0.45	E	GY	M	G			C	F				N	N :			
401.	A010	04	7	0.20	E	GY	M	S	SM		C	F A				N	N :			
402.	A010	04	8	0.15	E	GY	F	G			C	F				N	N :			
403.	A010	04	9	1.40	C	GY	M	S	SM		C	F A				N	N :			
404.	A010	05	1	5.00	C												:			
405.	A010	05	2	0.40	C	GY	C				C	F A				N	N :			
406.	A010	05	3	1.20	E	GY	M	S	SM		C	F				N	N :			
407.	A010	05	4	0.05	E	GY	C	S	SM		C	F A				N	N :			
408.	A010	05	5	1.40	E	GY	M	G			C	F				N	N :			
409.	A010	05	6	0.10	E	GY	C	S	SM		C	F				N	N :			
410.	A010	05	7	0.10	E	GY	C	S	SM		C	A				N	N :			
411.	A010	05	8	0.90	C	GY	M	S	SM		C	F				N	N :			

Fig. 3. Print-out of data table.

(f) Formation of basic data files

To reduce the cost of repeated processing of the data during our subsequent research, the text has been reconverted to binary form (four computer words per set) and filed on magnetic tape, indexed by station. Fortran and assembly language subroutines were written to provide random access to the file by station and to deliver data on successive sets expanded to a 51 word record suitable for processing using Fortran. This filing package was partly written for us by J. L. Cutbill.

It might appear that the steps outlined above are unnecessarily complex. Certainly some of them are due to the mixture and location of hardware we are using. But the provision of the intermediate text form (Fig. 3), gives us an immediately usable listing of our data (within 48 h of the expedition's return), and in this form text editors are available to correct mistakes, insert omissions and generally tidy up the data base, before it is finally filed.

ANALYSIS OF MARK-SENSE DATA

(a) Use of computer graphics

An important step in the analysis of sedimentological data is the preparation of a visual display, conveying at a glance the variation of parameters such as grain size, colour, bottom structures, internal structures, occurrence of fossils, distribution of palaeocurrent observations, etc., up a vertical section. The sedimentary characteristics of a section are not obvious before logging and the mass of observations within a section can leave one confused afterwards. A visual display allows one to observe the section as a whole and to distinguish critical phenomena for further investigation and analysis.

We have adopted certain conventions in display format. For example, grain size is represented on a horizontal axis, a standard logarithmic scale being used (the Wentworth Grade Scale). Each bed becomes a rectangular unit, the dimensions of which are defined by grain size and thickness. The vertical succession of these units represents the grain size and bed thickness variation up the section. A line or column on the left-hand side of this framework represents colour (red or non-red); bottom structures can be symbolically shown at the base of each unit and, similarly, internal structures can be represented; other data such as fossil occurrence can be added to the right of the section. Gradations of grain size within or between beds are represented by sloping right edges to any unit (see Fig. 4).

We have found that it takes us an hour to draw up, by hand, a visual display of 30–50 m of one of our sedimentological sections. In a single field season of eight weeks we have logged 4000 m of section. The process of drawing-up would therefore take a tedious 80 h or more. In this manual drawing process, it is the retrieval of grain size and thickness data, either from the mark-sense forms or from a computer print-out, and the drawing of the basic framework of the diagram that

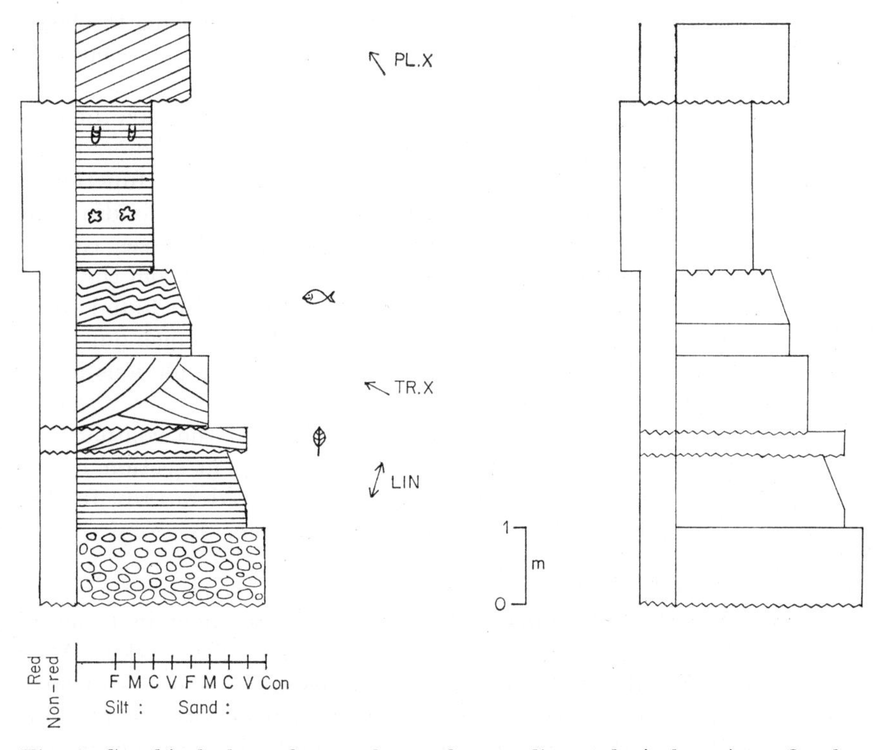

Fig. 4. Graphical plots of part of one of our sedimentological sections. On the left, full plot; on the right, as drawn by computer plotter.

takes the most time and effort. We can obviate a large part of these difficulties, however, by arranging for an on-line digital incremental plotter to draw out these sections straight from the data on file. Ideally we would like all the data, including internal structures, drawn mechanically in a simplified format but, apart from the considerable length of plotter time this would involve, a large amount of detail is noted on the stub of the mark-sense form and is not machine read. We have, therefore, written a program to draw only the framework of the diagram, all the rest being easily added by hand, as required (Fig. 4).

(b) Facies profiles

One method of reducing the data is to analyse them in terms of an independent parameter. We have based our analysis on the "facies profile" idea of Selley (1968). Our method involves the selection of thickness intervals of section, of an arbitrary extent, (e.g. 10 m), and the addition of the thicknesses (within that interval) in which certain features are found. The results are expressed as proportions of the different features, and characterize the interval. We are using the facies profile method to study the variation in proportions of internal sedimentary structures, colour, and grain size.

A facies profile for the entire section is a useful summary of that section, but details of facies change are averaged out and ignored, and in a sequence of great variation the profile could become a function of the length of section. This criticism was overcome by dividing the logged section into arbitrary thicknesses; these thicknesses can be varied to emphasize the scale of event to be considered. From the work attempted so far, intervals of 10 m and 60 m have been found to be informative. Such "reduced facies profiles" enable us to pick out facies changes where variation is too complex or subtle to be obvious in the field.

From the mark-sense data we have programmed the computer to calculate these "reduced facies profiles" in a 1000 m section in approximately a minute. This process can be repeated on any further data with only minimal further expenditure of time.

(c) Sequence of grain size variation

A second method of reducing data is to analyse them in terms of one of their own parameters (dependent variables). The work of Allen (1965) and Friend (1965) in similar, fresh-water, clastic sequences suggests that the grain-size properties are of great importance in interpretation. Both of these workers give great emphasis to the occurrence of fining-up cycles; that is, relatively short (2–25 m) sequences in which, in general, the grain size progressively decreases from bottom to top.

In this present discussion we do not intend to suggest a mode of formation for these cycles, and we do not intend to show that their occurrence is non-random. At present we wish to go no further than to suggest their presence in our data and to suggest a computer retrieval method.

Our first problem was to define cyclicity in such a way that a computer could recognize it. When considered in a rigorous way it becomes apparent that there are coarsening-up cycles as well as fining-up cycles. To avoid confusion, we have termed these "fining-up semi-cycles" and "coarsening-up semi-cycles", the two of which in combination make a cycle. A fining-up semi-cycle is a sequence of beds in which the grain size does not increase upwards; the converse is a coarsening-up semi-cycle. Thus we define a cycle as beginning at the initiation of an upwards coarsening trend and ending at the next initiation of a coarsening trend (Fig. 5). A program has been written to pick out these semi-cycles from the mark-sense data, and to compute a weighted mean

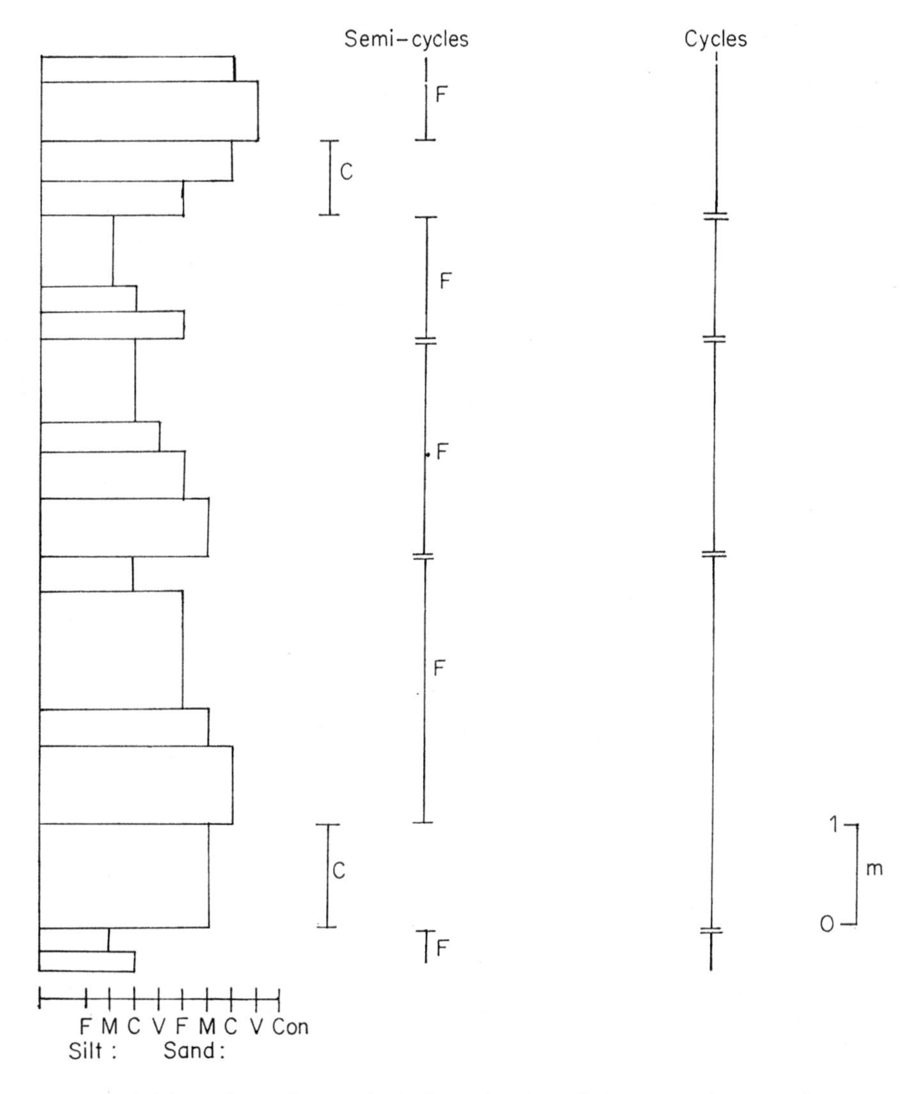

Fig. 5. Division of a sedimentological section into fining-up and coarsening-up semi-cycles (F and C), and cycles.

grain size for each semi-cycle. With this second type of reduction of data complete, the interesting questions that can be asked are endless.

Visual examination of a graph of weighted mean grain size against cumulative thickness of a section suggests stages of systematically increasing and decreasing grain size with peaks of maximum mean grain size every 70 to 100 m. We can define these as higher order cycles; however, much more work needs to be done to test the significance of these trends and to show that a similar cyclicity is visible in another parameter. It has been noticed that a similar wavelength cyclicity is seen in a group of several "reduced facies profiles" for a very long section.

(d) Sequence in internal structures

With this approach we are able to describe relations and infer the nature of processes, the results of which are preserved in the sediment, and compare them with those investigated in recent empirical work. The sequence of bed form roughness features occurring at the sediment water interface is considered to be as important as grain-size variation in the interpretation of fluviatile sequences (Allen, 1963).

Our approach to this problem is to express the probability of a particular sedimentary structure being followed by another in a transition probability matrix (Krumbein, 1968), for fining-up semi-cycles and coarsening-up semi-cycles. The matrix is found by keeping a tally of the structure sequence; if B follows A, it will be recorded in the AB position of a square matrix.

Some authors (e.g. Krumbein, 1965) have calculated transition probability matrices based on point-data up a sequence, and considered the transition from that point to the next. In our particular problem, however, we are concerned only with the transition of one structure to another and not with the relative abundance (in terms of thickness) of these structures; we therefore follow the method outlined by Carr *et al.* (1966). This could easily lead on to a consideration of stochastic processes in the fluviatile environment.

(e) Palaeocurrent analysis

One of the most important tools in the elucidation of the sedimentary history of a non-marine basin is the analysis of palaeocurrent patterns in space and time; the changing dispositions of erosional source areas and depositional basins can be deduced, and these suggest the mountain-building background to the sedimentation.

The recording of palaeocurrent data is dependent to a large extent on the type of rock-exposure in a section. Thus the amount of data collected in one field season by our three workers was 1000 readings, but could easily have been 10 000. Computation of vector means can be very laborious for this amount of information, so a program is being devised to retrieve palaeocurrent data from each mark-sense section, correct them for tectonic dip, and compute vector means and confidence limits (Curray, 1956) for indicators in that section. The results can then be plotted by hand on a section or on a map. The degree of variance for different types of current indicator can be examined and vector trend surfaces of current can be fitted for the basin at different time horizons.

(f) Correlation between various parameters

We are also investigating, in some detail, the relationships between internal structure, grain size and colour, on the lines of Friend (1965). The mark-sense data are particularly suited to this sort of study.

(g) Conclusions

Fluviatile sequences are monotonous and, to the casual observer, featureless; however, examination of the preliminary results especially the reduced facies profile, show considerable variation. With the methods outlined above, we are in a position to distinguish subtle variation in rock associations and to analyse the changing distribution of rocks in space and time.

The advantages and disadvantages of the mark-sense method are those inherent in all computer programming. We are able to list the information, collected in the field, very quickly, but the programs are all fairly complex because of the flexibility of the logging method, and therefore they take a long time to "debug". However, once the program is complete, it is possible to analyse the data rapidly and to use these programs on all information that may be collected in future field seasons. This means that progress can be made with one's research, because each season's observations can rapidly be processed by one's latest analytical techniques, leaving time for further thought and development before the next season.

REFERENCES

Allen, J. R. L. (1963). Henry Clifton Sorby and the sedimentary structures of sands and sandstones in relation to flow conditions. *Geol. en Mijnbouw.* **42**, 223–228.

Allen, J. R. L. (1965). Fining-upwards cycles in alluvial successions. *Liverpl. Manchester. Geol. J.* **4**, 229–246.

Carr, D. D., Horowitz, A., Hrabar, S. V., Ridge, K. F., Rooney, R., Straw, W. T., Webb, W., and Potter, P. E. (1966). Stratigraphic sections, bedding sequences, and random processes. *Science, N.Y.* **154**, 1162–1164.

Curray, J. R. (1956). Analysis of Two-Dimensional Orientation Data. *J. Geol.* **64**, 117–131.

Friend, P. F. (1965). Fluviatile sedimentary structures in the Wood Bay Series (Devonian) of Spitsbergen. *Sedimentology* **5**, 39–68.

Krumbein, W. C. (1965). FORTRAN IV computer programs for Markov chain experiments in geology. *State Geol. Surv., Kansas, Computer Contributions* **13**, 1–38.

Krumbein, W. C. (1968). Statistical models in sedimentology. *Sedimentology* **10**, 7–23.

McKee, E. D., and Weir, G. W. (1953). Terminology for stratification and cross-stratification in sedimentary rocks. *Bull. geol. Soc. Amer.* **64**, 381–390.

Selley, R. C. (1968). Facies profile and other new methods of graphic data presentation: Application in a quantitative study of Libyan Tertiary shoreline deposits. *J. Sed. Pet.* **38**, 363–372.

2. Recording of Geological Data in the Field Using Forms for Input to the IBM Handwriting Reader

DAVID J. W. PIPER, W. BRIAN HARLAND AND J. L. CUTBILL
Department of Geology, Sedgwick Museum, University of Cambridge, England

ABSTRACT

Data from measured stratigraphic sections has been successfully recorded in the field on forms readable by the IBM 1287 opitical reader. This reader will accept handwritten digits and the letters C, S, T, X, and Z. A tree structure was successfully used for the basic data structure, and alpha-numeric codes established for a very wide range of geological information. The coding system was relatively easy to learn and use, and significantly improved the quality and quantity of stratigraphic data recorded without reducing the speed of traversing.

INTRODUCTION

During July to September 1969 we, and our colleagues on the Norsk Cambridge Svalbard Expedition, made general stratigraphic studies in Spitsbergen. We aimed to collect as many data as possible in a form suitable for machine processing. Many data on stratigraphic sections were recorded on handwritten forms which can be read directly by the IBM 1287 handwriting reader. We believe this use in the field of hand-written forms readable by machine, and the basic structuring of our data, to be new. In addition we used systematic procedures for defining, numbering and recording data on localities, specimens and photographs. These are based on methods developed by one of us (W.B.H.) over many years for use by Cambridge expeditions to Spitsbergen (see Forbes *et al.*, this volume, p. 311) and have recently been revised for general use in the Geology Department at Cambridge. All the files are thus compatible with eventual machine processing.

We describe here only our machine readable recording techniques.

Systematics Association Special Volume No 3. "Data Processing in Biology and Geology", edited by J. L. Cutbill, 1970, pp. 17–38.

These have been thoroughly tested in the field during the summer, but it is still too soon for a full report on the subsequent machine processing of the data.

DESIGN OF THE DATA STRUCTURE FOR STRATIGRAPHIC SECTIONS

In recent years several workers have recommended the use of standard forms for logging stratigraphic data in the field (e.g. Bouma, 1962; Selley, 1968). Marrack *et al.* (this volume p. 1) have developed a mark sense sheet from earlier hand manipulated forms. All of these systems use the bed as the fundamental unit of description, and provide "boxes" in which specific items of data are recorded.

One advantage of such recording systems is that the data are collected in a standard format, suitable for later automatic processing. The form also acts as a reminder to look for and record certain types of data—this advantage has been stressed by many users. There is a converse danger that anything that cannot be recorded on the form tends to get overlooked. Some systems allow a visual representation of the measured section to be built up as the recording is carried out.

While bed by bed analysis is suited to sedimentological investigations for which most existing standard recording systems have been designed, it does not allow a sufficient degree of generalization for much stratigraphic work, where such detail is not always necessary. The standard recording systems mentioned above have been designed to record a restricted range of information from a single facies (turbidites by Bouma; littoral deposits by Selley; and fluviatile sequences by Marrack *et al.*) Our aim was to produce a data structure flexible enough for recording general stratigraphic sections, allowing varying degrees of detail in measurement and recording. The system had to be suitable for use in detailed sedimentological and palaeontological studies.

Restrictions in existing recording systems arise because they use an array as their basic data structure and this allows only one value to be entered for each kind of data. Most geologists recording in field notebooks use a tree structure which allows great flexibility in recording. This point is discussed elsewhere in this volume (Loudon, p. 135; Cutbill and Williams, p. 105; Cutbill, Hallan and Lewis, p. 255) and one extreme example will suffice here. In organizing the data from a stratigraphic section containing conglomerates, boulders are described as components of a bed, and the boulders themselves are then described. But since a boulder may well contain a stratigraphic section, to describe a component of a bed in a rock unit from a stratigraphic section may require one to restart the operation of describing a stratigraphic section.

Boulders containing boulders are not unknown. This kind of description is extremely difficult using arrays but presents no difficulties if one is using a tree structure which allows recursion (or if one is using a field notebook). While it is seldom necessary to provide for such extreme cases the organizing of stratigraphical data is easier if one uses a structure which provides for these extremes.

The basic structure

Our data structure is intended for sequences of sedimentary rocks, rather than isolated observations or the study of tectonic structure, or igneous and metamorphic rocks. The sequence is traversed and divided into a number of *rock units* whose thickness is measured either directly or by fixing the boundaries of the unit by survey and calculating the thickness allowing for dip. The unit is chosen by the geologist to suit his particular needs and can be anything from a thin bed, a centimetre or so thick, up to units of tens or hundreds of metres thick. The description of one rock unit is the basic record in the resulting files and a description of a section consists of a number of such records arranged in order.

Rock units are subdivided into bed types each having distinctive lithology, bedding structure or other character. Thus a rock unit might consist of alternating flaggy micaceous sandstones, cross bedded sandstones and shales. A rock unit may comprise only a single bed type. The position and thickness of individual beds of any bed type are not measured but the average thickness of beds and the proportion of the rock unit made up of a particular bed type are recorded.

The distribution of bed types within a rock unit is described as the *bed type arrangement*. We have recognized six basic *relationships* between bed types:

single bed type;
random sequence;
simple sequence (without repetition);
repeating sequence;
gradational sequence (without repetition);
repeating gradational sequence.

Any rock unit may comprise a nested hierarchy of bed types, with individual bed types having a certain relationship within a *bed type group*, and bed type groups having relationships within larger groups, until the whole rock unit has been described. A single bed type group contains only one bed type.

The structure used to accommodate this information in a record can be seen in the following diagram.

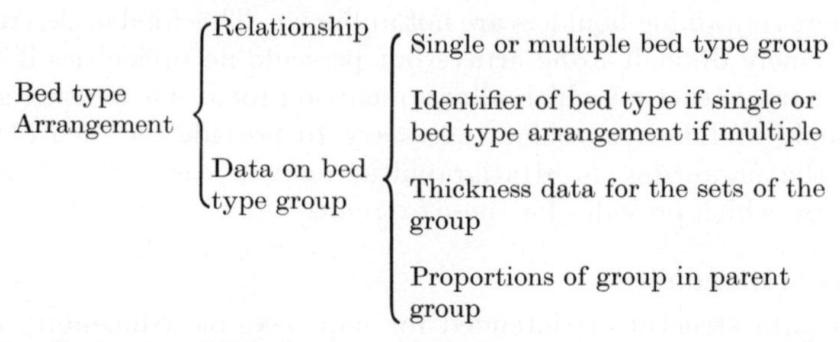

Thin distinctive beds in an otherwise monotonous sequence may be recorded as *point features* within a rock unit and described in terms of their position, thickness and bed type. Lists of specimens or photographs relating to a specific position within a rock unit can be recorded in an analogous manner.

The approach to describing a bed type parallels that for a whole rock unit. There are properties of the whole bed type, and the bed type may be subdivided into a number of lithological *components*. The properties of the whole bed type that can be recorded are:

nature of boundaries	lithification
colour	porosity/permeability
texture	grain size
stratification	sorting

The properties of a component that can be recorded are:

composition	colour
form	abundance
size	

When a component (such as a pebble) cannot be simply described in terms of composition because it is itself an identifiable rock type, then it can be described as a distinct rock unit. This recursive escape is not needed often, but makes possible the satisfactory description of extremely complex bed types.

In addition, any special features of the bed type may be given. The range of these is almost unlimited, but the most common are details of fossils and sedimentary structures.

Summary of record structure

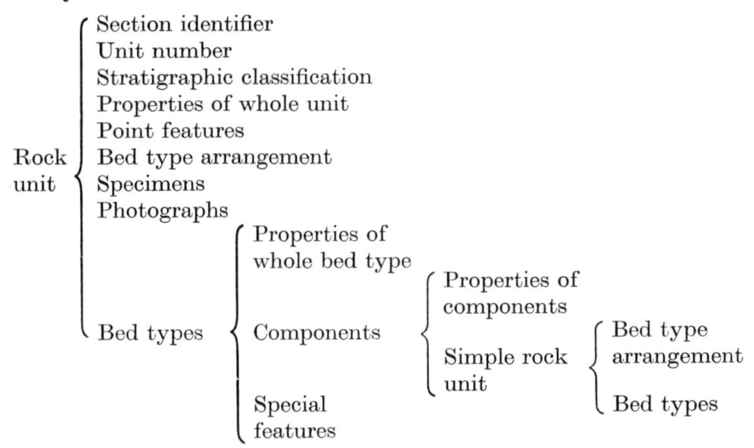

DESIGN OF RECORDING FORMS AND CODING SYSTEM

An advantage of using a machine readable document for field recording is the elimination of repunching of data before they can be machine processed. This reduces both costs and processing time. Present machine reading techniques are limited and impose considerable restrictions. Complex data are difficult to code using mark sensing (see Marrack *et al.*, this volume, p. 1). The IBM 1287 Optical reader can read handwritten digits and five capital letters, C, S, T, X, and Z, which is enough for more flexible recording. The reader can take documents ranging in size from 2·55 × 3·00 in to 5·91 × 9·00 in. The data must be written in fields of fixed length and in fixed positions on the document. Apart from X, letters may only appear in the last character position in a field. A field need not be filled, and the reader ignores blanks, right justifying the data. Unreadable characters are replaced by a special fault character.

Because the reading is performed under computer control, one can use complex reading sequences which depend on the actual data recorded. Also handwriting, sense marks, and printed characters may appear on one form. For our purpose a simpler approach was adequate. We used only handwriting and the first output consists of a magnetic tape on which each sheet has been translated into a fixed length record. This can be done using an existing IBM package and avoids the cost of special programming.

We do not have full costs available yet but the following figures are a rough guide:

(1) Form printing. £30 per 1000 sheets for the minimum order of 5000, to £3 per 1000 sheets for an order of 50 000.
(2) Registration of form—£40.
(3) Reading rate (depending on number of fields and average number of characters per form) about 120 sheets per minute. Hire of computer and reader about £80 per hour.
(4) Subsequent processing costs are small and it usually takes longer to load the magnetic tape than to run the program.

The limitations of the recording forms necessitate an alphanumeric code for entering the data. Its design is complicated as it must express a tree structure in a fixed length record and also allow the survey data, which will not form part of the final file to be recorded along with the stratigraphical description. As the first field use of the sheets was on upper Palaeozoic and Mesozoic sequences in Svalbard, we attempted to find the most satisfactory design for recording these particular data. With the experience we have gained, we hope to modify the system for more general use. We believe that very little modification of the details of the data structure and coding will be necessary. Codes for such rock types as evaporite, which we did not encounter in 1969, will have to be expanded. The specialist code section of the form is intended for codes designed for particular projects. Our codes for rock units, palaeontology and sedimentology were specially designed for use in Svalbard and we do not expect them to be suitable for other detailed investigations.

Forms and codes of this type are intended only as one method of adding data to a file of stratigraphical sections. One file could contain data originally recorded on our data sheets or on a tape typewriter or punch cards using a tagging system (see Cutbill and Williams, this volume, p. 105).

In constructing our code we have not attempted to make a rock classification and then provide codes (or names) for the classes. Rather we have provided codes for the language used to describe rocks and it has been necessary to restrict the language. This inevitably tends to force a classification on the rocks being described. We have yet to evaluate such effects.

We chose A5 (21 × 15 cm) size forms (Fig. 1) as these are convenient for handling in the field. They were punched on the left-hand side to fit a ring binder. A 3 cm wide margin on this left side of the page was used for recording non-codable information in longhand. A brief "crib" to the commonly used codes was printed on the back of each sheet (Fig. 2). In the ring binder the crib from the previous sheet was thus visible when filling in the current sheet.

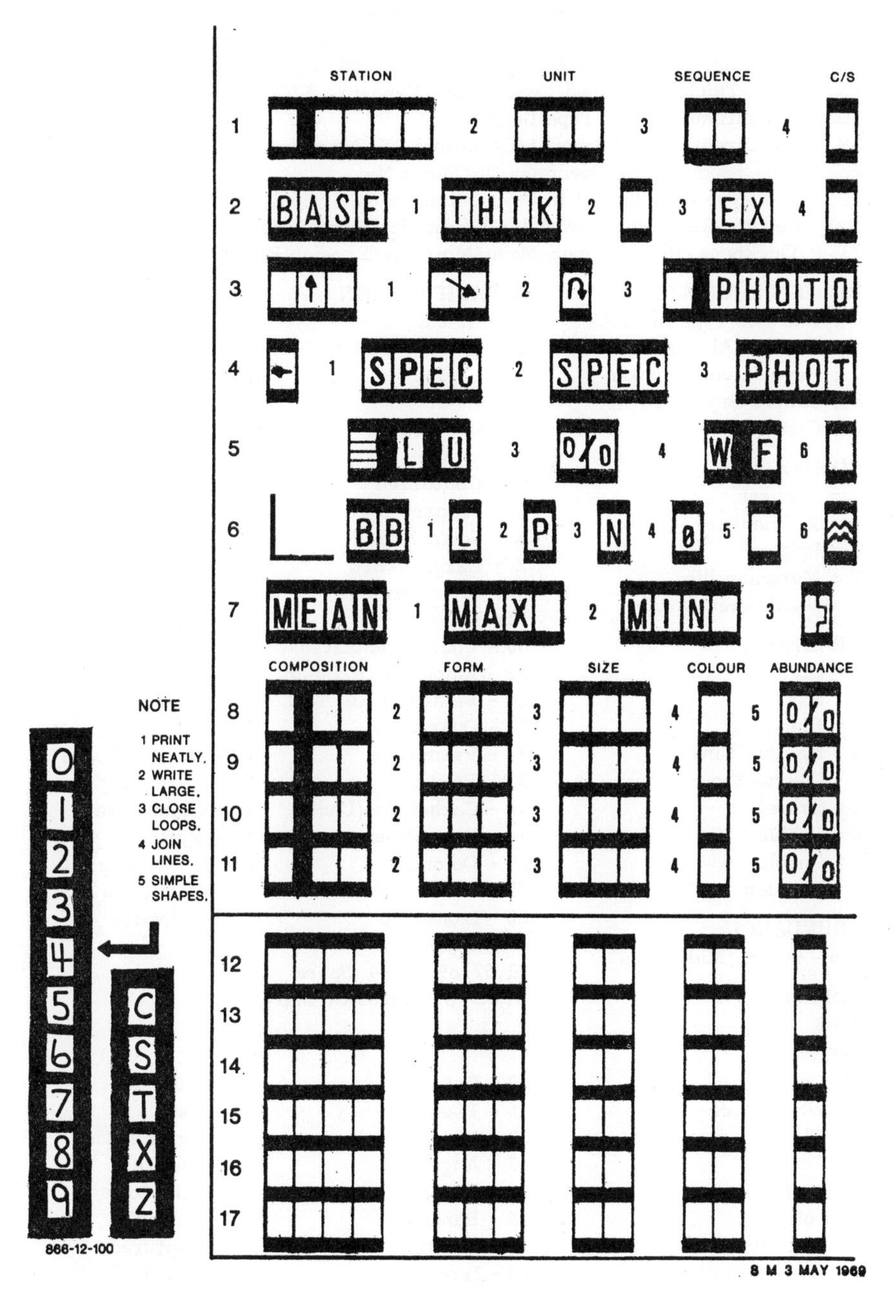

Fig. 1. The Cambridge Form for stratigraphic data. The colour is a pale green.

1. Letter/No/Unit/Sequence/Control Scientific
2C. Origin/Direction/Method/Inclination/ C=Uphill, X=Downhill
(=Taping, C=Aneroid, X=Vert. Int., S=Paced, Z=Direct Measurement)
2S. Base/Thickness/Taped, eStimated/Exposure %/Covered, eXposed, floaT
3C. Measurement Unit/Tape Length
3S. Dip Direction/Dip Amount/C=Normal, X=Overturned/Photo
4. Stub/Specimen No/Specimen No/Photo No
5. Bed Type Relations/% In Unit/Colour/Saccharoidal, Cruddy
6. Bedding/Lith/Por/C=Nodules/Grain Size/Sort/Deform=X
7. Thickness Of Bed Types-Mean/Min/Max/Erosion Resistance 0-9

BED TYPE RELATIONS
0 : One Bed Type In Unit
1 : Random Alternation
2 : Repetitive Sequence
3 : Simple Sequence
4 : Gradation

COMPOSITION
X = Mineral
S = Sedimentary
T = Igneous
Z = Metamorphic
C = Undifferentiated

GRAIN SIZE
0 = Clay
1 = Silt
2 = Fine Sand
3 = Medium Sand
4 = Coarse Sand
5 = Gravel
6 = Pebble
7 = Cobble
8 = Boulder

ABUNDANCE
% 0 - 99, X, Mid-Class

POROSITY
0 - 3

FORM
0 - Unspec
1 - Whole Rock
2 - Clast
3 - Fossil
4 - Pellet/Lump
5 - Oolite
6 - Intraclast
7 - Nodule/Concretion
8 - Cavity/Vein

STRATIFICATION
0 :< .2 cm.
1 : .2 - 1
2 : 1 - 5
3 : 5 - 60
4 : 60 - 120
5 :> 120
C : Contorted
S : Planar
X : Small X-Bed
Z : Large X-Bed
T : Structureless

ROCKS
00 - Undifferentiated
23 - Amphibolite/Basic
44 - Feldspathite/Acid
77 - Quartzite
22 - Psammite
33 - Pelite
66 - Carbonate (Undif)
11 - Limestone
99 - Dolomite
55 - Carbonaceous
88 - Ironstone
40 - Evaporite
12,57,89,48,94 Spare
STUB S = Notes On Margin

17 - In Situ
16 - Pseudo
32 - Round
67 - Angular
28 - Platy
75 - Banded
84 - Spotted
45 - Irregular
00 - Unspec

COLOUR
0 - Yellow
1 - Grey
2 - Green
3 - Brown
4 - Black
5 - Orange
6 - Purple
7 - Red
8 - Blue
9 - White

MINERALS
00 - Undifferentiated
15 - chert
29 - Quartz
54 - Feldspar
82 - Sheet Silicate
93 - Ferromag
65 - Glauconite
61 - Pyrite
31 - Phosphate
71 - Calc Carb
76 - Dolomite

64 - Intrusion
39 - Lava
51 - Pyroclastic
73 - Fissile
92 - Schistose
49 - Folded
53 - Ruptured
21 - Well Jointed

Fig. 2. Reverse of form, showing the field crib to codes.

FIELD EXPERIENCE

The machine readable forms were used extensively for measuring stratigraphic sections in Svalbard in 1969. Prior to this they had not been used in the field nor had the users been trained in their use. The possibility that poor writing would prevent the forms being read did not materialize. The actual error rate, 400 unidentifiable characters in 55 000 and 30 mis-identified, was acceptable. Indeed the simplicity and brevity of the characters used, compared with ordinary longhand, is an advantage when working with cold hands.

When used in heavy rain, the forms became too grubby and dog-eared to be read by the IBM 1287. Analogous problems are experienced with ordinary field notebooks. We are investigating the possibility of continuous roll recording paper, of which only a small part would be exposed to the elements at any one time.

However formidable it may appear on paper, the coding scheme was not too complicated. It can be learnt rapidly in the field and after a few days we seldom had to look up a code apart from the special features languages. The key on the back of the form is an adequate summary of the codes for normal lapses of memory and there is no need to use the detailed manual in the field. First year undergraduates, who were recording on the sheets at the dictation of a geologist, learnt the system sufficiently to be able to measure sections on their own (after a few weeks of direct participation) without any additional training.

The system of Marrack *et al.* has been criticized as monotonous to use in the field. This is more the property of their rocks than their recording system. We found our sheets interesting to use. There is a particular pleasure in discovering a subtle use of the codes to record an unusual feature or relationship.

We agree with other users of standard recording forms about their value as a reminder to look for and record certain items of geological data. In particular, we found that our description of rock units and their component bed types was far more complete than in free format recording in the field notebooks. The structure of our special feature coding, which is virtually a free format system, meant that this advantage did not extend to the recognition of "special features", but the form did ensure that their size, orientation, and abundance were noted.

It is not easy to express degrees of uncertainty with the system. Degrees of exposure can be recorded, and this is a useful guide. Methods of expressing doubt can be incorporated into specialist language, or can be put as a longhand margin note.

Where a section measuring party made observations not conveniently coded on the forms or put on the margin they were entered in notebooks, and related directly to position in the section by reference to the header information on the appropriate sheet.

We found our original coding for bed type relations to be too restrictive and the one described here is a revised version. We have also introduced a method of referring to previously recorded rock units and bed types; this will simplify recording in many cases. We had expected that our method of distinguishing rock units, rather than beds, would obviate the need for back referencing. In practise this was greatly reduced but it would have been very convenient on many occasions to be able to refer back.

CONCLUSIONS

1. Tree structures provide a satisfactory framework for recording data on stratigraphic sections.
2. Coding systems for recording variable data of this kind in the field can be designed and are relatively easy to learn and use.
3. The use of these techniques improves significantly the quality and quantity of stratigraphical data recorded on a traverse without reducing the speed of the traversing.

ACKNOWLEDGEMENTS

We must thank our colleagues on the expedition to Svalbard who acted as guinea pigs, and particularly David J. Batten and David G. Smith. The successful use of mark-sense techniques by Friend, Nicholson, Marrack and Yeats encouraged us to take a further step along this road. We must also thank T. Coleman and R. Jenkins of IBM who provided technical help at extremely short notice.

REFERENCES

Bouma, A. (1962). "Sedimentology of some flysch deposits". Elsevier, Amsterdam.

Selley, R. C. (1968). Facies profile and other new methods of graphic data presentation: application in a quantitative study of Libyan Tertiary shoreline deposits. *J. sedim. petrol.* **38**, 363–372.

APPENDIX: A CODE FOR STRATIGRAPHIC SECTIONS

The forms used are referred to as SM3 data sheet and are shown in Figs 1 and 2. The back of the form is a "crib" containing the main codes. The following notes on the code are taken from our technical manual.

1. SECTION MEASURING TECHNIQUES

Before discussing the coding system used on the data sheets it is necessary to review the methods of measuring stratigraphic thickness during a traverse. Three ways are commonly used—direct measurement, measurement of distance along surface of ground and measurement of vertical distance. The units of measurement may be divisible or integral. The practical distinction is between recording with a decimal point and recording an integral number of units.

Direct measurement is simple to record. In the other two techniques the traverse is divided into legs between substations and it is necessary to record the direction and inclination of the substation at the end of each leg from that at the beginning of the leg. The positions of points within each leg are fixed by simple proportions. Stratigraphic thickness can then be worked out allowing for dip.

The general case is most simply explained when a continuously recording instrument such as an aneroid is being used. The data which must be recorded at each substation are:

direction of next substation;
elevation of next substation;
unit of measurement;
instrument reading at substation.

For each point to be fixed between substations it is necessary to record the reading of the instrument at the point.

Any system of section measuring which produces equivalent data can be used on the forms. The conventions for the techniques for which computer programs exist are described below. In general certain values in the traverse data, such as measurement unit, remain constant for most of the traverse. Such items have default settings which may be altered by explicitly entering a new value. This new value becomes the default setting unless explicitly altered back to the original value.

2. FILLING IN THE FORM

The SM3 data sheet is divided into 89 fields, each with spaces for one or more characters. For reference in these instructions, the lines are numbered from 1 to 17, and the fields in each line are numbered 0, 1, 2, etc. Only the letters C, S, T, X, and Z and the digits 0 to 9 can be read by the IBM 1287, other characters are invalid. Further, the letters, C, S, T, and Z may only appear in the *last* character position in a field and if this position is reserved for a letter it must never contain a digit.

Characters must be written clearly in 2B pencil and as large as possible as shown on the bottom left-hand corner of each form. Pencil and other marks must not extend into the green printed border around each character space. The stub on the left-hand margin of the form is not read by the machine, and may be used for recording non-codable information in longhand. Since the machine right-justifies all characters in a field by the omission of spaces, least significant digits in all measurements must be entered. The top right-hand corner of the sheet must not become dog eared, since this is the leading corner for feeding into the IBM 1287. During reading, forms which are damaged or contain unrecognizable marks are rejected. Usually part of such forms will have been read correctly and only the actual fields containing errors need be altered on the magnetic tape. Extra care should be taken with letter S which sometimes the 1287 fails to recognize. Characters must be as large as possible. In particular for numbers like "2000" there is a tendency to write the zeros rather small. This should be avoided.

3. SM3 CODE

The data recorded on the SM3 sheets will be translated by a program into the computer format described earlier. The code is therefore temporary and the data are decoded after being read. Because the data are complex, the sheets must be used for multiple purposes. In particular it is necessary to distinguish between the survey data which do not go into the computer file, and the descriptive data which do. It is also essential that the sheets can be sorted if they get out of order. This is done through a system of identifiers entered on line 1 of each sheet. Figure 3 illustrates some examples.

Each complete traverse forms a *station* and is given a station identifier (a letter and number) which must be entered in fields 1/0 and 1/1 of every sheet. Every point fixed by survey on a traverse is a substation. There are three kinds:

control substations—defining the start of each leg of the traverse;
rock unit substations—defining the start of a rock unit;
point feature substations—fixing the position of a point feature within a rock unit.

Control and rock unit substations are given *unit numbers.* These numbers must be sequential, increasing in the direction of traverse, but need not be consecutive. Where a control and rock unit substation

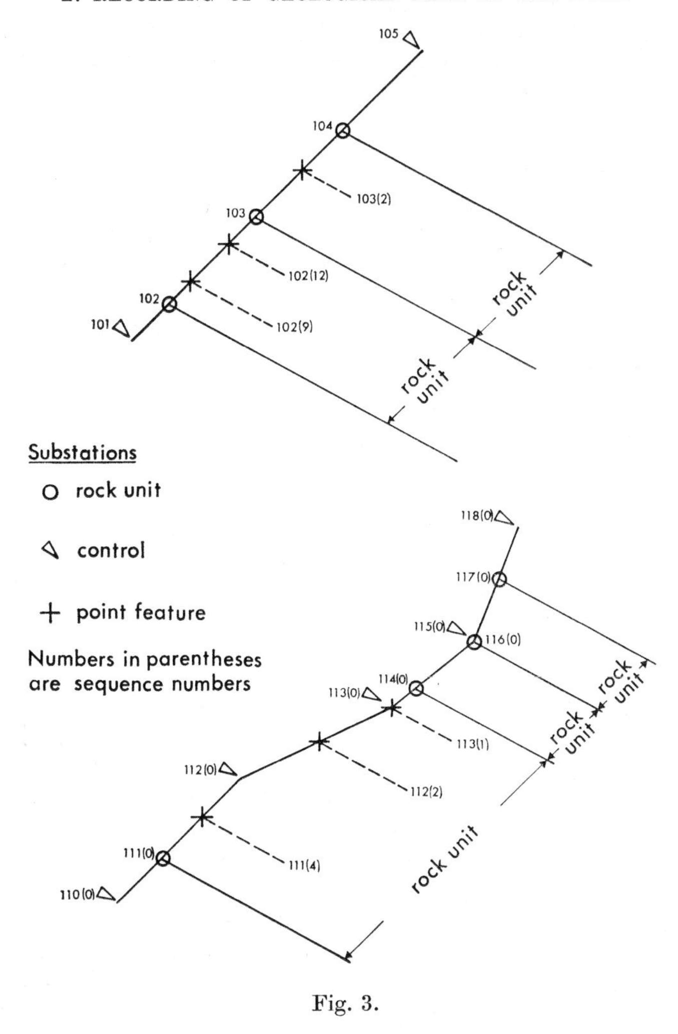

Fig. 3.

coincide, they are still given different unit numbers. The control substation is given the lower number. All sheets containing data associated with the numbered substation must carry the unit number in field 1/2. Different subsets of information with the same unit number are given different sequence numbers in field 1/3. The order of sheets with the same unit/sequence combination is not significant. Substations fixing a point feature are not given different unit numbers, just different sequence numbers. The sheets may be used for (1) survey data at a control substation; (2) fixing position of start of rock unit; (3) fixing position of point features; (4) describing bed type relations; (5) describing bed types. It is sometimes possible to use a single sheet for more than one purpose.

3.1 The rules for identification data

Sheets with survey data at a control substation have C in field 1/4. The start of a rock unit is indicated by sequence O. Sheets for bed type arrangements, bed types and types and point features all have non-zero sequence fields and the last unit number used (even if this was for a control substation).

3.2 Control data

This section deals with the survey data entered on sheets with C in field 1/4. The format depends on the survey method which is entered in field 2/2. The available methods are:

Z — Direct measurement of stratigraphic thickness
C — Continuous counting vertical intervals
S — Continuous counting of intervals along surface
T — Direct measurement of surface distances
X — Direct measurement of vertical intervals

The default setting is T. The first sheet for a traverse must always be a control sheet to establish the measurement method.

The length in centimetres of the unit of measurement used for survey distances or direct measurement should be entered in field 3/0. Default setting is 100. For measurements in feet enter 30 and the true conversion factor will be inserted by program. For pacing methods enter the pace length and for vertical interval methods the height to eye level of the observer. If the settings are 100 (metric) or 30 (feet) then all other measurements on the form must be the same. If any other setting is specified, and the measuring method is a survey one, then all measurements other than survey data must be metric. If any other setting is specified and direct measurement is being used, then all measurements, other than thickness (2/1), must be metric.

3.2.1 Direct measurement (Z)

The stratigraphic thickness is measured or estimated directly and entered on the sequence 0 sheet for each rock unit (field 2/1). A control sheet is needed only to establish the method and record if the traverse is from older to younger rocks or vice versa. The latter is done by entering in field 2/3 C for the former and X for the latter. C is the default setting. Once set this direction cannot be altered.

3.2.2 *Survey methods, general points*

The position of the next substation is fixed from the one before in terms of bearing, inclination and distance. Common parameters for all methods are:

(1) Bearing (field 2/1, degrees magnetic, 0–360) of the next substation along the traverse. Must be present.
(2) Inclination (field 2/3, degrees 0–90) from horizontal to next substation forward along traverse. Must be present.
(3) Vertical direction (field 2/4), C if traverse is uphill, X for downhill. Default setting is C, uphill.

3.2.3 *Continuous vertical measurement* (C)

A continuously recording instrument is used to measure vertical distances (e.g. aneroid altimeter, laying off equal increments with abney level at eye level). The instrument or count is not reset at substations. Record the reading of device at each substation as (field 2/0) (an integral number of units). Substations at start of rock units and at point features are also fixed by entering the instrument reading at substation in this field. A sheet must always be completed at the last substation in the traverse, giving the instrument reading.

3.2.4 *Counting of intervals along surface* (S)

Normally the method involves pacing. The rules are the same as for continuous vertical measurements.

3.2.5 *Direct measurement of surface distances* (T)

This is usually done by taping, the measuring device being reset to zero each time it is moved. The distance between substations is measured directly. The parameters are as follows.

(1) Reading at substation (field 2/0). Default setting is 0. Measurements are in multiples of 0·01 of the measuring unit (i.e. cm for a 1 m unit).
(2) Distance between substations (field 3/4). Default setting 50·00. Two digits after decimal point.

Rock unit and point feature substations are fixed by entering reading at the substation in field (2/0).

3.2.6 *Discontinuous vertical measurement*

This is the vertical equivalent of taping where the measurement unit may be subdivided and the count is restarted at each control substation. The rules are the same as in 3.2.5.

3.3 Coding convention for percentages

Where a two character field is used for recording percentages X is used to indicate 100%. Otherwise the number represents the mid-point of a 1% range, so that 23 indicates between 22·5 and 23·5%. 0 means less than 0·5%.

3.4 Fixing position of rock unit

The sheet must have new unit number and sequence 0. For survey methods the base field (2/0) must be filled in with the reading on the measuring device at the start of the unit. For direct measurement thickness (field 2/1) must be entered. If both fields are filled in then both thicknesses will be entered in the record. If the thickness is entered the method of measurement should be noted in 2/1. The codes are:

T — measured
S — estimated

The default setting is S.

Percentage exposure of unit may be entered in field 2/3 and the type of exposure in field 2/4. The codes are:

C — covered, scree, etc.
X — exposed
T — float, no exposure but surface material is believed to represent the unit.

The boundaries of the rock unit may be entered in fields 5/1 and 5/2.

3.5 Bed type arrangements

The codes for various relations are:

0 — single bed type
1 — several bed types
2 — repeating sequence of bed types
3 — non-repeating sequence of bed types
4 — continuous gradation between bed types. Those described represent points in the gradation.
5 — bed type group
6 — repeating gradation

The description of a rock unit consists of a sheet giving a bed type arrangement followed by sheets describing the bed types to which the relationship applies (see earlier discussion). If the relationship applies to a bed type group then instead of a bed type description there should be a sheet containing relation 5 in field 5/0. The format of this sheet is as follows:

2/0 and 2/1. Unit and sequence number of the sheet at which the description of the group begins. This and the following sheets must contain the description of all bed types belonging to the group. A further relation 5 sheet can appear among these.

5/3. Percentage of group in rock unit (or in higher group).

7/0 to 7/2. Thickness of sets of beds forming group. (See 3.6 for details.)

2/3. Bed type arrangement applied to a group.

The following example may help to clarify the use of bed type groups:

Seq. No.	0	1	2	3	4	5	6	7	8	9	10	11	12	13	14
Relation	4		5	5		2		5		1			3		
Field 2/1			5	9				12							

For relations 2, 3, 4, and 6 the order of the sequence numbers of the bed types is significant since this indicates the order in which the bed types occur. Note that this is the true stratigraphic order from oldest to youngest and not necessarily the order in which they are met in the traverse. For relation 1 the order is not significant.

3.6 Description of bed type

A description of a bed type consists of a number of parameters referring to the whole bed type, a list of components and a list of special features. Two or more sheets may be used if there is insufficient room on one sheet or if it is desired to associate a photograph or specimen with a particular component or special feature. The sheets must have the same unit/sequence number. There is no significance in the sheet order. If a general parameter has more than one value (e.g. two kinds of bedding present) this can be indicated using two sheets but there is no way of indicating relative abundance of the two types. There are no default settings. Blank fields mean that data were not looked for. The general parameters are as follows:

Percentage of bed type in rock unit or bed type group (field 5/3).

Thickness of sets of beds belonging to a single bed type. Field 7/0 mean, 7/1 maximum, 7/2 minimum. Note that there are three figures after decimal point.

Nature of boundaries of bed type, lower in 5/1 and upper in 5/2. Codes are:

S — sharp
T — gradational
C — covered (not observed)

Colour. Unweathered hand specimen colour in 5/5, weathered whole rock colour in 5/4. Follow the usage of the G.S.A. rock colour chart, abbreviating to the final name in the colour only. Pink should be included in red, or in white if very light.

Texture (field 5/6). Whole rock textures. General codes have not been developed.

S — saccharoidal texture in limestones
C — "cruddy" (applicable to any rock of indeterminate texture—use sparingly)

Bed shape (field 5/6). General codes not developed. Z means beds noticeably lenticular.

Stratification (field 6/0). This refers to primary sedimentary layering distinguished by abrupt changes in petrology. It does not refer to the fissility of the rock. The first character in the field describes the spacing of bedding planes. If two types are present (e.g. a fine lamination and bedding surfaces at 20 cm spacing) record one on a continuation sheet. However, should stratification represent the boundaries between bed types (e.g. interbeds of silt and sand) it will be recorded in fields (7/0 to 7/2) and should not be repeated here. The second character describes the internal structure of strata.

C — contorted beds due to slumping or preconsolidation deformation or very irregular strata, such as in reefs.
S — planar (or nearly planar) stratification.
T — structureless units, includes completely bioturbated units. No stratification between upper and lower surface of bed type unit.
X — small-scale cross stratification (cosets generally less than 5 cm thick).
Z — large-scale stratification (cosets generally greater than 5 cm thick)

A letter must always be present here if information is put into the 6/0 field.

Lithification (field 6/1). No precise definitions have been attempted for field use. The codes are:

S — soft, unconsolidated
T — semi consolidated, can be dug out with a trowel
X — consolidated, requires a hammer

Porosity (field 6/2). No precise definitions are used in the field. The codes are:

0 — tight (e.g. clay, dense limestone)
1 — poor (e.g. very small vugs, pores, etc.)
2 — moderate (e.g. medium-sized pores, partially cemented sandstone)

3 — good (e.g. well-sorted uncemented sandstone or saccharoidal limestone)

Nodules (field 6/3). C in this field indicates that post depositional concretions are present. Redeposited concretions should not be recorded here.

Grain size (field 6/4). Grain size refers to modal size.

0 = clay	$< 4\ \mu$; smooth to tongue, no grains visible with lens
1 = silt	4–62 μ; Grains visible only with lens
2 = fine sand	62–125 μ; v.f. sand on Wentworth scale
3 = medium sand	125–500 μ; f + m sand on Wentworth scale
4 = coarse sand	0·5–2 mm; c + v.c. sand on Wentworth scale
5 = gravel	2–4 mm
6 = pebble	0·4–6·4 cm
7 = cobble	6·4–25·6 cm
8 = boulder	$>$ 25·6 cm

These are defined precisely, but only estimates are possible in field.

Sorting (field 6/5).

0 = noticeably poorly sorted (e.g. greywacke) sigma phi $>$ 2·5
1 = average sorting
2 = noticeably well sorted (e.g. beach or dune sand) sigma phi $<$ 0·5

The components are described in lines 8–11. Each line can be used to described one component of the rock type. The significance of the codes varies with context and association. The available codes are shown in Fig. 2.

Field 0 identifies the general nature of a component.

Field 1 identifies the chemical composition of the component or rock.

Field 2 describes the physical form of the rock or component. The first digit gives a general form. The second and third digits qualify the first.

Field 3 identifies the size of a component, using the grain size code in field 6/4. The first digit is modal size, the second maximum and the third minimum. Either the third, or both the second and third, may be omitted. If used for veins, indicates width.

Field 4 identifies the colour of component.

Field 5 indicates the abundance of the component using the percentage conventions.

Two approaches are available for applying these codes. First a single component can be used to describe the whole rock. In this case the abundance field contains X for 100%. In the second method the rock is described in terms of a number of components whose abundances sum to 100%. The two approaches may be combined by using one component with abundance X for the whole rock and then listing one or more components (e.g. a limestone with 5% quartz pebbles). The abundances of these subordinate components need not sum to 100%.

When a component is used to describe the whole rock, the significance of the fields is as follows:

Field 0: S — sedimentary rock
T — igneous rock
Z — metamorphic rock
X — mineral bed not formed from clasts
C — undifferentiated

On this basis massive evaporites, recrystallized limestones and dolomites are all X. C is used for rocks whose chemical composition is known but which cannot be fitted into any of the other four classes.

Field 1: chemical composition, the rock codes should be used.

Field 2, digit 1: Since whole rock is already implied by context there is no need to repeat it and this digit can be used for a significant feature of the rock, e.g.

S11 5 — oolitic limestone
T23 8 — veined basic igneous
S33 3 — fossiliferous shale

Field 2, digits 2, 3: Further qualification where applicable, e.g.

X99 775 — banded nodular dolomite
S33 073 — fissile shale

Note that this information qualifies the rock and not the special feature noted in digit 1.

Field 3: The size in this context could only be the grain size of whole rock and is therefore redundant, but can be used to indicate a second mode on a range.

Field 4: Colour can only be that of whole rock, so can only be used for a second colour.

Where a line is used to describe a separate component the key factor is the first digit in field 2 which identifies the nature of the component. If it is identified as a clast, then fields 0 and 1 identify the rock or mineral type from which they are formed. If the component is a fossil, pellet, oolite, nodule or vein then field 0 is generally not significant and field 1 reflects the composition. If the form is identified as whole rock then the bed type consists of two or more rock types mixed but not stratified such as partially dolomitized fossiliferous limestone. Mineral cements, sparry calcite, etc. are given form 9.

The second two digits of field 1 are used for a variety of qualifiers most of which are self-explanatory. However, "*in situ*" is intended for fossils in position of growth. "Pseudo" means that the form is descriptive but not "true" and is intended for such things as pseudo breccias.

The size field refers to the component. There is no way within the line of indicating the grain size of a rock type forming a component.

The colour field refers to the fresh colour of the whole component.

Because of the restrictive nature of the codes for describing components in terms of their rock types an escape code system has been provided (see below).

3.7 Short cuts

Normally the bed type relation and base of rock unit go on one sheet. If there is only a single bed type in the unit, its description can also appear on the same sheet. Otherwise the first bed type must be described on a new sheet.

Constant data. Properties applying to all bed types in a unit or bed type group can appear on the sheet establishing the unit or group. These become default settings within the unit or group and will be assumed for all bed types unless an alternative value is given subsequently.

Escape code for components. If component field 1 contains XX then fields 2 and 3 contain unit/sequence cross reference to a bed type group. The sheet indicated can be used exactly like a sequence 0 sheet and lead on to a complex rock description.

Cross reference. Anywhere where a bed type or bed type group description can start it is possible to refer to one described previously. The convention is to put 9 in bed type relation field and the unit/sequence concerned in 2/0 and 2/1. The referenced data can be overwritten by data on the referencing sheet. The distance back that can be referenced is limited by the computing requirements (about 20 sheets in practice). However, each time a set is referenced it gets put to the head of the list.

If no composition is given for components, then terrigenous is assumed.

3.8 Point features

A point feature is indicated by an entry in the base field coupled with a non-zero sequence number. This is the reading on the measuring instrument if surveying, and the thickness from the start of the unit if measuring directly.

A point feature may have specimens or photographs associated with it. Alternatively it may have a description of a rock unit. The method of measurement for this unit is direct and the thickness is entered in field 2/1. The normal bed type relations apply. Note that the sheets following that starting the point feature are assumed to belong to it until either another point feature is started or a sheet already noted as belonging to a bed type group appears. Therefore a rock forming a point feature cannot itself include a point feature.

3.9 Photographs and specimens

Specimen numbers consist of a letter and number but the letter must be the same as the station identifier. The number is entered in 4/1 and, if several consecutive numbers are to be entered the last goes in 4/2. These letter/number combinations must appear on specimen labels.

Photograph numbers consist of a letter (3/3) and a number (3/4) consisting of a 3-digit film number and a 2-digit frame number. If a sequence of photos is to be entered the last two digits of the film number and the frame number can be put in 4/3.

Photographs and specimens can be associated with particular features of a rock unit by placing them on the same sheet. If they are recorded on the first sheet of a rock unit no information about the position in the unit of the specimen or photograph is implied. The actual position of a specimen within a unit can be fixed by establishing a point feature for it. If this feature is not a rock unit (nothing in field 5/0) a cross reference to a bed type or bed type group can be made in fields 7/0 and 7/1.

3.10 Specialist codes

The last six lines on the form are available for specialist codes. The code is identified on the form by a digit in the last field on the line. Codes needing more than one line per statement may use 0 in this position to indicate a continuation line. The association between the digit and the decoding subroutines is done each time the decoding program is run. There is thus no limit to the number of codes which may exist but only nine may be used in any one batch of forms. Decoding routines involve about two weeks' work to prepare and test so programming support should be assured before designing new specialist codes. The actual codes now available were designed for use by N.C.S.E. in Svalbard but are not repeated here.

3. The Implementation of an Automated Cartography System

IAN S. EVANS

Experimental Cartography Unit, NERC/RCA, Oxford, England

One of the major projects of the E.C.U. is to perform research on automated cartography and to develop an operational system suited in particular to environmental information. For this purpose, the Unit was established in October 1967 with a major grant from the Natural Environment Research Council (U.K.). The basic aspects of such a system have now been implemented, in the course of experiments with information from a variety of government agencies concerned with mapping. There are many additions and improvements to be made, but the Unit is already producing maps of various kinds.

The system involves use of a digitizing table, a small control computer, and a flat-bed automatic plotting table of high accuracy. These are illustrated in Bickmore (1968b) and in Roberts and Evans (1969). The Unit is structured into three sections: data capture and editing; programming and hardware; and output and design. The data capture section arranges data so that they are suitable for digitizing; this includes coding, edge matching and the resolution of ambiguities. It seems appropriate to open with a discussion of hardware, before moving on to programs and data handling procedures.

HARDWARE

(a) Digitization

While an ideal system would be automated from the recording of field information through to the plotting of the final map, it is necessary to face the fact that for years to come much information will be derived from pre-existing maps. Many of these will be rough field sheets. Hence, the digitization of maps will remain a major mode of "input" to

Systematics Association Special Volume No. 3. "Data Processing in Biology and Geology", edited by J. L. Cutbill, 1970, pp. 39–45.

an automated cartography system, and the process used will have to cope with some rough material, unsuited to automatic line-following or scanning.

For such work, manual intervention in the digitizing process is necessary, or at least economic. The only appropriate equipment currently available is the D-Mac Pencil Follower. There are many other manually operated digitizers, which can be used for points; but only the D-Mac, with its relatively free-moving pencil, is convenient for the digitization of lines. The Unit has a Mark 1B Pencil Follower, with a table 1m square, outputting to a 7-track, IBM-compatible incremental magnetic tape unit. Output to an 026 card punch, and to an electric typewriter, is used less frequently, but the typewriter is used for recording identifiers on the magnetic tape.

A stylus ("pencil") is used to follow lines, and it leaves a trace on lightly coated material (masonscribe) to indicate what has been digitized. A coil centred on the point of the stylus generates a magnetic field which is picked up by a follower under the table. The follower moves in the y direction, up and down a gantry which moves in x. Taut wires transmit the x and y positions of the follower to shaft encoders which digitize them with a resolution of 0·1 mm.

Several other pencils are available, but we favour a new type developed for us by Dr Bessant and Mr Jebb of Imperial College, London. This involves a magnifying glass and a larger mount to permit a steady, two-handed grip.

In producing accurate plots of what has been digitized, the Unit is testing the D-Mac table more than most users, and this does reveal a number of defects. The most obvious is an inevitable consequence of the table's advantage in utilizing both sides of the working surface; the need to maintain exact correspondence between the position of the pencil and that of the follower. In practice, this is only a minor source of error: when a sharply curving line is followed too quickly there is a slight overshoot effect, but at normal following speeds coupling is satisfactory.

Larger errors can be caused by the winding of the wires on the shaft encoders, which is affected by temperature and wire age; by any deformation of the gantry, which can produce skewing of the axes; and by variations in the magnetic field, the adjustment of which is a compromise. The latter effect seems particularly serious near table edges, where distortions of up to 1 mm in absolute accuracy may occur. Some of this distortion is repeatable, but parts of the table can be affected by oscillation of the follower, producing completely unacceptable lines. For this reason only a 70 cm square of the table can be used at the moment. (We understand from the makers that this problem has been eliminated in more recent models.) The best results are obtained by using a smaller area, say 40 cm square. This should produce a repeatability of about 0·2 mm and an absolute accuracy of 0·3 or 0·4 mm. A scale reduction of 2 or 3 brings such work up to the exacting accuracy standards of the Ordnance Survey. It should be possible to improve the table until its absolute accuracy approaches its resolution (0·1 mm).

The Unit has also used the A.E.G. geameter, an automatic lock-on follower, which is attached to the plotting table described below. Though very accurate (resolution 0·01 mm), this has proved unsatisfactory because:

(i) It occupies the computer on-line, as well as the table, and therefore it needs to be several times faster than the D-Mac to justify the much larger capital investment which has to be dedicated to the geameter.
(ii) In fact, digitizing on the geameter takes over twice as long as on the D-Mac; because of the need for frequent operator intervention it rarely reaches its full speed of 2 cm/s.
(iii) Manual steering of the equipment from one line to the next is not easy.
(iv) Positioning on a point is difficult and inaccurate.
(v) The geameter is sensitive to background noise; for example, it easily moves off onto symbols which overlap the line being followed, even if this involves some change of direction. Hence, it is often necessary to clean up the document to be digitized, and most field sheets are unsuitable.
(vi) There is difficulty in rounding sharp corners and in following lines running close to other lines; this necessitates manual intervention, or at least a considerable reduction in speed.
(vii) Some pecked lines can be followed, but most cannot.
(viii) No record of what has been digitized is generated, so that the operator has to keep account of this by marking up a dyeline copy.

At the moment, then, the geameter is useful only for simply curving lines, and major improvements are necessary before it can become a cartographic tool.

New data derived from a stereoplotter can be digitized from the start by recording the revolutions of the x and y lead screws. This necessitates a procedure for erasure of errors but is obviously preferable to creating a map and then going through the extra stage of digitizing it. The Ordnance Survey will shortly install a trial system of this nature for use with contours, but it would also be useful for any cultural detail readily identifiable in the stereoscope.

A completely different mode of digitizing is by line-scanning, which is very rapid but requires a high-quality input map. The main difficulty is in sorting out the different lines and labelling them; this requires sophisticated software with an interactive CRT display. For this reason line-scanning seems more suited to map reproduction than to map compilation or data bank formation; however, several North American organizations are well advanced in the development of such systems.

For the immediate future, the Unit has ruled out line-scanning, which involves expensive hardware as well as software. Nor does it seem likely that the geameter can be improved sufficiently, although a new generation of lock-on line followers might change the situation. The manual intervention involved in using the D-Mac table is more of a help than a hindrance when digitizing rough or complex documents; for the near future our main hope is for improvement in the absolute accuracy of the table.

(b) Computing

A PDP-9 computer was installed in June 1968, with a 16k extended core of 18 bit words. It is interfaced to the geagraph/geameter, and also to a teletype, three DEC tape drives, and two IBM-compatible tape drives. Though otherwise suitable, our particular machine has suffered an undue number of breakdowns.

The original intention was to perform major processing work off-line on the Oxford KDF-9, and for this reason the magnetic tapes were packed at the low density of 200 characters per inch. There is, however, little spare capacity on the KDF-9, and progress is much slower than with an on-line computer. At the present, most processing is done on the PDP-9, and packing density is being increased to 556 characters per inch.

Since there is no time-sharing on the PDP-9 it is used either to drive the geagraph/geameter, or for program development and processing work. The Unit has now reached the stage where the PDP-9 has become overloaded, and could be fully employed by either of these tasks. Hence, there is a great need for further computing capacity. This will be essential when the Unit has more digitizing tables, and is undertaking regular map production to exploit the system it is developing.

(c) Plotting

A high accuracy flat-bed plotter, the geagraph (utilizing a 1·5 m square Aristomat table) was installed by A.E.G. in August 1968. This has proved very satisfactory and generally reliable in use. It plots to a resolution of 0·01 mm, and at slow speeds can approach this repeatability. Accuracy can be sacrificed to improve speed, the maximum speed being 6·7 cm/s. Actual plotting speeds of around 1 cm/s on intricate lines produce adequate accuracy.

Three different attachments are mounted on the gantry for different modes of plotting. A 0·2 mm rapidograph pen quickly provides a positive. For reproduction, however, a 0·1 mm cone scriber is used, producing a negative on coated plastic. Good quality scribing of broader lines requires either a hot point or a chisel point with tangential control to keep it normal to the direction of movement.

Greater flexibility is now provided by a light-spot projector, fitted early in 1970. This projects a beam of light through one of 48 symbols selected from a rotary disc. The symbols may either be placed as point symbols, or used to draw lines or double lines of various thicknesses. Tangential control maintains constant line thickness or double line spacing. Any symbol used for drawing a line may be combined with one of nine different patterns of pecked line generated by hardware. The nine possible patterns are formed by having the light on or off for any of sixteen basic units, the length of which can be varied (manually) between 0·18 and 1·2 mm.

BASIC SOFTWARE

Though the existence of appropriate hardware is a *sine qua non* of automated cartography, production of useful software is equally important. Most of the Unit's programs are written in Fortran IV, but frequently used tape-processing routines are in machine code. Though most software would ideally be independent of hardware, considerable rewriting is in fact necessary to transfer programs to other hardware, because of varying formats and systems requirements. The geagraph system involves drawing straight lines between closely spaced points, and is significantly different from those employing arcs of circles. We do, however, have a program to write tapes in standard format, for export to non-DEC computers. Under the heading "Basic Software" I will treat several features and programs which are essential to the production of any map by the Unit's system.

(i) Feature coding

Each line or point is identified by up to 50 5-digit feature codes which are typed in at the time of digitization. If no new code is given, a new line automatically receives the feature codes of the previous line. Since code typing is a distracting and error-prone operation, it is best to digitize all lines with the same code together. Obviously feature coding is essential for the production of a map differing in content from that digitized. A new map can be compiled at plotting time by inputting the selected codes through the teletype.

All features that it might conceivably be desirable to separate at a later stage, should be distinguished by different codes at the time of digitization. It is possible to change codes later on a line segment basis, but this is more troublesome than if the appropriate decision were made prior to digitization. The Unit has a tentative system of coding for topographic and thematic information covering both O.S. and U.S.G.S. maps, but broad agreement is necessary before a rigorous system for data banking is adopted.

Multiple coding is necessary not only for relating statistical information to locations, but also for labelling boundaries according to the areas on each side, e.g. the formations of geology maps or the series of soil maps. In this way it is possible to draw, for example, all lines which have chalk on one side, for the production of a colour mask. Figure 1 demonstrates the results of utilizing this facility. Multiple coding is even more necessary for administrative boundaries, which might need to be coded with not only two county names, but also two rural district and two parish names.

(ii) Coordinate transformation

Another capability of fundamental importance is to change from arbitrary digitizer coordinates to an absolute coordinate system. Before the digitizing of any sheet commences, the corner points are digitized. At the computer processing stage, their desired coordinates in a stated absolute system are typed in. The program then makes a least-squares fit of the first set to the second, to establish the correspondence between coordinate systems. The coordinates of all detail on the map are transformed to the absolute system by bilinear interpolation between the established corner-point positions.

This program permits maps digitized as several separate sheets to be reassembled automatically, with the sheets in their correct relative positions. It does not ensure that lines cut by a sheet edge meet exactly, for if the sheets have been digitized independently there may be operator error in opposite directions on the two sides. Hence, before digitizing commences, it is very necessary to check correspondence along sheet edges, especially for lines which have been sketched in the field, as in soil and geology maps. It seems that the best procedure is to place the original and the masonscribe of an adjacent previously digitized sheet alongside that being digitized, and to continue lines from the exact point where their digitization on the adjacent sheet ended.

Coordinate transformation rectifies any linear distortion in the original document; this is especially useful where there is no alternative to digitizing from paper field sheets. It also reduces similar distortion which may arise from the digitizing table, though such distortion should be dealt with by servicing. All uses of corner points depend on the accuracy with which they are digitized, which therefore requires particular care. The corner points of large sheets will be in marginal parts of the table and likely to suffer error in digitizing: their use for rectification may then increase rather than reduce distortion. Hence, it is particularly important to start with small sheets when using the coordinate transformation facility.

(iii) Tape editing

It is, of course, crucial to be able to correct errors after the digitized material has been drawn and checked. Exactly the same facilities are required to update tapes later on, if deletion is involved (addition is simpler). The requirement to edit tapes for these two purposes dictates many features of their format, and dominates the design of an automated cartography system. It would be very irritating to have to discard a whole tape because a few crucial errors could not be corrected. The retention of unchanged information while intermixed information is updated is fundamental to the philosophy of an automated system.

The Unit has now developed an editing system which is useful and easy to operate, but rather slow. After a line has been plotted, the gantry stands back so that the line may be inspected, and the operator can either accept or reject the line, or input a new code. Inspection of the accuracy of each line on the table is possible if the original is placed underneath a plot onto tracing paper, fitted to the corner marks. However, to inspect each line as it is drawn wastes too much table and computer time; it is better to produce a complete plot, compare it with the original off the table, and perform the editing on the computer more rapidly. For this purpose, a line may be accepted or rejected as soon as its drawing has commenced; the program will then move on to the next line. Finally, it is possible to accept "all remaining lines" when correction has been completed. New lines may be added to replace those deleted. While operator accuracy is checked by the masonscribe, machine accuracy can be checked only by producing an accurate, complete plot; this also checks that the tape deck was recording as intended.

Checking the accuracy of line work is less onerous than checking feature codes. There is plenty of room for error in 5-digit codes typed by the digitizer operator. Codes could be inserted at a later stage by an editor, but this would involve at least as much work, and still not prevent errors. For checking, frequently used codes can be called separately and drawn in different colours, but where many codes are represented by a few lines each it is necessary to revert to a line by line check. In the special case of contours, we have found that the production of anaglyphs (discussed below) shows up mislabelling, though a few mistakes may be overlooked.

Codes may alternatively be checked by allotting them different symbols, pecking patterns or pecking unit lengths, in a light-spot projector plot, or by listing them with the coordinates of the first and last points of the line, so that the line can be identified manually and the codes checked off-line. The former method is easier but involves plotting multiply-coded lines several times over. The latter method is laborious, but minimizes the use of time (which is often at a premium) on the plotter and computer.

Code checking would be greatly facilitated by the use of an interactive CRT, or even a rough but rapid drum plotter; use of a slow, but accurate, plotter for this purpose is wasteful. Hence, our present techniques represent a temporary stage in the development of a more economic system. It should be possible to locate a line by indicating a point close to it, for the purpose of deleting the line or changing its code.

(iv) Data compression and generalization

The fact that ten B.C.D. characters, or six frames of binary, are required to store the coordinates of a point digitized on the D-Mac, makes the use of magnetic tape essential for the recording of curved lines. The spacing of points along lines depends on the speed with which the follower is moved, which depends in part upon the intricacy of the line. Usually there are from 1 to 3 points per mm. There are more points on sharp bends and fewer on straight lines; in fact if a line is known to be straight the operator need only record its two ends. Hence, the digitization process itself performs a useful data compression, in relation to the total number of points which could be digitized.

Accordingly, it is difficult to perform data compression without some elimination of data, i.e. generalization. The use of circular arcs does not compress data sufficiently to compensate for the computing time and later decoding involved. However, to plot maps at greatly reduced scales it is necessary to discard many points, i.e. to generalize, if a reasonable drawing speed is to be maintained. The original A.E.G. software attempted this by omitting intermediate points provided that they lay within a specified distance of the line drawn. This was unsatisfactory because the computing time involved, especially for extreme generalization, took so long on the PDP-9 that it was slower than drawing every point.

Following experiments with different techniques (Lang, 1969), it was found that satisfactory results were obtained by either (a) retaining every *n*th point; or (b) retaining only those points more than a certain distance from the previous retained point. (b) has been adopted since it is rather faster, for what appears to be a similar result with oceanographic contours. The program calls for a speed which is proportional to the distance between selected points, with a minimum of 13·3 mm/s, i.e. 20% of the maximum. Actual speeds are somewhat lower, because of acceleration and deceleration for changes of direction. A point spacing of about 0·3 mm at the scale of plotting produces a map which is neither too detailed, nor too angular. The time taken is at least 30% less than with the original software, and more time may be saved by greater generalization. There is, however, plenty of further room for experiment with automatic generalization.

ADDITIONAL SOFTWARE

The above software is fundamental to the production of almost any map. Additional programs providing various special facilities are also essential for flexibility of output; these are described below.

(i) **Projection change**

Drawing small-scale maps on new projections has always been a tedious procedure, and since it has involved manual interpolation the results have not necessarily been accurate. Even large-scale, topographic maps suffer from being specific to the projection on which they were originally drawn. Digitally stored information does not have this limitation; it can readily be transformed to any mathematically specified projection. The ease of conversion between projections, and of producing oblique and exotic versions, is one of the most obvious advantages of automated cartography, and programs to achieve conversion have been implemented on large computers by various earth scientists.

The E.C.U. has recently developed a program to convert between any of the following projections; equal-area or equidistant cylindrical or zenithal; Mercator; orthomorphic zenithal; Gall; modified Gall; any rectangular grid. The pole can be changed to give transverse or oblique variants (Fig. 2); further projections are being added as necessary. Output can be either on-line to the plotting table, or to a magnetic tape for later use. Since the program is quite slow on a PDP-9, it is possible to select every *n*th point of the input ($10 > n > 0$), to produce a quicker and more compact result.

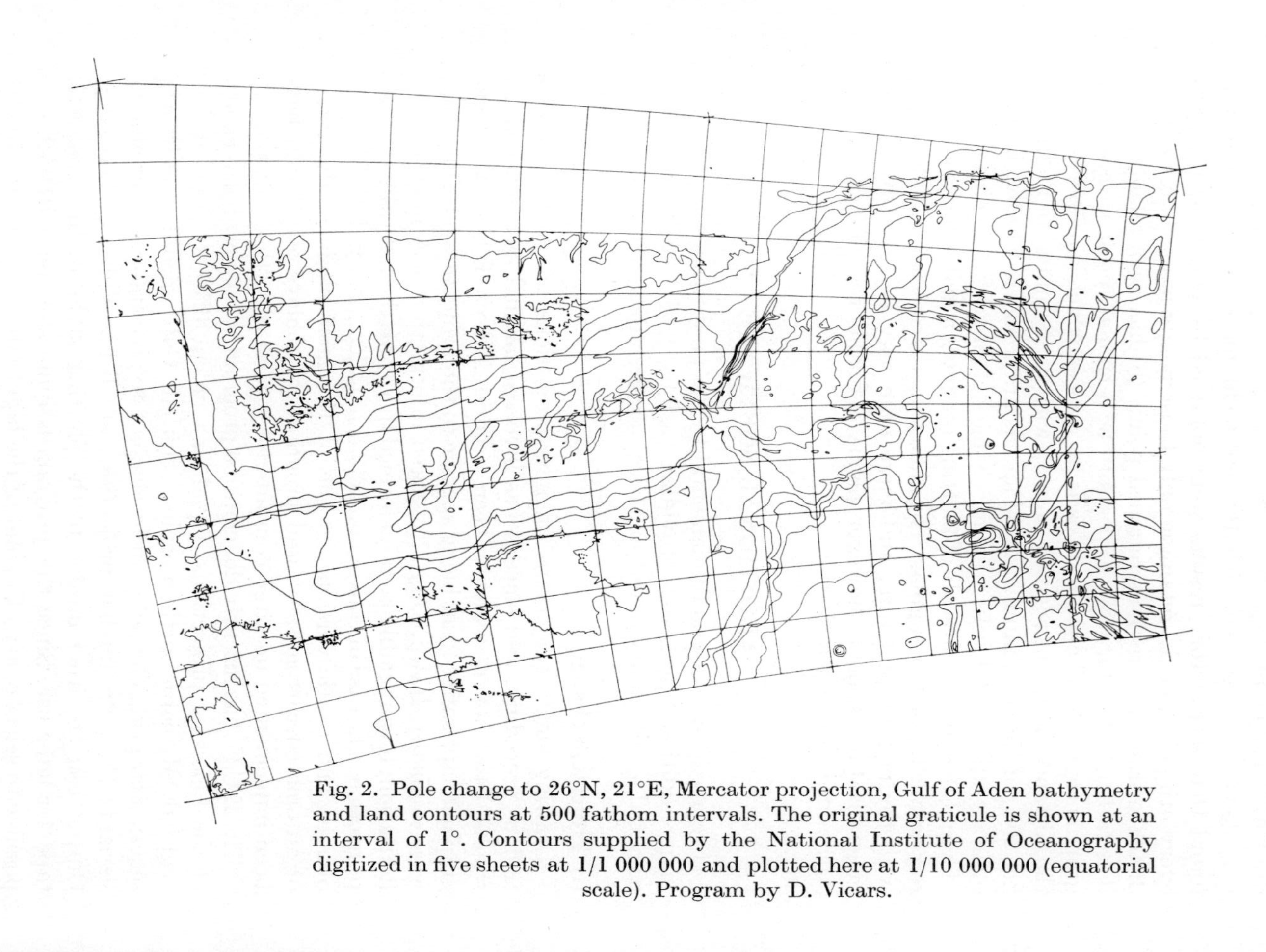

Fig. 2. Pole change to 26°N, 21°E, Mercator projection, Gulf of Aden bathymetry and land contours at 500 fathom intervals. The original graticule is shown at an interval of 1°. Contours supplied by the National Institute of Oceanography digitized in five sheets at 1/1 000 000 and plotted here at 1/10 000 000 (equatorial scale). Program by D. Vicars.

(ii) Three-dimensional views

Another field in which automated cartography has great advantages over conventional methods is in the production of anaglyphs or stereo pairs to represent the third dimension. Topographic surfaces, geologic boundaries, air or water mass boundaries, or more abstract surfaces, may be represented in this way. Given digitized contours labelled with the appropriate heights (or "z-values") the program is relatively simple. For each half of the pair, each contour is displaced in proportion to its height. The left and right displacements are set side by side for a stereo pair, or superimposed in two colours (usually red and green) and viewed through spectacles which allow each eye to see only its corresponding view. It is necessary to specify the proportionality constant between height and displacement, and the level of zero displacement. The proportionality constant must be chosen carefully to produce an appropriate relief at a given viewing distance (Adams, 1969). Relief should not exceed some 3–5 cm over a short horizontal distance (the field of view), for a viewing distance of 35 cm. With automatic plotting, a good three-dimensional effect can be achieved, whereas manual redrafting of contours produces lack of correspondence which makes it more difficult for the eye to fuse right and left images. Alternatively, if a separate plate of each contour were duplicated photographically, there would be difficulty in combining them with exactly the right displacement. Hence, it might almost be said that automation makes the large-scale production of anaglyphs and stereo pairs from contours feasible for the first time.

(iii) Hill shading

The third dimension may also be represented by calculating the degree of shade produced by a given light source, on a hillslope segment of known gradient and aspect (Yoëli, 1967). This requires a grid of spot heights, which may be interpolated manually from a contour map or automatically from digitized contours or scattered spot heights. The E.C.U. has produced grey-scale output of this type on the SC–4020 plotter at Chilton.

(iv) Contouring

The program which the E.C.U. has applied to geologic, geophysical and oceanographic data can accept up to 1500 irregularly distributed data points. These would normally be digitized on the D-Mac, but points digitized manually could be input when converted into the appropriate format. Height values at points on a square grid (90 × 90 or smaller) are interpolated by applying a weighting function to the known values at the nearest four to ten data points. Data points within a radius likely to include seven points on average are examined. If there are more than ten such points, the nearest ten are used; if there are less than four, the radius is repeatedly increased by $\sqrt{2}$ until four points are encountered. These points are then averaged after being weighted by the function $(1 - R^2)/R^2$, where R = (distance from the grid node in question to the data point)/(radius of area examined). R varies between 0 and 1. The use of this weighting function produces a continuous surface without necessitating examination of all data points for each grid point.

In a second phase, the resulting square figure field is scanned once for each altitude for which a contour is to be drawn. When two adjacent grid values bracket the desired altitude, the position of the contour between the two points is estimated by linear interpolation, and stored. When the scan is complete, these interpolated contour positions are arranged into the drawing sequence, and joined by smoothly curving cubic functions in x and y which are plotted by the geagraph as a series of short straight chords.

The results are very good for a smooth surface (Fig. 3) but can be poor for a rough surface especially if the data points are distributed irregularly. In the latter case, the surface will fit extreme points only if a fine grid mesh is used. The spacing for grid points should be comparable to the minimum spacing of data points. Further variants of the program, introducing either a shadowing effect or a moving local quadratic trend surface, are being tested. The program is currently being transferred to a larger computer with much faster computation, and it will then be able to handle more points and work through larger grids.

(v) Length and area measurement

A simple program to calculate the lengths of lines and the numbers of points on them has provided very useful measurements of the spacing of points in large sets of digitized data. The lengths of lines with different feature codes are cumulated separately. In this way the amount of detail on Ordnance Survey large-scale plans is being sampled, and it should soon be possible to estimate the total amount of magnetic tape required for a data bank of this information.

The program for measuring areas performs a Simpson's Rule integration for unequal intervals. It is not necessary to use the digitizing table simply as a planimeter. If several lines bounding an area are digitized separately in the normal way, the areas between each line and the x-axis can be calculated by program, and totalled outside the computer. Due regard must then be paid to sign; area is given as positive for a line digitized in the direction of increasing x, negative for decreasing x. We hope to be able eventually to tie such segments together automatically by a program which would compare the coordinates of termini.

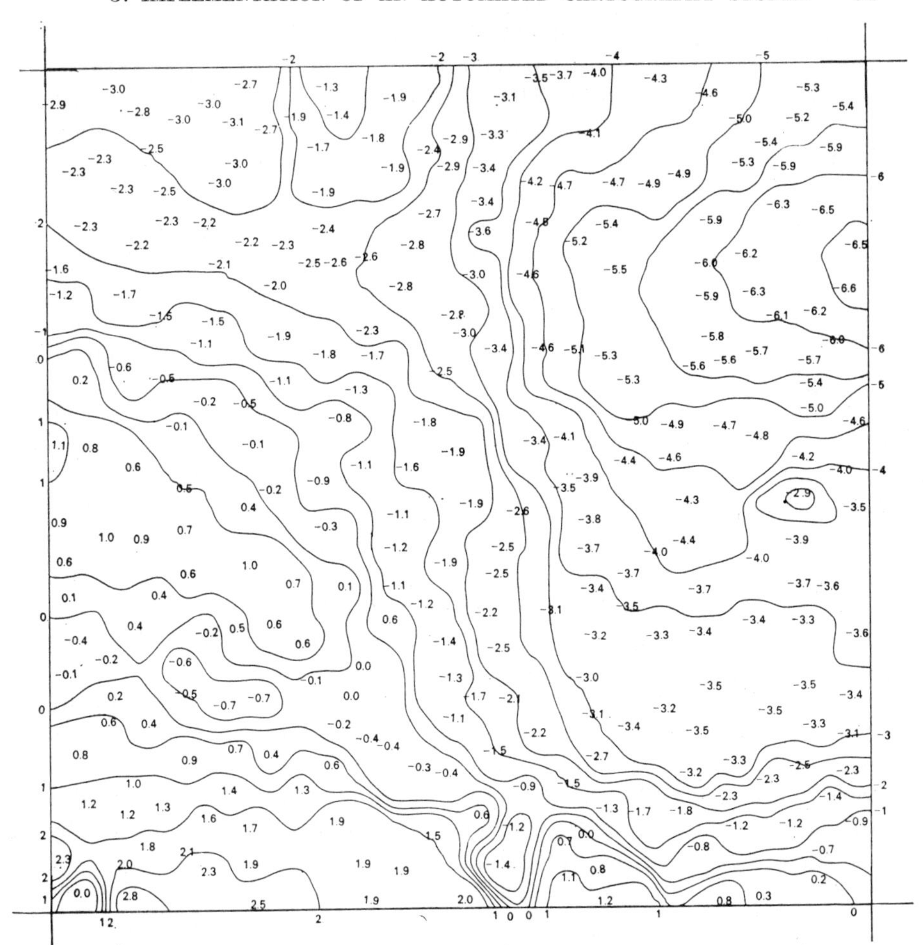

Fig. 3. An example of computer contouring applied to scattered point data. Bouguer Gravity Anomaly has been contoured automatically from data at 287 fairly evenly distributed points, with a program by D. Connelly. On the PDP-9, it took 30 min to interpolate values on a square grid, 45 min to calculate the positions of contours and 20 min to scribe the contours on the geagraph. Data point values, in milligals, were added manually; the decimal point marks the position of the data point. Contour interval 0·5 milligals, scale 1/126 720.

CARTOGRAPHIC PROJECTS

In the experimental work of the E.C.U., it is our policy to work on real problems—potential practical applications—rather than hypothetical data. Accordingly, the Unit collaborates with a number of different government agencies concerned with mapping, especially those concerned with the natural environment. Our major project to date has been the processing of bathymetric data supplied by the National Institute of Oceanography. The Gulf of Aden 1/2 000 000 sheet with metric isobaths, published in October 1968 by the Royal Geographical Society, demonstrated the ability of the D-Mac/geagraph system to produce linework at least equal in quality to good manual scribing. This and four adjacent sheets form the U.K. contribution to an international atlas of the Indian Ocean, now being compiled in Moscow. More recently, a large anaglyph of the Gulf of Aden, and a layer-coloured map, both with fathom contours, have been produced (Laughton *et al.*, 1969).

The digitizing of a 1/10 560 geologic field sheet for the Institute of Geological Sciences produced promising results (Bickmore, 1969), including a colour map at 1/25 000. For this area, plotting at 1/63 360 was satisfactory without generalization, when formation boundary lines were drawn 0·1 mm thick. The next stage is to produce a whole 1/63 360 sheet. Point data for gravity, and for depth of the upper surface of the chalk, have been contoured. Aeromagnetic data have been converted from British National Projection to a Universal Transverse Mercator (3°E) projection, and to the Lambert conformal conic projection used in France. There is a large potential demand for such work, so that international maps may be produced.

Another important sphere of collaboration is with the Nature Conservancy, especially for the production of vegetation and ecological evaluation maps. The Soil Survey has appointed Dr R. Webster to work with the E.C.U. and study the implications of cartographic automation and computing. Conventional soil maps can be handled by the same techniques as geologic maps, and the E.C.U. has drawn the western half of the Reading 1/63 360 map after digitizing the 1/10 560 field sheets. We plan, however, to go further and process field measurements located on a square grid, to produce various types of soil maps, general, special and applied.

A report on the implications of automated cartography and cartographic data banks for planning has been prepared for the Department of Economic Affairs. A major phase of this project was the preparation of a catalogue of the availability of located information, including its form and geographic resolution. Various maps dealing with point and area data in Norwich demonstrate the feasibility and flexibility of an automated system.

Work with the Ordnance Survey has provided us with valuable checks on the accuracy of our system. Comparison of various scales of digitizing suggests that it is best to digitize small areas at the largest available scale, but with care taken over the matching of lines at sheet edges.

The next stage of development is to bring together information from all these subjects, plus hydrology, geochemistry and geomorphology, for a single area. This will provide an embryonic cartographic data bank of environmental information, located with various degrees of precision, for the testing of data banking routines and the interrelation of different sets. When it is operational, statistical and analytical facilities for handling cartographic data will be built up, permitting for example the filtering and correlation of distributions. It was recently decided that the E.C.U. will develop such a data bank for the Bristol Channel area.

OUTSTANDING PROBLEMS

Several problems touched on above need to be tackled before a fully operational, economic system for general cartography can be established. One is the reliability of equipment; the number of difficulties suffered by the Unit could not be tolerated in a routine production environment. Equipment is, however, being improved, and the most recent generation of computers is more reliable.

The E.C.U. is working on the improvement of its checking and correction system, and now possesses a cathode-ray storage tube which produces a considerable increase in speed. We will investigate whether an editing system based on coordinates is adequate, or whether it is necessary to label each line segment. It will be useful to correlate lines at sheet edges to ensure the production of continuous lines when sheets are plotted together. This could be done by sorting into order the lines crossing each edge, and then correlating them with the corresponding list for the adjacent sheet; coordinates of points near the edge would be modified so that the lines each side joined exactly and there were no sudden changes of direction.

The problem of generalization is more difficult than implied above, since if adjacent lines are generalized independently their qualitative, topological relationship may change, that is the generalized lines may cross where they should not, or cease to cross where they should. This problem is particularly applicable to contours at greatly reduced scales. Fortunately, alternative methods of contour generalization are available if height data can be transformed to square figure fields of point values, so that new contours may be derived automatically. Yoëli (1968) has demonstrated the effect of contouring from thinned subsets of a figure field, while Tobler (1966) has shown how figure fields can be smoothed, producing contours suitable for smaller scales, without loss of data; it is even possible to retain maxima and minima by normalizing a smoothed map. On the other hand, displacement of cluttered detail at small scales would be very difficult to perform automatically, except perhaps by cathode-ray tube, and the answer to this problem may lie largely in symbolization and design, rather than computing. Generalization of built-up areas is a particularly difficult problem.

Little has been said, above, of the cost of these maps. To give reliable estimates, it would be necessary to repeat an operation several times; but the maps discussed above are the first of their kind. Digitization is an important cost, but it is several times as fast as manual scribing. Editing is as yet expensive in computer time. Hence the production of a single map is usually more expensive than by conventional methods. However, once a corrected tape has been produced, it is available for the production of further maps, differing in scale, projection, or content, or transformations such as anaglyphs. Such maps are cheap to produce even on our present plotter, and already accurate plotters over five times as fast are being advertised. It is likely that the cost of machine operations will continue to fall, while the cost of labour rises: we can at least say that less labour is required for automated cartography than for current methods, even for the first map from a given set of data. Where survey data are received in machine-readable form, an automated system is almost certain to be more economic. Other benefits of automation are speed, and the availability of cartographic data in digital form for statistical manipulation.

ACKNOWLEDGMENTS

The results described here have been achieved by the whole E.C.U. team. I would particularly like to thank A. Kemp for work on the illustrations, and J. M. Adams, D. P. Bickmore, D. Connelly, T. Lang, D. Rhind and W. Trevett for commenting on parts of this paper; the opinions expressed are my own.

REFERENCES

Adams, J. M. (1969). Mapping with a third dimension. *Geogr. Mag.* **42**, 45–49.

Bickmore, D. P. (1968a). Cartographic data banks. *Proc. 12th Conference, Fédération Internationale des Géodésistes*, **509**, 1–5.

Bickmore, D. P. (1968b). Maps for the computer age. *Geogr. Mag.* **41**, 221–227.

Bickmore, D. P. (1969). Computers and Geology. *Geogr. Mag.* **42**, 43–44.

Lang, T. (1969). Rules for Robot Draughtsmen. *Geogr. Mag.* **42**, 50–51.

Laughton, A. S., Whitmarsh, R. B. and Jones, M. T. (1969). The evolution of the Gulf of Aden. Red Sea Discussion. *Phil. Trans. R. Soc. A.*

Roberts, D. G. and Evans, I. S. (1969). The production of bathymetric and other oceanographic charts by automatic methods. *Proceedings of the Oceanology International '69 Conference* (Brighton, February 1969), Vol. 2.

Tobler, W. R. (1966). Numerical map generalization; *Michigan Inter-University Community of Mathematical Geographers, Discussion Paper* 8, 26 p.

Yoëli, P. (1967). The mechanization of analytical hill shading. *Cartographic J.* **4**, 82–88.

Yoëli, P. (1968). Reliefdarstellung durch Höhenkurven mit Rechenautomaten und Kurvenzeichnern und deren Genauigkeit (Relief representation by contours with computers and plotters and its accuracy). *Z. Vermess-Wes., Stuttg.* **93**.

4. Recent Advances in Source Data Automation

NICHOLAS J. SUSZYNSKI*

Information Systems Division, Smithsonian Institution, Washington D.C., U.S.A.

ABSTRACT

Much has been written concerning the role of computation and data processing in the operations of our technical society. All fields—from atomic energy to zoology, and now museology—have been targets for computer applications, but to what extent are the end users in these areas actually aware of the evolution, characteristics, and capabilities of these applications? In an effort to explain some of these aspects, this paper will review the history of source data automation, discuss some of the more recent advances, and demonstrate several of these applications as they are implemented at the Smithsonian Institution. Finally, one of the recently announced data entry terminals and its capabilities are discussed in depth to provide an overview of what a new family of source data automation equipment offers today.

KEYBOARD TO PUNCHED CARD EQUIPMENT

For the past 50 years, the 80-column punched card has provided the standard means of information transfer in data processing systems. Punched cards have been used, not only to enter data into the file, but also to maintain the data in the file (Bourne, 1963). The redundant handling required to process the punched cards introduces continuous opportunities for errors and is unnecessarily time-consuming.

Since all of the data are contained in the punched cards, to rearrange the information, it is necessary to rearrange the cards by physically sorting and collating them into various sequences, grouping and regrouping them until the desired sequences and subsequences are established. In the process, all of these cards are handled by the operators

Systematics Association Special Volume No. 3. "Data Processing in Biology and Geology", edited by J. L. Cutbill, 1970, pp. 57–68.

* Now Assistant Director of Research, The Federal Deposit Insurance Corporation, Washington, D.C.

of the tabulating equipment, introducing opportunities to misarrange the decks of cards and thus affect the validity of the final output. In addition, translation of the data from the basic documents into punched cards almost always involves at least one middle man, and often two or more. Normally, before the data can be captured for data processing, a hard-copy document, usually prepared on a typewriter, is required. This document as a rule is designed to transact business and not to prepare the data for keypunching; hence, to increase the reliability of the data a clerk recopies parts of the original document onto a collection form which, when completed, is transmitted to the centralized keypunching section. If the precision of the data content is important, it is first keypunched and then verified by another operator to insure that all information has been correctly entered onto the cards.

Thus far three redundant operations have been required subsequent to the typing of the initial document. Finally, the cards are often visually verified by the unit responsible for the initial transcription of the document. The costs in terms of money spent to perform these redundant operations and in terms of time wasted on the part of the professional staff can be very substantial in such procedures; yet, until the advent of computers some 25 years ago, most of the procedures associated with the data entry into the processing system paralleled the procedures described above. What is unfortunate is that in many installations such procedures are still followed today.

Why not have the first individual responsible for the hard-copy preparation, record and capture the data in machine readable form? If this could be done, annoying time delays and the money needed to transcribe, keypunch, and verify the data contained in the original document could be greatly reduced. An early solution to this problem was provided by the advent of keyboard-to-paper tape equipment.

KEYBOARD TO PAPER TAPE EQUIPMENT

Keyboard-to-paper tape equipment generates perforated paper tape as a by-product of a hard-copy preparation. These devices in main combine the functions of a keypunch with a typewriter. The depression of a key on a typewriter-like keyboard prints the character and punches a combination of round holes across the paper tape, each sequence of perforated holes representing a particular keyboard character.

Paper tape offers several features not present in cards. Probably the most important advantage is that for practical purposes, paper tape provides a record length of any size. The other major benefit of paper tape equipment is that the same length of tape which was just punched can be read back into the device to generate additional copies ("original" copies of the document just typed), except that the equipment when driven by the tape, types at the rate of 10 characters per second, an equivalent of 120 words per minute. This is two to three times faster than an average typist. In addition, for practical purposes automatic typing introduces no errors or erasures. Form letters of high quality are produced in this manner. For example, the U.S. Senate is filled with Friden and Dura equipment on which typists insert the name and address of the constituent and the equipment, driven by punched paper tape, types the rest of the letter at 120 words per minute.

The Smithsonian Institution also makes wide use of paper tape equipment in its bureaus and offices (Squires, 1968). For example, the National Portrait Gallery uses Dura equipment to process data associated with the Catalogue of American Portraits, an application that we are installing on the Honeywell H-1250 computer. The Smithsonian Oceanographic Sorting Center employs Smith-Corona Marchand equipment to prepare data for inventory control purposes and to maintain control over the distribution of specimen to cooperating institutions. The U.S. National Museum of Natural History relies on both Friden and SCM equipment to produce a computerized data bank for the information retrieval of geological and biological data associated with the Smithsonian Institution's Information Retrieval System (Creighton, 1969). The Smithsonian Office of the Treasurer uses Friden equipment to gather fiscal data for the computerized accounting system and the Smithsonian Archives experiments with Friden equipment to develop indexes to its files.

The use of source data automation equipment is widespread throughout business and educational communities. Many members of the academic community are familiar with the work of Dr James H. Soper of Canada who wrote on the subject as far back as 1964 (Soper and Perring, 1967). When source data automation principles are applied, the computer readable data are prepared by the same person who creates the hard-copy. Such a person is infinitely more qualified to judge the contents of the data than a keypunch operator located in a remote facility to whom the data on the document represents little more than a stream of alphanumeric characters.

Source data automation offers decentralization of data capture and guarantees reduced error rates by eliminating the intermediate clerks and keypunch operators, and by assuring that the computer readable material is prepared by a person better qualified to make decisions about the accuracy of the data. Add to these advantages, the time and money saved when the redundant re-transcription, keypunching, and key-verifying are removed from the data preparation cycle and one can readily see why source data automation is becoming so widespread.

KEYBOARD TO COMPUTER EQUIPMENT

Recent advances in computer technology, coupled with more reliable and less expensive telephone switching systems bring direct data entry a step closer to the computer by making it possible to use remote terminals that are both operationally and economically reasonable. Remote terminals were initially developed for military applications, rapidly followed by airline reservation systems, stock quotation systems, bank and brokerage account status information, credit and real estate information, manufacturing and merchandising inventories, and numerous other applications where speed of response is of the essence.

Initial market penetration by the terminal devices was extremely slow. This slowness can be attributed in the early stages to the high cost of equipment, and later to the fact that no general purpose software was available to link the terminal device to the computer. The high costs of these devices are diminishing very rapidly. Four years ago it was difficult to lease a quality terminal employing a cathode-ray tube (CRT) for less than $250·00 per month. Today the Viatron Company advertises such a terminal for $39·00 per month.

The other problem, that of the software required to support a terminal in communicating with a data base, remains rather spotty. There are computer systems, especially those with "time-sharing" features, which provide remote computational capability with relative ease. These time-sharing systems, however, are limited to very large computers, to specialized computers, or to specific applications, mathematical computation for example. Most of today's computer installations, at least those in the low- to medium-price range, do not have a time-sharing capability. This in itself need not be an insurmountable problem if a definite need exists for the submission of data from remote locations since most of today's computers have the capability for data transmission, if not for time-sharing itself.

In the context of this paper, a time-sharing environment suggests a simultaneous ability for more than one terminal user to interact with the computer from a remote location in such a way that, while connected with the computer, each user has an illusion of an exclusive dialogue with that computer. The "on-line" system suggests that the facility exists to enter and transmit data from point A to point B, but no conversation is implied. In either case, if one selects his remote terminal judiciously, he may acquire a device possessing the functions of a typewriter, a keypunch, a teletype, and in addition, some of these devices may possess a certain amount of programmable logic which can be developed to respond to individual needs.

This facility for interacting with the preprogrammed logic, stored within the confines of a remote computer, and in some cases, within the terminal device itself—such as Burroughs L200 or TC500—offers a great potential for an error-free capture and manipulation of data. Some of these terminal devices can be operated only when interconnected with a computer, others can be operated "on-line" as well as "off-line", or as it is sometimes called, in "stand-alone" mode with the data captured on it, batched, and submitted at a later time to the computer. In some instances, the data can be recorded using a cassette or a reel of magnetic tape, in others it is done on the surface of a magnetic disc, and still in other cases, in the form of punched paper tape.

All of these devices are able to accept data by means of a typewriter-like keyboard. All of those that are connected to a computer can recall and display remotely stored data on a printed page or on the cathode-ray tube. In addition, terminals that have access to a computer can be used either in a conversational mode or they at least can transmit data to and from the computer. It should be stressed that these various capabilities exist only to the extent that they have been programmed and developed for a particular application on a particular computer. This capability then is not so much a function of a terminal device, but rather of the executive operating system associated with the computer.

For today, at least, uses of a terminal device in conjunction with a time-shared computer, cannot be justified across the board as a substitute for a keypunch, although such applications are coming to light. One of the more notable of these is currently being conducted by the Internal Revenue Service which has been experimenting with the equipment of various manufacturers. In one of the Internal Revenue Service's experiments, a dedicated, small computer with many terminals is used for one application only—the conversion and initial edit of individual income tax returns into a computer readable media.

Successful applications of entering data directly into a computer have shown that, if the initial data capture, edit, update, and subsequent data purification are made a part of the source data automation procedure, economies over keypunching can be attained and substantial savings in time can be effected to say nothing of greater purity of the data itself. These savings are even greater when dealing with textual expressions rather than columnar data, the former being more often encountered in museums. Since such direct-entry, conversational data systems are based on a human-to-computer interaction, a speed of 10 to 20 characters per second is quite satisfactory for data transmission, and therefore, normal telephone lines make a very serviceable communication link between the terminal and the computer (McPherson, 1967).

One system developed at the Smithsonian Institution for the "Center for the Study of Man" will result in the alphabetized directory of some 4000 anthropologists associated with the Center. What is also important is that this data will remain in a computer readable media, and in the future, it will be a relatively simple and inexpensive matter to produce up-to-date directories just by entering the changes.

The input device used in the Current Anthropology project is an IBM 2741 terminal, leased from the vendor of the text-editing system for $150·00 per month. Two-thirds of this price represents the fee for the use of the text-editing software. This particular terminal is a modified selectric typewriter containing its own circuitry which provides teletype capabilities and also offers the user diacritic marks and upper- and lower-case letters. To reduce costs, instead of making special interconnection with a dedicated telephone line, an acoustical coupler is used to connect an ordinary telephone set to the IBM 2741 data terminal. The computer is reached by dialing the appropriate telephone number. After the passwords are exchanged, the operator of the terminal device is able to access and react with a computer program which will aid him in the establishment of a data base within the disc file attached to the computer. Text-editing programs of this type provide the user not only with a formatted data base, but also offer various other aids such as hyphenation and pagination of the text, and the ability to rearrange the data in various sequences for subsequent printing.

In this particular case, a newly graduated anthropologist using a survey questionnaire operates one terminal device while the second one is operated by a clerk typist. The objective of this activity is to produce completely formatted pages for publication, containing three columns of justified text on each page. The final product looks very much like a newspaper page, with the names of the anthropologists, their addresses, affiliation, interests, and special areas of competence indicated on each entry. A number of indexes will also be produced to provide cross-references to the publication. This latter portion, of course, is not a function of source data automation, but of subsequent sorting and manipulating of data captured on the computer.

Past experience indicates that an individual, knowledgeable of the data content, can process four to six questionnaires of this type in an hour. In the Current Anthropology Project, 35 to 75 words of information are abstracted from each questionnaire. Since the hourly charge for the use of this system is approximately $5·00 an hour, the cost per questionnaire processed is roughly $1·00. This cost can be misleading if compared only to the cost of keypunching, which obviously would have been lower. Nevertheless, when a comparison of costs is made, it should include not only those costs connected with the keypunching of data, but also the cost of preparing edit and audit programs, rental costs of computer time to process the data through these programs, and the costs associated with the establishment and maintenance of the master file. Any subsequent costs of sorting and printing of such a file are not included since these would be incurred no matter how the data were prepared.

Another recently announced terminal with a great deal of promise is the Viatron System 21 with which its manufacturer hopes to capture a large portion of the data entry market.

Viatron introduced several innovations into the manufacturing of terminal devices. One of the more important ones is the use of a new technology known as Large Scale Integration (LSI), using the Metal Oxide Semiconductors (MOS), which offers great reliability and cost reduction possibilities. To a large extent, it is this reliable MOS/LSI technology which packages Viatron's microprocessor, cathode-ray tube, keyboard, and two tape cassettes into a desk top console. This technology greatly increases reliability. In the Viatron 21, it eliminated some 40 000 soldered connections, and substantially reduced the total number of circuit boards required for a given device. If the conventional circuitry were used in the Viatron, the microprocessor would have contained over 750 separate circuit boards, but when using MOS/LSI techniques, only 25 are required. This technology, initially developed for the space effort, is beginning to make definite inroads in the area of computers and computer terminals.

Viatron's System 21 is organized around its microprocessor. Each microprocessor is in fact a "hard-wired" program. At present Viatron offers only one set of hard-wired program steps which include all of the functions of a keypunch, a verifier, and a teletype.

The most basic recording media offered with this terminal device is a cassette of magnetic recording tape much like those available with portable tape recorders. Such a cassette can contain 50 000 characters or 625 records of 80 characters, an equivalent in storage of 625 punched cards.

One of the shortcomings of a keypunch device is that the operator cannot immediately see the results of her keystrokes as she would see them on the typewriter. A device equipped with a cathode-ray tube overcomes this shortcoming by displaying each character as it is accumulated in the buffer, before being transferred to the tape in the cassette. Since the data rest in temporary storage before they are transferred to tape, the same operator can visually examine the cathode-ray tube display and change any incorrect information before it is written out as an 80 character record on the magnetic tape.

The keyboard has the outward appearance of an IBM selectric

typewriter keyboard. In addition, under the control of a microprogram, this keyboard can be used either as a standard typewriter keyboard or as a keyboard of the IBM 029 keypunch with the numeric digits condensed into a 3×4 matrix analogous to the keyboard of an adding machine.

To make the data entry almost foolproof, the cathode-ray tube is divided into four sections. Each section can display up to 80 alphanumeric characters in four lines of 20 characters each. The top section is called the Read Record and functionally corresponds to the card that has just been punched and is ready to be scanned by the read brushes of a conventional keypunch.

The second section of the video screen is called the Write Record. Functionally, it corresponds to the card in the process of being punched. While the operator depresses keys, typing the data into the Write Record, she can also observe the previously punched 80 characters located in the Read Record area and any subsequent characters which she has typed into the Write Record. Accordingly, at any point in time the operator can see at least 80 characters of information and possibly up to 160 characters, depending on the amount of data typed into the Write Record.

The third successive section of the video display contains the Master Record, which corresponds to an optional feature available on keypunching equipment. The Master Record provides the operator with an alternative to the Read Record from which fixed information can be duplicated into the Write Record.

Viatron uses the Master Record for still another function. By typing in the titles of successive data elements as they should appear in the Write Record into the Master Record, the typist is provided with a visual aid to further assure error-free data entry.

The fourth and last section of the video screen displays the Control Record. This record corresponds to the control program of a conventional keypunch and may perform such functions as:

selective duplication of fixed information from either the Read Record, or the Master Record, into the Write Record;

the determination of which of the 80 columns of the Write Record are to be skipped over or zero filled;

the determination of whether a given field of data is written in numeric or alphabetic mode, etc.

All together, there are eight such functions corresponding to a keypunch which, when typed into the Control Record, automatically provide control functions over the data entry, thereby further assuring correct completion of each new Write Record.

In most applications it is necessary for the operator to see only two of the four sections on the video screen just described. These two sections are: the Write Record and the Master Record. The Control Record, once established, provides really no information in processing successive records, nor for that matter does the Read Record need to be displayed continuously.

To minimize potential errors in data entry, it is possible to focus the operator's attention on that portion of information that varies from record to record. To accomplish this, the 80 characters of the Write Record are interleaved with 80 characters of the Master Record, in eight alternating lines of 20 characters each. In such an operation, the Master Record contains the instructions for completing the Write Record and the Write Record contains the blank spaces to be filled out.

To convert data from a cassette of magnetic tape to a computer readable media, Viatron provides three choices:

An on-line cassette to punched card converter which incorporates a modified IBM 029 keypunch and converts the data from a cassette into punched cards at the rate of 48 000 strokes per hour.

A cassette to computer compatible magnetic tape drive which converts data at the rate of one cassette every 9·6 min.

Transmission of data contained in the cassette directly over the telephone line into the computer. By adding a communications adapter, Viatron becomes a remote terminal device capable of interacting with a computer in a time-sharing environment.

One remote terminal can transmit data over the telephone network and store it in a cassette of another terminal device located at the collecting station (the computer room perhaps), and does not require computer time to collect data from the transmission. Once the data are transmitted they can be converted into punched cards or into a computer readable spool of magnetic tape using one of the previously described procedures.

It can be argued with success that it is cheaper to transmit from remote terminal to remote terminal and then convert the cartridge into a computer readable media, rather than to go directly into the computer in a time-sharing environment. Nevertheless, it is pointless to press these arguments in a vacuum. The more complex the decisions that must be made and the higher paid the operator, the more likely it is that an on-line diagnostic editing and auditing of data will be desirable; conversely, simple and straightforward data can be captured with less costly off-line methods.

For those who do not have access to the computer, or whose limited amount of data processing does not require arithmetic operations,

Viatron offers a printing robot. This device has an array of 50 solenoids which, once clamped on an IBM selectric, fit over the keyboard and actuate the typewriter. The terminal device takes control and transmits the data to the robot at the rate of 12 characters per second.

This detailed discussion of Viatron was introduced to provide a clear picture of what a new family of source data automation equipment offers today. Although Viatron was selected as an example, it should be noted that this is not the only equipment capable of providing the functions described above. There are numerous manufacturers with other variations of the above mentioned capabilities. Computer Access Systems, IBM, Honeywell, Sanders Associates, Mohawk, Sycore, and many others provide equipment that not only captures data, but also allows the user to edit, re-format, and search for a given record on the output media.

CONCLUSIONS

It would be a simple matter to conclude by suggesting that all source data automation should be performed on-line to the computer, preferably under the control of an interactive dialogue. Assuming a good supervisory operating system exists, one would need only to write compact programs for selective editing, rearranging, updating and, in general, maintaining a file current and in the desired order. Although this cannot be done economically in many cases, it would be beneficial for the museum community to experiment with terminal devices on a small scale to gain detailed knowledge of what they can thereby accomplish. By so doing, museums can gain invaluable experience that will provide them with the necessary information to better assess the potentials of computers and electronic data processing.

This experimentation, more than anything else, will determine the real demand for on-line systems, and will also provide vendors of these systems with the information they need to respond to specific requirements.

Computers and computer technology, as all new work-saving devices, will be evaluated in most cases on the basis of their utility and their economic value. It is obvious that if an assignment can be accomplished in a week by a steno-typist for x dollars while the same work can also be accomplished in several days on a computer for twice that amount of money, there probably would be few interested in the second approach. On the other hand, if the computer can perform this task in a day with some increase in costs and the basic data are also retained in machine readable form for subsequent processing, some may decide that the rapid turn-around plus the retention of data for the future is worth the extra expense. There is much to be said for the convenience of having immediate access to the computer, for having data instantaneously returned to the user upon his demand, and for the retention of this data within the computer for future needs.

It has been said that modern man is convenience-oriented and that which at one point in time is a luxury item, often at a later time becomes a necessity. An often quoted example of this points out that if the production of an automatic gear box for the automobile depended on a cost-effective analysis, it would never have materialized; however, since some buyers were willing to pay a substantial premium for it and as the manufacturing know-how was perfected and the economies of mass production began to pay off, the cost of the luxury item began to approach the cost of the standard item. This type of progress most certainly will occur in the on-line access to the computer.

A survey conducted by the Diebold Group, in the winter of 1968 representing some 1700 users, indicated that improved customer service and rapid access to data were the most important factors in the acquisition of terminal devices (Diebold Group, 1968). It is interesting to note that providing computer capabilities to remote users was considered fifth in importance on this survey, indicating that the users generally had to have a better reason to go to on-line equipment than the mere fact that they were remote to the computer. In other words, the decision to acquire terminal equipment is heavily affected by economic factors and therefore to justify a terminal, some sort of a cost effective analysis must be performed.

Predicting what the future will bring is always a difficult task; however, industry sources indicate that by 1975, 60% of all computers will have teleprocessing capabilities and the rental revenue from these terminal devices will reach $300 million. This growth probably is grossly understated since the advances in software development and in manufacturing technology are moving at a very rapid rate.

If companies such as Viatron now can provide an advanced terminal device for $39·00 a month, in two years someone else may offer it for $25·00. At that time, people may begin to use them in their homes for local information such as special advertising of department store sales, theatre programs, civic affairs, etc. Obviously then, software systems will have to be developed with enough flexibility to respond to all of these individual needs. Thus, it follows that the greatest unfulfilled need existing today—that of a meaningful interaction between the user in quest of a solution, and his computer—will have to be met. In the meantime, there is no reason not to take full advantage of what today's technology offers, especially since this technology is moving in "seven-league boots".

As an example of this rapid development, in late August IBM announced its newest computer 360/Mod 195 which uses MOS technology very extensively (EDP Daily, 1969). In this computer, each memory chip, $\frac{1}{8}$ in square, contains up to 664 discrete components (such as transistors, resistors, etc.), and stores 64 bits of data in binary form. Two such chips are mounted on a $\frac{1}{2}$ in square ceramic substrate which is used as a basic building block for the buffer memory. This extensive use of monolithic circuits brought the basic machine cycle down to 54 ns. A staggering speed when one realizes that there are as many nanoseconds in one second as there are seconds in 30 years.

All of those who have spent their professional life using computers know that our appetites will never be fully satiated with what is even now being projected for the future. Each year that passes offers continuously expanding possibilities and promises of greater developments. That is why it is so important for museologists to take advantage of these current developments, and to grow with them thus insuring intelligent utilization of what the future brings forth, both for themselves and for those serviced by the museum itself.

REFERENCES

Bourne, C. P. (1963). "Methods of Information Handling". John Wiley, New York.

Creighton, R. A. (1969). The Smithsonian Institution's Information Retrieval System (SIIRS) for Biological and Petrological Data. Sixth Annual National Information Retrieval Colloquium.

Diebold Group. (1968). Requirements for terminal devices among computer users.

EDP Daily. (1969). August 21. Vol. 1, No. 32, p. 238. EDP News Services, Washington, D.C.

McPherson, J. C. (1967). Data communication requirements of computer systems. *I.E.E.E. Spectrum* **4** (12) 42–45.

Soper, J. H. and Perring, F. H. (1967). Data processing in the herbarium and museum. *Taxon* **16**.

Squires, D. F. (1968). Collections and the computer. *Bio. Science* **18**, No. 10.

Viatron Computer Systems Corporation. (1969). Viatron System 21 System Description.

5. Optical Processing of Microporous Fabrics

JOHN C. DAVIS

Kansas Geological Survey, University of Kansas, Kansas, U.S.A.

ABSTRACT

Characterization of a system consisting of solid elements and attendant pore structure is difficult using conventional techniques. Vectors describing size, shape, and orientation of each element must be determined. The problem can be simplified by considering only element projections on a plane through the solid and then directly measuring intercepts along parallel traverses. Unfortunately, most relevant information is lost in this approach.

All information contained within a plane through the porous system can be retained and treated by optical processing techniques. Using a laser optical bench, the Fourier transform of the image of the porous system can be created at the focal point of a simple lens. In the transform, elements of the image are resolved into their spatial and angular components; relative frequency is expressed as light intensity. With proper photographic techniques, the transform can be recorded and digitized. The result is equivalent to a two-dimensional histogram of all possible traverses at all possible angles across the image. Specific components of the transform may be isolated using appropriate pass-filters.

The technique is being evaluated as a method of analyzing the pore structure of petroleum reservoir rocks. Studies presently are being conducted to correlate engineering properties with details of the reservoir fabric.

INTRODUCTION

The data that scientists receive come from many sources. Some are generated in the form of quantitative measurements created by complex machines or scientific instruments. Other information is perceived directly through the human senses of touch, taste, and smell. A geologist, however, receives most of his information as visual impressions

Systematics Association special Volume No. 3. "Data Processing in Biology and Geology", edited by J. L. Cutbill, 1970, pp. 69–87.

through his sense of sight. The human eye is an optical processor which projects a two-dimensional real image of the world onto the retina. The retina is composed of cells which convert light energy into electrical energy which is then transmitted along the nerve system to the brain. The human eye is sensitive to both wavelength and intensity of light and because of the retina structure forms are perceived. The three-dimensional nature of the real world is ascertained by combining two two-dimensional images, one being relayed to the brain by each eye.

If visual impressions are the most important source of information to geologists, it seems likely that pictorial representations must be the most efficient manner of storing and transmitting geologic information. Although maps and diagrams are important for conveying conclusions or interpretations, most raw data are transmitted with photographs.

PHOTOGRAPHS AS DATA ARRAYS

A photograph can be considered simply as an information storage device, a photochemical "black box". However, some knowledge of the physical characteristics of photographic materials is necessary because these influence optical processing capabilities and limitations. At its simplest, photographic film is a sandwich consisting of a coating containing photosensitive chemicals, a sheet of transparent supporting material, and a layer of dye or opaquing (Fig. 1). The opaque material is an antihalation coating and prevents stray light from entering the film. The supporting material may be glass or, more commonly, acetate or plastic film; its purpose is to provide a physical support for the

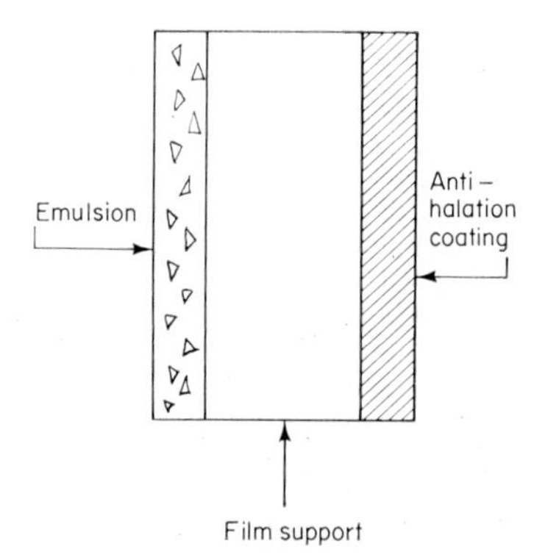

Fig. 1. Diagrammatic cross-section through typical photographic film or plate. Films usually are 0·1–0·2 mm thick; glass plates are about 1 mm.

emulsion. Emulsion is composed of fine grains of silver halides dispersed in hardened gelatin. Silver halide compounds are unstable and light quanta striking the grains cause release of silver ions. In the photographic development process, developing agents diffuse through the gelatin of the emulsion and convert silver ions to metallic silver. During fixing, the remaining unexposed silver halides are dissolved by the fixing agent and removed. The final image is formed of small grains of metallic silver dispersed through the emulsion. Variation in the number of silver grains per unit area of negative determines the optical density and transmission characteristics of the image.

Usefulness of a photographic device as a memory unit depends upon the nature of its response to an input stimulus and the density to which information can be packed into it. Darkening of a photographic emulsion is dependent on the number of quanta of light which fall on the surface. This is in turn a function of light intensity and length of time the photographic emulsion is exposed. Photographic material ideally should respond in a linear fashion to energy of the incident light and should be capable of recording a wide range of light intensities. Unfortunately, emulsions are strongly nonlinear in normal photographic situations (Todd and Zakia, 1969, p. 57). This imposes some serious limitations on use of optical processors.

Efficiency of photographic films as data storage devices is impressive from the viewpoint of information density. With data encoded in a binary state (i.e. opaque or transparent), conventional photographic emulsions can easily record 100 lines in a millimeter, a density of approximately 10 microns between successive bits (Reich and Dorion, 1968, p. 577). At this packing a square image on 35 mm film contains almost six million bits, roughly the equivalent of a 320 K digital storage device. Resolving powers of 100 lines per millimeter are classified as "high"; however, special films may exceed this resolution many times. Certain spectrographic films, for example, have resolving powers of 2000 lines per millimeter. It is theoretically possible, using these films, to pack two billion bits into a single frame of 35 mm film. Unfortunately, achieving this packing density and being able to effectively recover the recorded information is dependent not only on film properties but also on the optical system. Although elaborate optical devices are necessary to achieve maximum resolution, relatively unsophisticated equipment can record photographic images or information arrays at extremely high densities if compared to conventional digital storage devices. The image in Fig. 2 consists of the pattern of pores and grains in a thin section of an oil-reservoir rock. This image has been digitized using a flying spot scanner having an equivalent optical resolution of 40 lines per millimeter. The digitized record consists only

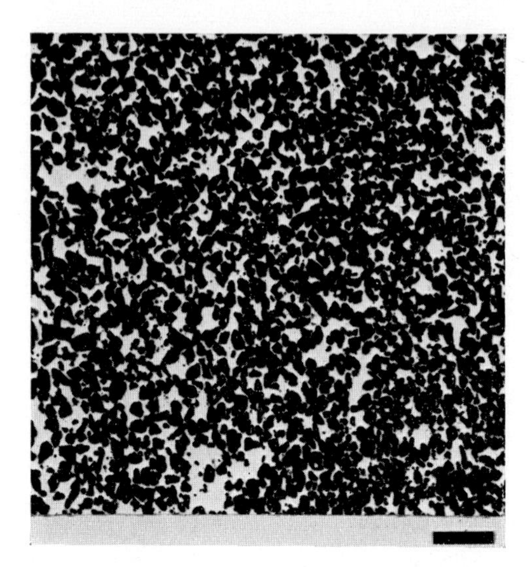

Fig. 2. High contrast image of porous sandstone. Sand grains are black, pores are white. Bar represents 1·0 mm.

of a series of binary "ones" or "zeros", denoting the existence of clear or opaque areas at successive positions across the film. The film was scanned to produce 1024 lines of 1024 data points, arranged as a square array (Fig. 3). The magnetically recorded binary record occupies about 300 ft of 800 BPI (high density) 7-track computer tape and contains over one million items of information. All of this originated in one piece of photographic film 24 mm square. By packing each word bit by bit, the data can be squeezed into a digital storage or processing device having a capacity of 64 K words. These comparisons indicate that photographs are extremely efficient storage devices for recording information in a binary fashion.

Photographic images, of course, are not necessarily discrete nor binary. In fact, most ordinary photographs are neither; they are continuous recordings and contain information about magnitude as well as location. Unfortunately, it is extremely difficult to measure variations in optical density and to control light intensities with precision. Additional problems arise because photographic film responds to light in a nonlinear manner unless special precautions are taken. Also, many natural phenomena encompass a wider dynamic range of light intensities than film can record, introducing tonal distortions in the image. For these reasons, non-binary recording on photographic film is fraught with difficulties and recording and processing in an analog mode is subject to uncertainty.

Fig. 3. Computer listing of output from a flying spot scanner, which made over one million density measurements on a segment of photographed thin section. The man on the right is holding original thin section against listing.

DIRECT PROCESSING OF PHOTOGRAPHS

Once information has been stored in the form of an image in photographic emulsion, the problem of retrieving and processing arises. This may be done in a variety of ways; for example, by sampling with a scanning densitometer and operating on the array of sample values with a digital computer. This is the process employed with the digitized record shown in Fig. 3. Alternately, the image may be processed directly using electro-optical techniques. These techniques are based on physical laws governing wave optics and allow creation of a number of useful transformations of the input data. All methods are closely related to time-series analysis, and are based upon the ability of lenses to produce Fourier transforms (Goodman, 1968, p. 77). Among the techniques are auto- and cross-correlation, power spectrum analysis, and spatial frequency filtering. The methods can be used for discrimination, comparison, image enhancement, and classification.

Because the film image used as input is continuous rather than discrete, output is likewise continuous. Digital methods produce numeric results. In contrast, optical methods produce output similar in form to the input; that is, as patterns of light intensity. Therefore, it generally is necessary that output from optical processors be recorded and displayed in some manner approximately equivalent to conventional digital results. This usually is in the form of a video contour display or photographic recording followed by isodensity contouring.

The technology of optical processing is highly advanced. Several journals are devoted exclusively to optical analysis and technology and the literature of the field contains thousands of entries. Unfortunately, almost all of this is unfamiliar to geologists as most of it is couched in the terminology of electrical engineering.

Optical processing methods enjoyed a brief flurry of popularity in geophysical seismic prospecting (Jackson, 1965), but have been supplanted for the most part by new digital methods. Interest presently is being revived by prospects of optically processing satellite photography to reveal subtle geologic features (Brody and Ermlich, 1966; Fontanel and Grau, 1968; Chevallier *et al.*, 1968). Again unfortunately, much of this material is military-sponsored and classified; only a few geologically pertinent examples have been published. Some possibilities of optical processing in earth sciences have been pointed out by Preston *et al.* (1966), Dobrin (1968), Smith (1969) and in a series of articles published by Pincus (1966, 1969a, 1969b, 1969c; Pincus and Dobrin, 1966; Pincus and Ali, 1968). The collection of papers presented at the Symposium on optical and electro-optical information processing technology (Tippett *et al.*, 1968) contains many descriptions of devices and techniques potentially useful in the earth sciences. Although, as the extent of the title suggests, the scope of optical processing is wide, the remainder of this paper will be devoted to one particular optical processor and one particular technique. This is based on a research project currently under way at The University of Kansas and partially funded by the American Petroleum Institute.

SPATIAL FREQUENCY ANALYSIS OF ROCK PORES

The simplest form of optical processor consists of a parallel beam of coherent monochromatic light impinging on a simple lens. The lens focuses the incident beam to a point in the back focal plane. If an object such as a photographic film is placed in the incident beam, destructive and constructive wave interference occurs in a complex manner that is dependent upon the pattern in the film. The focal plane will no longer exhibit a single point of light at the focus but rather a complex design of varying light intensity. It is obvious, however, that all information induced into the light beam by the film is present at the focal point because at twice the focal length an inverted real image of the object can be seen. If the input film is placed in the incident beam at the front focal point, the intensity of light across the back focal point is described by the equation

$$f(x', y') = \left| \iint f(x, y) \exp^{-j(\omega_x X + \omega_y Y)} \, dx \, dy \right|^2$$

Because of the physical action of the lens, the light intensity in the back focal plane is a Fourier transform of the image in the front focal plane. The transformation which occurs in a simple optical processor is shown

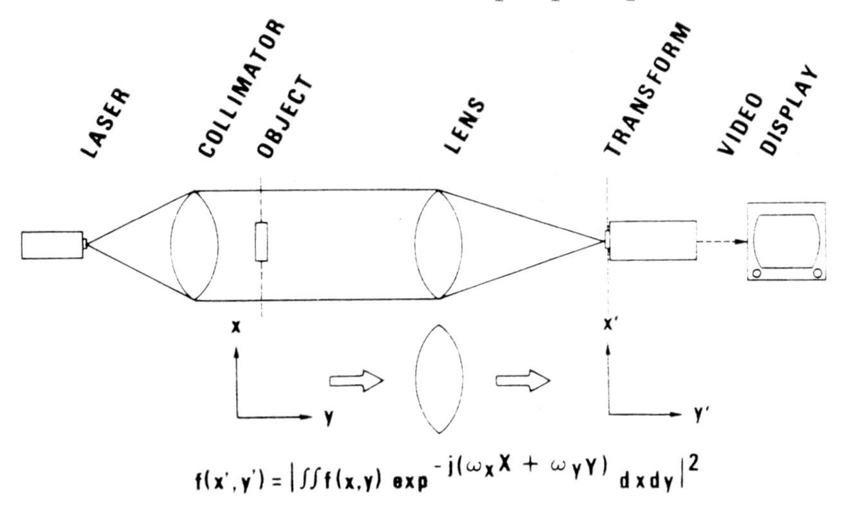

Fig. 4. Arrangement of optical elements for producing power spectrum of an object in transform plane coincident with television detector. Lens performs the transformation indicated at bottom. X, Y, are coordinates on the image, X', Y' are coordinates on the power spectrum, j is a complex-valued constant, and ω_x and ω_y are phase angles.

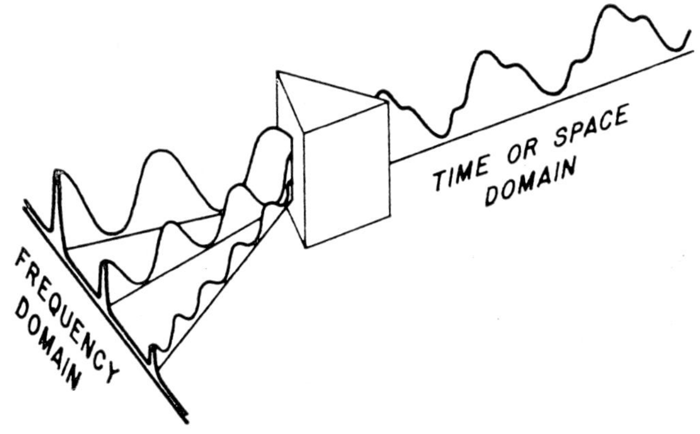

Fig. 5. Complex signal (white light) is transformed from time or space domain by glass prism to frequency domain (colored spectrum). Each component of complex signal is separated according to its frequency.

in Fig. 4. A familiar example of the same phenomenon is shown in Fig. 5. The wave form on the right is a complex signal resulting from the superposition of several varying wavelengths or colors of light. A glass prism will resolve these wavelengths into a spectrum, creating the display shown on the left. One side is the Fourier transform of the other and the time-variant light wave has been changed into the frequency domain in which each wavelength or color is separated from the others. The same transformation can be made with a simple lens, except that the frequency separation occurs as colored rings, familiar as chromatic dispersion.

The simple optical processor uses coherent monochromatic light, so any wave-form disturbances induced into the incident beam are the direct result of interference caused by spatial characteristics of the image on the film. Because the wave pattern in the incident beam is a function of spatial frequencies of spacings in the input film, the display at the back focal plane represents the spatial frequencies in the object. The transform contains both a phase component and an amplitude component. Changes in phase of light waves are extremely difficult to detect and almost impossible to record with conventional methods. Fortunately, in this particular optical technique, phase information is not required and can be disregarded. Interest is centered solely on the amplitude of the light which is expressed as its intensity or power. Intensity, of course, is something which can be recorded directly either by video techniques or by photography.

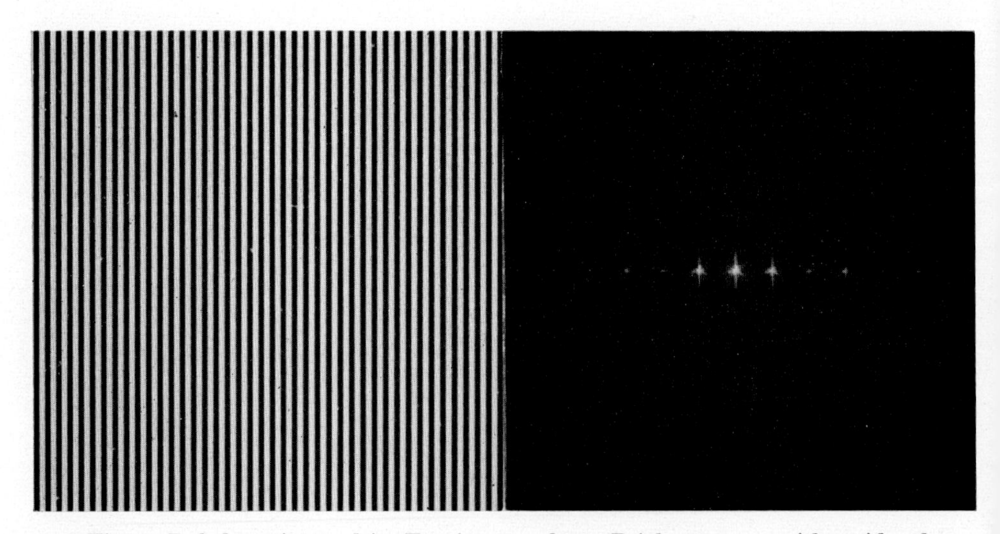

Fig. 6. Ruled grating and its Fourier transform. Bright spots on either side of center spot represent fundamental spatial frequency; successive less-intense spots are harmonics.

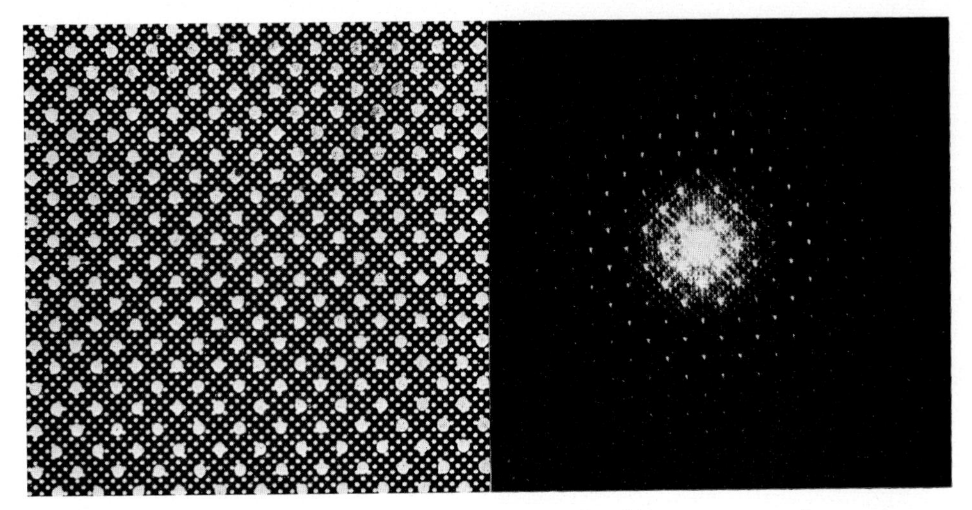

Fig. 7. Complex periodic input of superimposed grids creates complex power spectrum on right. Spectrum resembles single-crystal x-ray diffraction pattern.

If output of the optical processor is captured on photographic film at the back focal plane, the resulting image is the power spectrum of the input object (Goodman, 1968, p. 5). There are certain characteristics of the power spectrum which potentially make it useful for interpretation and classification of images of various types of geologic phenomena. It should be emphasized that the power spectrum obtained optically is a direct transformation of the image itself and contains all information originally in the image with the exception of phase relationships. If the input image consists of a series of parallel ruled lines, the output transform is a series of dots perpendicular to the input ruling (Fig. 6). The brightest spot is in the center; its intensity is a function of the average light transmission through the image. The distance from the center point to the first spot on either side is a function of the spacing between successive transparent areas on the ruling. If spacings on the original grid are small the distance of this point from the center becomes greater. Successive spots outward from the center are harmonics of the fundamental spatial frequency. This provides an easy way to calibrate power spectra empirically. If, for example, the input image had rulings 1 mm apart, the power spectrum would contain points representing spacings of 1 mm, $\frac{1}{2}$ mm, $\frac{1}{3}$ mm, and so forth. More complex periodic inputs such as Fig. 7 will produce patterns of repeating and interacting harmonics. These transforms are similar in appearance to x-ray diffraction patterns of single crystals and arise from exactly the same phenomenon. In fact, the use of power spectrum techniques to

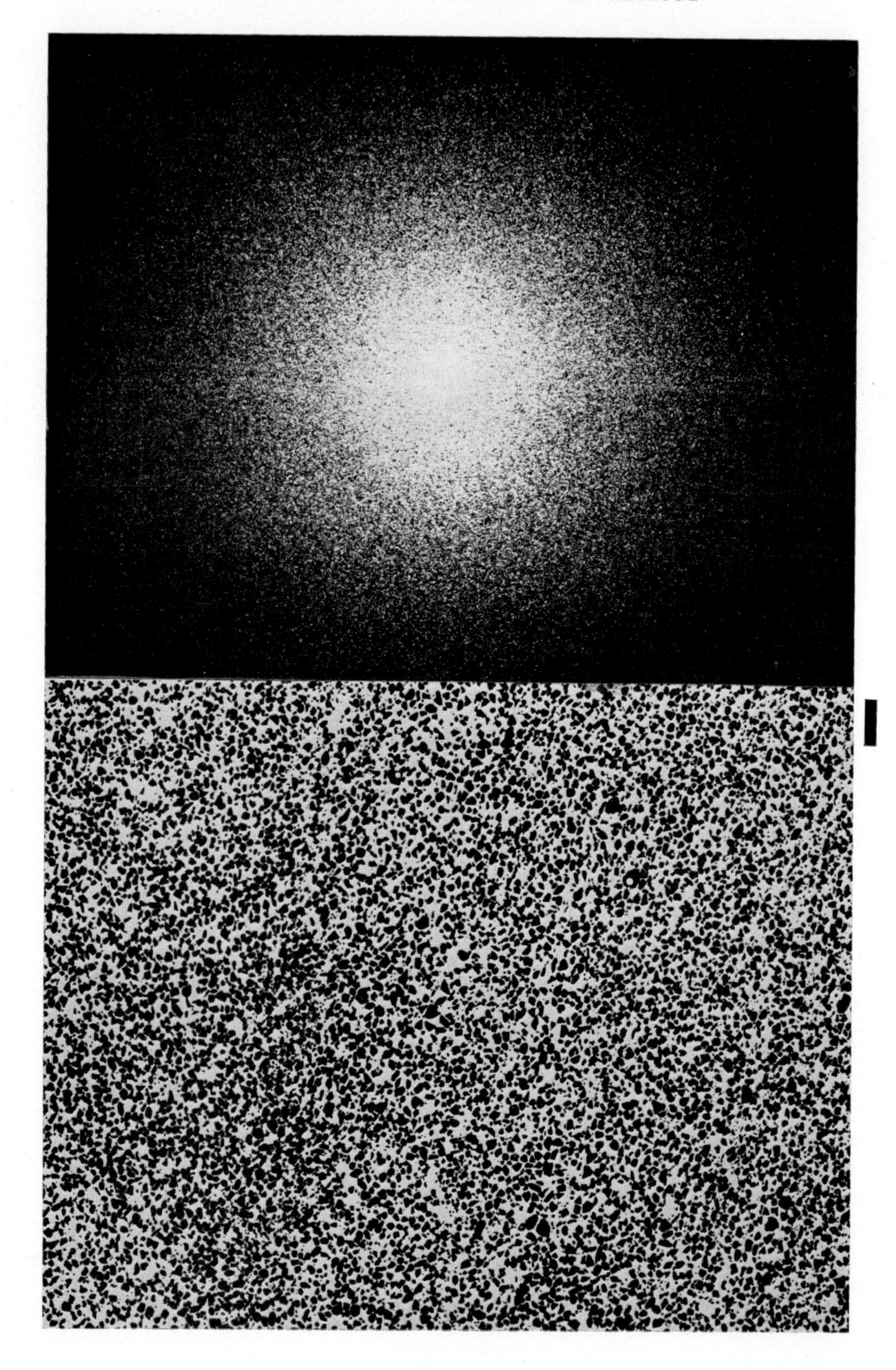

analyze complex molecular structures is described in detail by Taylor and Lipson (1964).

If the input image is not periodic in nature, the power spectrum is not discrete but forms an area of varying light intensities. High light intensities in the power spectrum occur at positions representing spatial frequencies abundant in the original image. Low intensities occur in areas corresponding to spatial frequencies which occur less abundantly. In a photographic recording, the power spectrum usually appears as a large, diffuse bright area (Fig. 8). If isodensity contours are constructed on the recording, the power spectrum can be used to determine the relative abundances of spacings of specified size in the original image. Experiments to date have failed to disclose any evidence of periodicity in pore spacings of reservoir rocks, suggesting that the pore network is the result of a random process. The power spectrum of a random pattern represents the contribution of each spatial frequency to the total image (Otts, 1968, pp. 15–18). If the power spectrum is divided into classes of spatial frequencies, its magnitude forms a histogram of the relative abundance of different spacings within the image. This is analogous to conventional histograms of grain size prepared by sedimentary petrologists, except that the power spectrum is three-dimensional. That is, it is a graph of the spacings encountered along all possible traverses taken at all possible orientations across the image.

Orientation information is preserved in the power spectrum. The frequency distribution of spatial sizes present at a specified orientation can be determined by examining the power spectrum along the perpendicular diameter. As would be expected, the power spectrum is symmetrical about its center point and when contoured shows radial symmetry. Contoured spectra obtained from the optical processor resemble circular or elliptical bell curves in three dimensions.

POWER SPECTRA OF POROUS MEDIA

There are compelling economic and scientific reasons for attempting to analyze the pore network of sedimentary rocks. Behavior of contained fluids such as petroleum or water is directly dependent upon details of this network. It is suggested that the fabric of clastic sediments may reflect conditions of deposition. A detailed pore analysis would of necessity describe the particle framework as well and would allow quantitative appraisal of fabric. Finally, analysis of the rock pore system poses interesting classification problems. Solutions to these problems may be equally applicable to other networks—geological, biological, and geographical.

Fig. 8. Continuous spectrum produced by random (non-periodic) pattern such as high contrast photograph of sandstone. Grains are black and pores are white. Bar represents 1·0 mm. Faint vertical rays in power spectrum are caused by edge of input film.

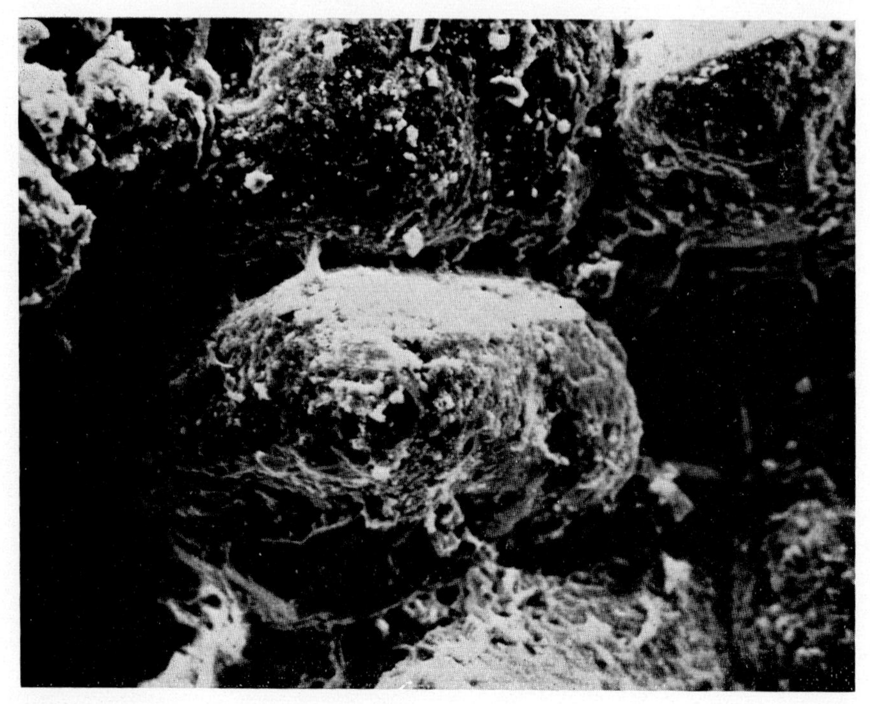

Fig. 9. Scanning electron (SEM) photomicrograph of Ordovician St Peter Sandstone. Although sand is well-sorted and rounded, pores are highly irregular. Field of view is approximately 0·4 mm long.

The pore network and reciprocal particle framework occupy three-dimensional space. For a variety of reasons, at the present time it is essentially impossible to investigate more than a thin slice through this space. Figure 9 is a scanning electron (SEM) photomicrograph of an Ordovician sandstone. Although the grains in the rock are extremely well rounded and highly sorted, pores are irregular and relations between grains are complex. Figure 10 is an SEM photomicrograph of a Cretaceous sandstone from a producing oil reservoir in central Wyoming, and has a fabric typical of many reservoir rocks. These photographs indicate the difficulty of determining and describing shapes of pores more than one grain deep within the rock.

Although the pore network cannot be directly examined in three dimensions to any significant depth, the problem can be simplified by considering only an infinitesimally thin slice of the rock. If necessary, three-dimensional analysis may be approximated by series of successive slices or by a box work of slices set at angles. However, useful information can be obtained by examining only the cross section of the pore network which appears on two-dimensional planes. This approach is expressed in the ordinary petrographic thin section as in Fig. 11. Extraneous details arising from differences in composition of grains can be suppressed by reducing the image to a continuous binary form. A typical pattern of grains and pores in a plane through sandstone is shown in Fig. 12. The image was produced by contact printing a thin section of rock impregnated with red plastic onto process-type orthochromatic film. This image is an ideal form for input into the optical processor.

The following illustrations consist of power spectra from samples chosen to show contrasting features. Parts of the original photographs used to construct the power spectra also are shown. In order to evaluate the transforms in a quantitative manner, it is necessary to measure relative brightness (intensity) of the spectrum, which is directly proportional to power. This was accomplished by the IDECS processor, a television-like device (Dalke and Estes, 1968), that contoured the power spectra at seven density levels.

Fig. 10. SEM photomicrograph of Cretaceous Muddy Sandstone. Grains are angular and moderately sorted; pore network is extremely complex. Field of view is approximately 3 mm long.

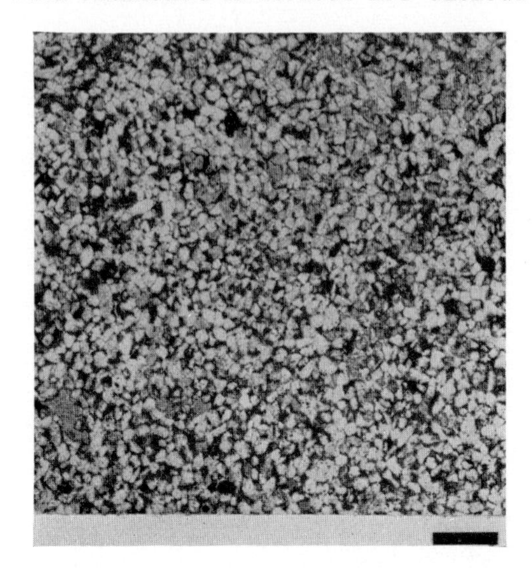

Fig. 11. Photomicrograph of Cretaceous Woodbine Sandstone from Texas. Quartz grains appear light grey. Pores are filled with red plastic and appear dark grey. Bar is 1·0 mm long.

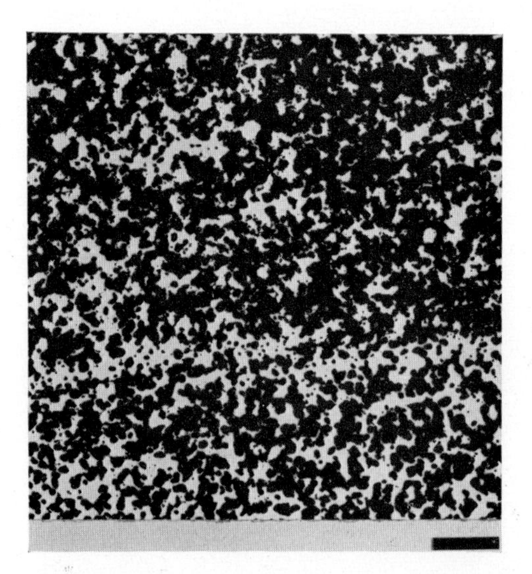

Fig. 12. High contrast print made from thin section similar to Fig. 11. Film is not sensitive to light passing through red plastic, so pore areas are not exposed and remain white. Grains transmit white light so become black in print. Bar is 1·0 mm long.

Figure 13 shows the pore patterns of a sample of the Gaskell sand (Lower Eocene) from southern California, and its power spectrum. Although the pore volume of the sample is relatively low, most individual pores have large cross-sectional areas. This is a consequence of poor sorting in a coarse-grained sandstone. The resulting power spectrum shows that almost all power is concentrated in spatial frequencies greater than 0·2 mm.

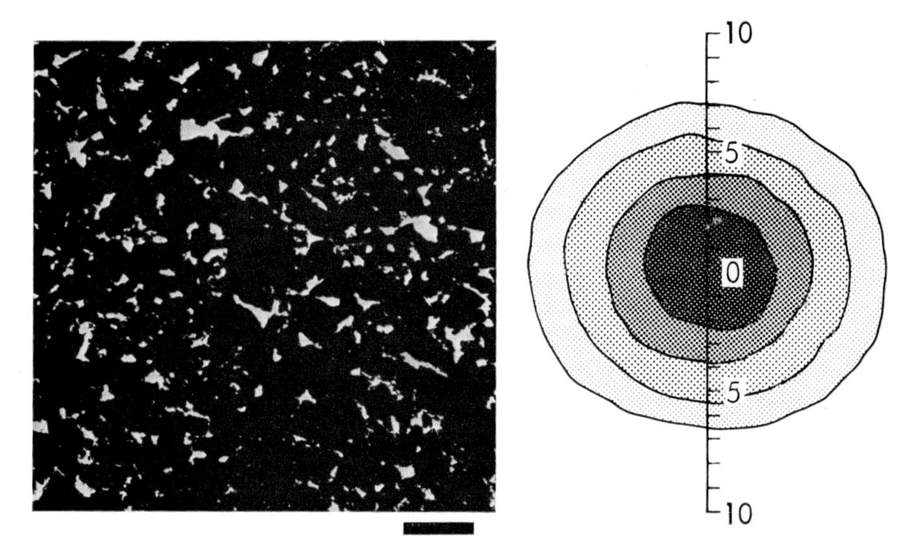

Fig. 13. Pore pattern and isodensity contoured power spectrum of Gaskell sand (Lower Eocene), Southern California. Contours are percent opacity and are $>80\%$, 80–60%, 60–45%, and 45–35% outward from center. Scale divisions represent $1/N$ mm. Bar represents 1·0 mm.

A sample of Muddy Sandstone (Lower Cretaceous) from eastern Wyoming presents a contrasting pore pattern and power spectrum (Fig. 14). The sample is composed of well-sorted fine sand, and has a uniform distribution of pore spacings extending from about 1 mm to the resolution of the optical system. The power spectrum shows that gradually decreasing contributions are being made to the total porosity by the finer spacings. Concentration of power in larger spatial frequencies is not evident as in the Gaskell sand, and significant power exists in spatial frequencies below 0·1 mm.

Both examples are homogenous sandstones having no significant directional patterns in their pore structures. In contrast, Fig. 15 is a sample of Dakota Sandstone (Cretaceous) from Kansas, having pronounced microbedding. Pores are highly elongated in the horizontal

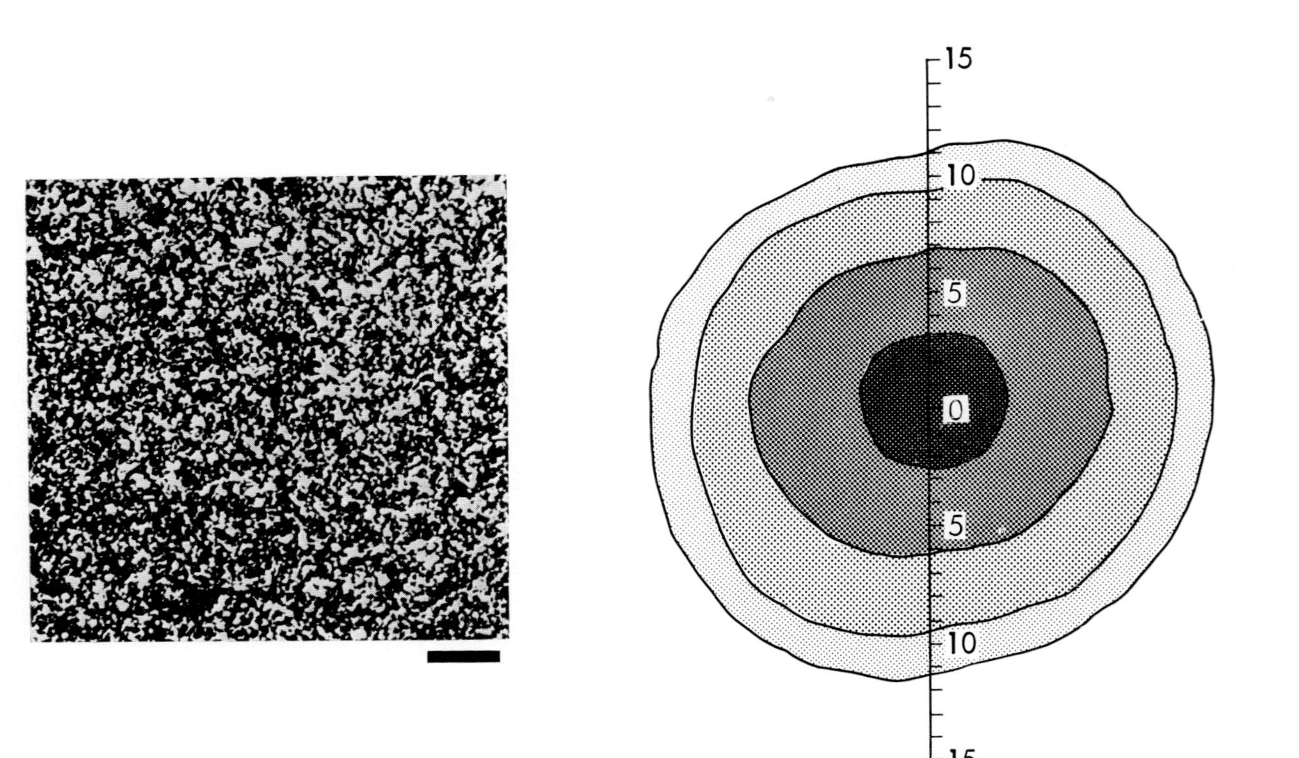

Fig. 14. Pore pattern and isodensity contoured power spectrum of Muddy Sandstone (Cretaceous), Wyoming. Conventions are same as on Fig. 13.

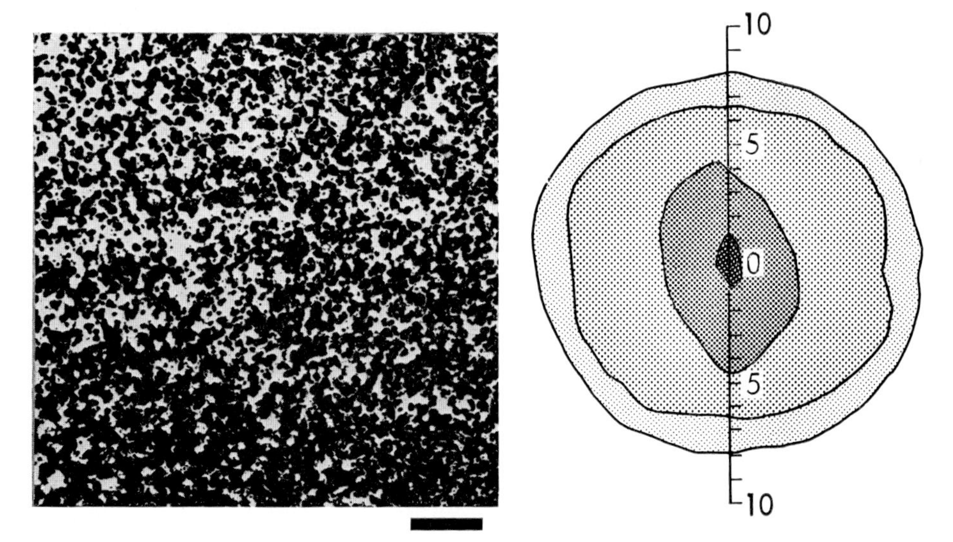

Fig. 15. Pore pattern and isodensity contoured power spectrum of Dakota Sandstone (Cretaceous) from Kansas. Note elongation of spectrum caused by inhomogenous pore structure. Conventions are same as on Fig. 13.

direction, the result of differential cementation. The power distribution is strongly ellipsoidal in the size range larger than about 0·3 mm, but is uniform at small spacings.

Although these illustrations suggest that the power spectrum may reveal significant features of sedimentary pore networks, research is continuing into the potentials of optical processing. Current activities are concentrated in three areas: (1) technological development of rapid methods for isodensity mapping of spectrum photographs; (2) establishing relationships between power spectra and more conventional parameters used to describe pore structure; (3) theoretical and model studies of the spectra of two-dimensional random patterns. At this point it is not possible to assess the usefulness of optical processing in reservoir analysis. However, rapid advances are being made in many other disciplines using optical techniques, and the potential in the earth sciences seems equally great.

ACKNOWLEDGEMENTS

This report is part of API Project 103, Numerical Characterization of Porous Media, a joint study by the Kansas Geological Survey, the Department of Chemical and Petroleum Engineering at the University of Kansas, and the American Petroleum Institute. Several organizations have assisted the project, including Pan American Research Corporation, Mobil Geophysical Laboratories, the Center for Research, Inc., Technical Operations, Inc., and Metals Research Instrument Corporation. In addition, many individuals have contributed material, analyses, and advice.

REFERENCES

Brody, R. H. and Ermlich, J. R. (1966). Fournier analysis of aerial photographs. Proc. 4th Symp. Remote Sensing of Environment, Ann Arbor, Mich., p. 375–392.

Chevallier, R., Fontanel, A., Gran, G. and Guy, M. (1968). Application du filtrage optique a l'etude des photographies aerienness. 11th Cong. Intl. Photogrammetrie, Luasanne, France, Comm. 7, pp. 1-15.

Dalke, G. W. and Estes, J. E. (1968). Multi-image correlation systems study for MGI, final report. Center for Research Inc., Univ. of Kansas, Contract No. DAAK02–67–C–0435.

Dobrin, M. B. (1968). Optical processing in the earth sciences. *IEEE Spectrum* **5**, 59–66.

Fontanel, A. and Grau, G. (1968). Traitement optique de l'information en geophysique et dans le domaine de la photographie aerienne. *Onde élect.* **48**, 1–10.

Goodman, J. W. (1968). "Introduction to Fourier optics". 287 pp. John Wiley and Sons, New York.

Jackson, P. L. (1965). Analysis of variable-density seismograms by means of optical diffraction. *Geophysics* **30**, 5–23.

Otts, J. V. (1968). Random vibration. U.S. Dept. of Commerce, Clearinghouse PB 184 113.

Pincus, H. J. (1966). Optical processing of vectorial rock fabric data. Proc. 1st Intl. Cong. Intl. Soc. Rock Mechanics, Lisbon, vol. 2, pp. 173–177.

Pincus, H. J. (1969a). Sensitivity of optical data processing to changes in rock fabrics, Pt. 1, Geometric patterns. *Int. J. Rock Mech. Min. Sci.* **6**, 259–268.

Pincus, H. J., (1969b). Sensitivity of optical data processing to changes in rock fabric, Pt. 2, Standardized grain patterns: *Int. J. Rock Mech. Min. Sci.* **6**, 269–272.

Pincus, H. J. (1969c). Sensitivity of optical data processing to changes in rock fabric, Pt. 3, Rock fabrics. *Int. J. Rock Mech. Min. Sci.* **6**, 273–276.

Preston, F. W., Green, D. W. and Aldenderfer, W. D. (1966). The use of statistical communication theory for characterization of porous media. Pennsylvania State Univ., Mineral Industries, Publ. 2–65, vol. 1, p. B1–B20.

Pincus, H. J. and Ali, S. A. (1968). Optical data processing of multispectral photographs of sedimentary structures. *J. Sedim. Petrol.* **38**, 457–461.

Pincus, H. J. and Dobrin, M. B. (1966). Geological applications of optical data processing. *J. Geophys. Res.* **71**, 4861–4869.

Reich, A. and Dorion, G. H. (1968). Photochromic, highspeed, large capacity, semirandom access memory. *In* "Optical and electro-optical information processing". pp. 567–580. M.I.T. Press, Cambridge, Mass.

Smith, A. R. (1968). Techniques for obtaining fabric data from coarse clastic sediments. *Brigham Young Univ., Geol. Stud.* **15**, 13–30.

Taylor, C. A. and Lipson, H. (1964). "Optical transforms". 182 pp. Cornell University Press, New York.

Tippet, J. T. Berkowitz, D. A., Clapp, L. O., Koester, C. J. and Vanderburgh, A., Jr. (1968). "Optical and electro-optical information processing". 780 pp. M.I.T. Press, Cambridge, Mass.

Todd, H. N. and Zakia, R. D. (1969). "Photographic sensitometry". 213 pp. Morgan and Morgan, New York.

6. Serial Sections of Fossils Prepared by the Annular Sawing Technique

K. A. JOYSEY AND J. L. CUTBILL

University Museum of Zoology, University of Cambridge and Department of Geology, Sedgwick Museum, University of Cambridge, England

INTRODUCTION

The annular sawing technique was first developed in Britain by the Post Office Engineering Department for the economical slicing of quartz crystals for use as piezoelectric vibrators (Thwaites and Sayers, 1955). The saw itself consists of a thin annulus of metal which is clamped round its outer edge to tension the blade against deflection, and the inner edge of the annulus is armed with diamond dust as a cutting agent. The object to be cut is mounted so that it protrudes through the central aperture of the annulus, and during cutting it is fed against the inner edge of the rotating annular blade.

As described by Thwaites and Sayers (1955), the experimental saws constructed by the Post Office Engineering Department were capable of producing 20 successive slices of quartz of an area 14 × 18 mm, each 0·35 mm thick and separated by saw cuts of the same width. We understand from Mr C. F. Sayers that some years after the Post Office Engineering Department had developed their annular saw, it was discovered that a small saw working on a similar principle had been invented previously by a dental surgeon in Holland, and used for cutting teeth.

The Post Office annular saw was subsequently developed and manufactured by the Caplin Engineering Co. Ltd, and is now available as the Capco Q-35 Annular Saw, which is widely used for cutting semiconductor substances for the manufacture of transistors.

Systematics Association Special Volume No. 3. "Data Processing in Biology and Geology", edited by J. L. Cutbill, 1970, pp. 89–95.

SERIAL SECTIONING OF FOSSILS

Credit is due to Mr C. F. Sayers himself for recognizing the potential of the annular saw for cutting serial sections of fossils. Mr Sayers is an amateur geologist and he happened to be a friend of the late Mr W. N. Croft, who developed the Croft Parallel Grinder for the investigation of fossils by serial section (Croft, 1950).

It is well known that grinding techniques suffer from the disadvantage that the specimen itself is totally destroyed, and even though the ground surface may be recorded at intervals by means of cellulose acetate replicas, drawings and photographs, it is nevertheless impossible to refer back to the original material. When fossils are cut with the annular saw, although a proportion of the whole is lost in the width of the saw cuts, the sections remain as a permanent record.

In 1958, Mr Sayers exhibited some serial sections of fossil brachiopods in London, and at that time he kindly agreed to cut some sections of the blastoid *Pentablastus* on behalf of one of the present authors. The results of the study of these serial sections were subsequently published (Joysey and Breimer, 1963), and this paper included an account of the application of the annular sawing technique to fossil material. In some regions of the blastoids the material was suitable for cutting into slices of a thickness equivalent to the width of the saw cut, and so serial sections could be studied at regular intervals by examining both sides of each slice. While working on these *Pentablastus* sections it was found that by shining an intense light through the thickness of the slice, it was possible to trace plate sutures within the slice and observe the direct connection between the two surfaces. Nevertheless, it was possible to obtain separate photographs of the two sides by using a short depth of focus, partly relying on the scattering of light within the slice to blur the image of the opposite side (Fig. 1a). A continuous sequence of photographs of the serial sections was obtained by photographing both sides of each slice, and reversing the negative of alternate pictures, so that those surfaces actually photographed from below were printed, and later drawn as though viewed from above (Fig. 2).

Although serial sections of fossils cut on the annular saw were exhibited more than ten years ago, and the results of the *Pentablastus* study were published six years ago, we have been surprised that relatively few palaeontologists have taken advantage of this method. This may be a result of the capital expenditure involved, but we are

Fig. 1. (a) Section through an ambulacrum of the blastoid *Pentablastus*, select from a sequence of serial sections cut with the annular saw, demonstrating t quality of information available on the untreated cut surface. Thickness section 0·35 mm. Photographed by transmitted light (Reproduced from Joys and Breimer, 1963.) Compare with Fig. 2c.

(b) Surfaces at 0·25 mm intervals through a specimen of *Schwagerina* (magn fication × 10) and a section of one of the surfaces.

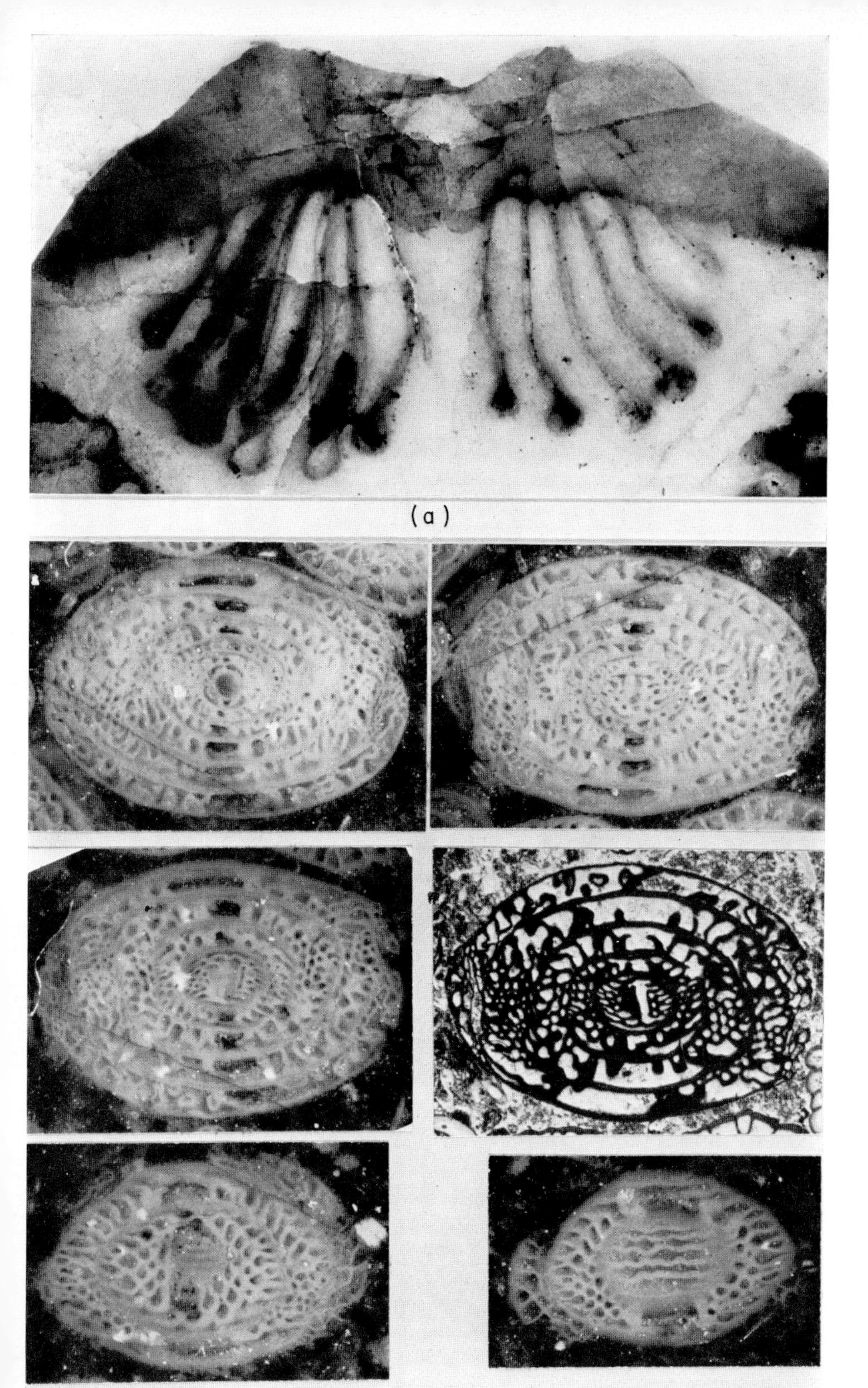

(b)

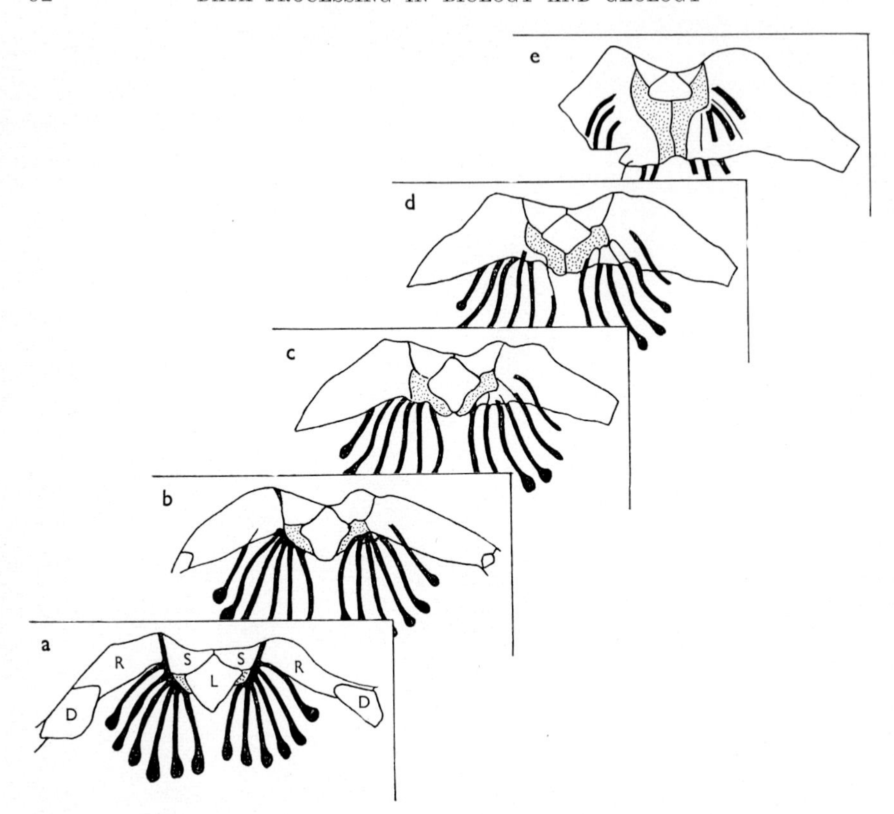

Fig. 2. Drawings of serial sections through an ambulacrum of the blastoid *Pentablastus*, cut with the annular saw, demonstrating the nature of the information revealed by this technique. The sequence shows the hydrospire-plates (stippled) coming into contact in the mid-line, and forming an under-lancet plate in the region aboral to the hydrospire-clefts. Deltoid (D); lancet (L); radial (R); side-plates (S).

Fig. 2c was drawn from the photograph on Fig. 1a (Reproduced from Joysey and Breimer, 1963.)

convinced that this technique can produce a great deal of new information, and our present purpose is to give it more publicity.

An annular sawing machine was installed in the Department of Geology, Cambridge in 1964 and since that time it has been used by several workers. Dr M. J. S. Rudwick and Dr R. Cowen have used it in their studies of brachiopods; Dr C. L. Forbes has used it in his own research on corals, and he has also cut serial sections of the Silurian coral *Syringaxon* and the Devonian coral *Spongophyllum* on behalf of Dr J. S. Jell of the University of Queensland, Brisbane; Dr C. R. C. Paul used the annular saw in his studies of cystoids, and one of us

(J. L. C.) has used it extensively in studies of Permian fusulinids from Spitsbergen (Fig. 2).

As a result of the work in Cambridge we have been pleased to learn that a Capco Q–35 annular saw has been installed by Dr J. S. Jell in the Department of Geology, University of Queensland, and another by Dr K. A. Kermack in the Department of Zoology, University College, London, for use on fossil vertebrates. At the 15th Symposium on Vertebrate Palaeontology and Comparative Anatomy held in London in September 1967, Dr D. M. Kermack demonstrated some serial sections of the skull of the reptile *Endothiodon* cut on this machine. We have little doubt that other machines, of which we are unaware have been installed elsewhere, and we understand that annular sawing machines of American manufacture are now available.

PRACTICAL INFORMATION

A specimen to be cut is mounted on a brass bar, the movements of which are controlled by a precision carriage, enabling the specimen to be offered up to the cutting edge of the annulus at known intervals. The specimen is usually fastened to the brass bar by a hard wax (e.g. Tanwax, from Haesler Sales, 4, Grange Street, St Albans, Herts).

For irregular specimens, which cannot be mounted easily on the brass bar, it is advisable to embed the specimen. This also facilitates orientation of the specimen. A mixture of equal parts of plaster of Paris (dental quality) and synthetic resin has been found suitable (e.g. "Ceemar" resin, from E. M. Cromwell and Co. Ltd, Galloway Road, Rye Street, Bishops Stortford, Herts).

If the specimen is particularly friable, it may be given additional strength before embedding by impregnation with "Ceemar" resin, and if hollow spaces are likely to be encountered within the specimen this impregnation is most effective if conducted under partial vacuum.

The Cambridge machine is a Mark 1 model which takes a blade $8\frac{1}{8}$ in outer diameter, $3\frac{1}{4}$ in internal diameter. Blades 0·006 in thick (from Impregnated Diamond Products Ltd, Tuffley Crescent, Gloucester) have proved perfectly satisfactory. Alternative blades are available from several sources (e.g. Joseph Electronics Ltd, 23/24, Warwick Street, London, W.1 are agents for "Greenlee" blades, made in U.S.A.). Cutting oil is essential as a lubricant (e.g. Sternol Neatcut 262, from Sternol Ltd, Royal London House, Finsbury Square, London, E.C.2).

Various improvements have been effected since Sayers's pioneer work on the Post Office machine and in suitable material such as a compact limestone, it is possible to cut sections as thin as 0·2 mm, while the width of the saw cut, using the thinnest blades on the market, is 0·22 mm. In more friable material a wider sectioning interval must be employed but, with a slow rate of cut at about 20 min/in, slices 0·3 mm or 0·4 mm thick may be cut from relatively difficult material such as calcite filled cystoids. At this rate of work, the operator will have time to perform other tasks between resetting the machine, and if the fully automatic model is not available it is advantageous to have an electric bell arranged so as to ring at the end of each cut.

The "Capco Q–35" High Precision cutting machine is manufactured by Caplin Engineering Co. Ltd, Elton Park, Hadleigh Road, Ipswich, Suffolk, and the Roditi Purchasing Co. Ltd, 12A, Golden Square, London, W.1 are agents for sales and servicing.

REFERENCES

Croft, W. N. (1950). A parallel grinding instrument for the investigation of fossils by serial sections. *J. Paleont.* **24**, 693.

Joysey, K. A. and Breimer, A. (1963). The anatomical structure and systematic position of *Pentablastus* (Blastoidea) from the Carboniferous of Spain. *Palaeontology* **6**, 471.

Thwaites, J. E. and Sayers, C. F. (1955). A novel type of saw for the economical cutting of quartz crystals or other materials. *P.O. elect. Engrs' J.* **47**, 1–3.

ADDENDUM

The foregoing information was prepared for the symposium in September 1969. Since then the following paper has appeared: Kermack, D. M. (1970). True serial-sectioning of fossil material. *Biol. J. Linn. Soc.* **2**, 47.

This paper gives an account of the use of the annular saw to cut a small skull of the therapsid reptile *Endothiodon* into 104 serial sections each 0·6 mm thick. It includes details of an embedding technique and a description of a method of supporting the sections with adhesive polythene tape while they are being cut. The author describes the use of the annular saw for cutting serial sections of fossils as a new technique, apparently unaware that the machine was installed at University College, London, because it was known to have been used successfully for this purpose. In this context we must repeat that credit is due to Mr C. F. Sayers himself for recognizing the potential of the annular saw for cutting serial sections of fossils.

We understand from Mr Sayers that the first fossils to be cut on the annular saw were some specimens of the brachiopod *Spirifer* given to him by Mr Croft. Slices of these brachiopods were exhibited at the

Centenary Reunion of the Geologists' Association at Burlington House in November 1958. A catalogue describing the exhibits was issued at the time, but it is unfortunate that when a report of the Centenary Reunion was published later (*Proc. Geol. Assoc.* **70**, 345) the exhibits were only listed in title.

One of the present authors was present at the Reunion, and it was on this occasion that Mr Sayers kindly agreed to cut some serial sections of the blastoid *Pentablastus* on his behalf. Mr Sayers also exhibited some serial sections of fossils cut with the annular saw at a Demonstration Meeting of the Palaeontological Association on "Palaeontological Techniques" in October 1959. One of us then had the pleasure of including the serial sections of *Pentablastus* as part of an adjacent exhibit on techniques for fossil echinoderms, with grateful acknowledgements both to Mr Sayers and to the Post Office Engineering Department.

7. The Use of the D-Mac Pencil Follower in Routine Determinations of Sedimentary Parameters

DAVID J. W. PIPER

Department of Geology, Sedgwick Museum, University of Cambridge, England

ABSTRACT

Thin sections or photographs of rocks can be projected onto the reading table of a D-Mac Pencil Follower, or viewed using a camera lucida, and digitized. Granulometric parameters and measures of grain "roundness" can be computed from these data. The approach requires new, more rigorous definitions of grain "roundness". Advantages over traditional methods are in the amount of time saved, and in the capture of data in machine readable form at the earliest opportunity.

INTRODUCTION

Routine measurements of the size, shape, and orientation of sedimentary particles can be made much more quickly than hitherto by using digitizing techniques. Methods have been evolved for analyzing planar sections of consolidated rocks. Both standard thin sections and photographs of planar outcrops or polished slabs can be used. Many of the techniques could also be applied to unconsolidated sediments.

Grain outlines are digitized and standard parameters computed from these data. If the petrology of the grains is recorded at the same time, a very large amount of data on a rock—far more than is usually collected by conventional means—can be machine stored and processed.

The programs used in this study have been written in Titan Autocode and run on Titan at the University Mathematical Laboratory, Cambridge.

Systematics Association Special Volume No. 3. "Data Processing in Biology and Geology", edited by J. L. Cutbill, 1970, pp. 97–103.

THE DIGITIZER

A D-Mac Pencil Follower type P.F. 10 000 Mark 1A† in the Department of Geology, Cambridge, has been used in this study. The digitizer has a 1 m square reading table. Material to be digitized is traced around with a reading pencil, which is accurately followed beneath the table by a servo-detector, whose position in terms of x and y coordinates on the table is passed to an electronics console, and output on punched paper tape. Output rates of over 400 coordinate pairs per minute can be achieved. The accuracy of the digitizing is dependent on the speed with which the shape is traced: at speeds of less than 1 cm/s, the accuracy is about 0·1 mm. This falls off to 2·0 mm at 10 cm/s.

A Leitz projection microscope fitted above the digitizing table permits thin sections and photographic negatives to be projected onto the table with magnifications up to $\times 30$, using polarized light where necessary for petrologic identifications.

Higher magnifications can be achieved using a camera lucida attachment on a Leitz SM-pol microscope. This allows magnifications of up to $\times 1000$, but the field of view is much restricted; the camera lucida image on the digitizing table has a diameter of only 11·2 cm. The standard D-Mac projected image following pencil has been modified for use with the camera lucida by blacking out the surface of the pencil and filling in the cross lines in white, so as to reduce the light intensity passing through the camera lucida. The servo-detector beneath the table is affected by magnetic objects but the effect of the microscope is negligible.

MEASUREMENT OF GRAIN SIZE AND ORIENTATION

Digitizing techniques have been used by the author to study variations in the size and orientation of pebbles in conglomerates in the Gowlaun Member of the Silurian of Western Ireland (Piper, 1969). Photographs were taken of approximately plane outcrops of conglomerate, and the attitude of the outcrop plane, and of a scale placed on it, were measured. Photographic negatives were projected onto the digitizing table. The ends of the scale, and the outline of the area to be analysed, were then digitized.

In the programs analysing these data, no attempt has been made to interpolate between digitized points. The long axis of the grain is determined as the longest possible line between two digitized points on the grain perimeter. The accuracy with which parameters such as long axis, short axis, and orientation of a grain are determined is thus depen-

† D-Mac Ltd, Queen Elizabeth Avenue, Glasgow, S.W.2.

dent on the density of digitized points around the grain perimeter. With only 20 coordinate pairs per grain, the computed orientations of a set of parallel grains showed a standard deviation of 4°; there was also a difference of about 4° between the mean orientations when the grains were digitized in a clockwise direction, and when they were digitized anticlockwise. Long and short axis determinations had a standard error of $\pm 2\%$.

The program for the primary analysis of the data computes the long axis, short axis, cross-sectional area, orientation of the long axis, and position of the midpoint of the grain, and stores the data on magnetic tape. Further programs have been written to print out histograms of orientation and size distribution of grains, and to produce a cumulative size frequency distribution. The input format allows a code indicating petrology (or a roundness estimate, or any other desired attribute) to be punched onto the output tape while digitizing, and size and orientation histograms for each grain petrology can be produced.

No attempt has been made to transform the granulometric data from plane exposures to a form comparable with that obtained by sieving unconsolidated samples. The general problems of such transformations are discussed by van der Plaas (1962).

The advantages of digitizing over traditional hand measurement are two-fold. Much time is saved in the measurement process. A semi-skilled operator can digitize a photograph of 100 pebbles in about 30 min. A similar amount of time would be required to measure and record by hand only the long and short axes; the measurement of orientation is a far more time-consuming operation. (Should only the long and short axes be required, they may be recognized visually, and their ends digitized—this allows several hundred pebbles to be measured in 30 min.) Granulometric data are almost always subsequently analysed statistically. If a digitzer is used, the data are stored in machine readable form at the earliest possible opportunity.

MEASUREMENT OF GRAIN "ROUNDNESS"

Some preliminary studies have been made on the problem of measuring "roundness" in sedimentary particles. "Roundness has to do with the sharpness of the edges and corners of a clastic fragment" (Pettijohn, 1957, p. 57). Roundness is an unhappy choice of a word to express this concept, and roughness would have caused less confusion. "Roundness" to a geologist has nothing to do with the approximation of the gross shape of a particle to a sphere.

The concept of roundness or roughness in sedimentary particles is loose. "Roundness" is usually visually estimated by comparison with standards (Krumbein, 1941; Powers, 1953). Many precise definitions relate the radius of curvature of edges or corners of the grain to some form of average diameter for the whole grain, Cailleux (1947), followed by most French geologists, used only the sharpest corner. Wadell (1932, 1935), followed by most American geologists, used the average angle of curvature of all corners. Lees (1964), in studying very angular material, used the angles between faces. A rigorous definition of this type must define the scale on which "roundness" is to be recognized, how corners (as opposed to sides) are to be recognized, and precisely how the radius of curvature of a corner is defined. This is quite easily done in the idealized drawings of grains in the definitive papers on the subject—but the author has found it impossible to formulate rigorous definitions for the general case.

Some authors, perhaps recognizing these difficulties, have used the proportion of convex to concave parts of a grain as a measure of "roundness". The method of Szadeczky–Kardoss (1933) is the most widely used. Since there is no direct measurement of the sharpness of corners, this index of "roundness" does not completely accord with Pettijohn's definition, but Luttig (1956) and Mohr *et al.* (1968) have demonstrated a close empirical relationship between the two types of "roundness" measure.

The Szadeczky–Kardoss "roundness" can be easily measured from the digitized trace of a grain, and programs have been written which use either a regular string of digitized points defining the grain perimeter, or requiring that the points of inflection be recognized by eye and digitized individually.

Stone and Dugundji (1965), in a study of terrain micro-relief, point out that roughness is a vector quantity, with several independent components. These include the amplitude of oscillations, the steepness of oscillations and the degree of periodic repetition of similar roughness elements. They used Fourier analysis of terrain cross profiles to define several roughness components.

Such a technique could easily be applied to digitized grain outlines derived from the pencil follower. Any detailed study of grain "roundness" or roughness must examine the several components of roughness.

The immediate geological need, however, is for a rather simpler definition of "roundness", requiring less complex programming and analysis than the work of Stone and Dugundji. One major problem is to achieve a satisfactory definition of the scale at which "roundness" is to be studied. If digitized grain outlines are used, the scale can be selected by analysing only those points with a certain minimum distance between them (the spacing may be absolute, or related to the size of the grain). A two-dimensional measure of "roundness" can be obtained from the distribution of angles between digitized segments of a

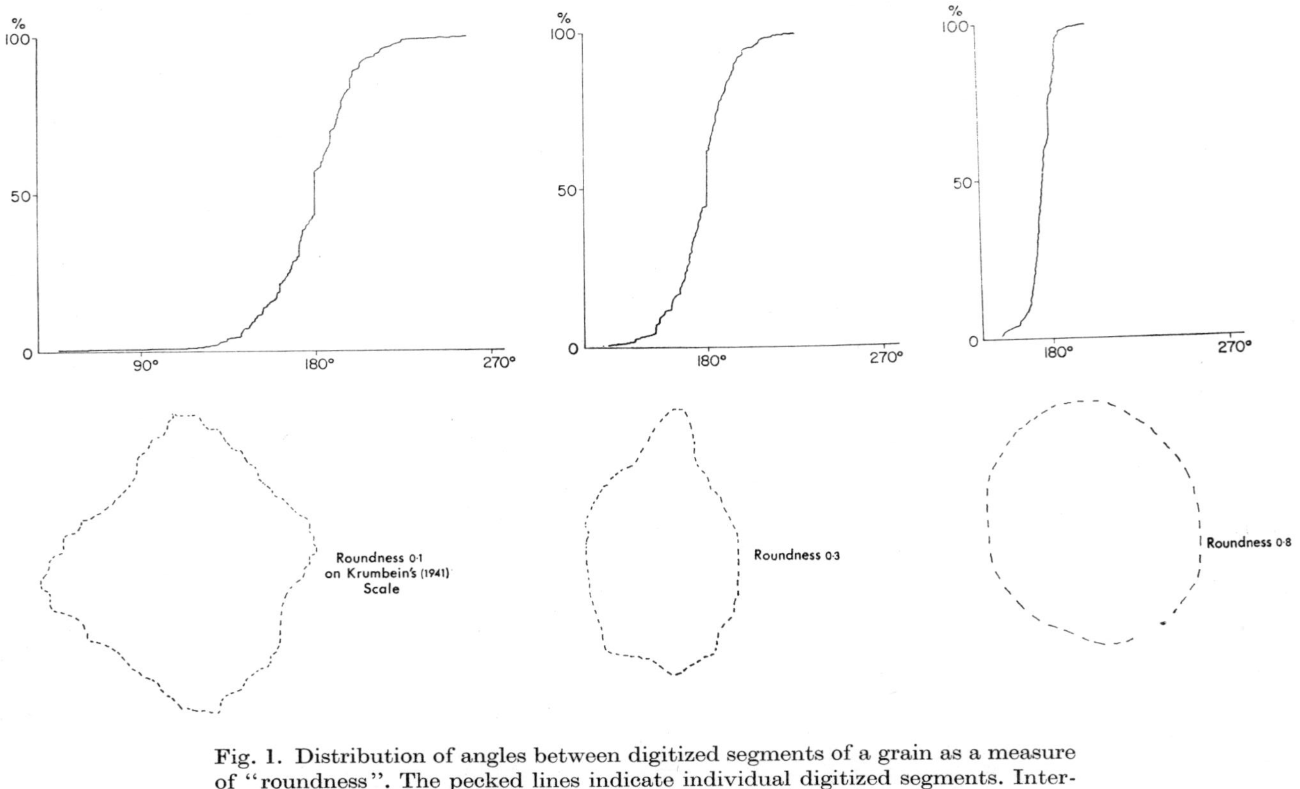

Fig. 1. Distribution of angles between digitized segments of a grain as a measure of "roundness". The pecked lines indicate individual digitized segments. Inter-segment angles plotted on a cumulative frequency diagram.

grain (Fig. 1). The range of very small and very large angles is a measure of the "sharpness component", while the number of such angles is a measure of the number of irregularities. Angles around 180° indicate almost smooth outlines. Another technique that has been studied is to calculate a generalized shape, using conventional smoothing techniques, and to determine the difference between the original and progressively generalized shapes.

It is not the author's intention to add yet another definition of "roundness" to the dozens already in the literature. The purpose of this note is to point out that once a grain outline is accurately digitized in detail—and this can be easily accomplished using a camera lucida attachment—the data can be analysed in a variety of ways, attempting to reveal different aspects of "roundness". Future studies of the geological significance of grain "roundness", of its use in distinguishing geological environments, or of changes of "roundness" with the action of certain geological processes, should not rely solely on visual estimation of gross "roundness"; the most significant components should be sought.

ACKNOWLEDGEMENTS

This work was carried out during the receipt of a N.E.R.C. Research Studentship. I should like to thank John Cutbill and David Williams for their advice at various stages in this study.

REFERENCES

Cailleux, A. (1947). L'indice d'émousse: definition et premiére application. *C. r. Somm. Séanc. Soc. géol. Fr.* **17**, 250–252.

Krumbein, W. C. (1941). Measurement and geological significance of shape and roundness of sedimentary particles. *J. sedim. Petrol.* **11**, 64–72.

Lees, G. (1964). A new method for determining the angularity of particles. *Sedimentology* **3**, 2–21.

Lüttig. G. (1956). Eine neue einfache geröllmorphometrische Methode. *Eiszeitalter Gegenw.* **7**, 13–20.

Mohr, A., Mrazek, J., Richter, K.-H., and Vogel, W. (1968). Ein neues Darstellungsverfahren in der morphometrischen Analyse zur Kennzeichnung der Kornrundung. *Geologie*, **16**, 660–675.

Pettijohn, F. J. (1957). "Sedimentary rocks". Harper and Bros., New York.

Piper, D. J. W. (1969). Geosyncline margin sediments in the Silurian of west Connacht, Eire. *Mem. Am. Ass. petrol. Geol.*, **12**.

Plaas, L. van der (1962). Preliminary note on the granulometric analysis of sedimentary rocks. *Sedimentology* **1**, 145–157.

Powers, M. C. (1953). A new roundness scale for sedimentary particles. *J. sedim. Petrol.* **23**, 117–119.

Stone, R. O. and Dugundji, J. (1965). A study of microrelief—its mapping, classification and quantification by means of Fourier analysis. *Eng. Geol.* **1**, 89–187.

Szadeczy-Kardoss, E. von (1933). Die Bestimmung der Abrollungsgrades. *Zentbl. Miner. Geol. Paläeont.* **34B**, 389–401.

Wadell, H. (1932). Volume, shape and roundness of rock particles. *J. Geol.* **40**, 443–451.

Wadell, H. (1935). Volume, shape and roundness of quartz particles. *J. Geol.* **43**, 250–280.

8. A Program Package for Experimental Data Banking

J. L. CUTBILL AND D. B. WILLIAMS

Department of Geology, Sedgwick Museum, University of Cambridge, England

INTRODUCTION

Computers have been used in geology for long enough for a pattern to emerge. A geologist starts to use statistical and analytical methods. He finds that more and better data are needed and so improves his collecting methods. Finally he discovers that he has a storage and retrieval problem on his hands. By this time all but the persistent have retired hurt. The remainder spend most of their time computing and do little geology. To get any return for their efforts these geologists have been forced to acquire a knowledge of computer programming and operating systems more appropriate to computer specialists. Museum curators whose problem starts with storage and retrieval lack an easy way onto the slippery slope and fortunately perhaps few get started.

We both followed this path and finally decided that the only way forward would be for a few geologists to start at the collection, storage, and retrieval end of the problem and attempt to write widely applicable programs. These could be used for data management by geologists whose main interest was in analysis, and for storage and retrieval by those responsible for larger data archives. General packages have a way of being not general enough and so unusable. Our aim therefore has been to construct a programmer's tool which will enable us to produce programs and systems faster, and therefore cheaper, than was possible before. This paper outlines the technical side of our approach.

The difficulties in the way of generalization are considerable. The data types and structures are varied. File sizes range from a few thousand to 10^8 characters and even small projects may have 50–100 files in existence at one time. It is essential to support a variety of methods of data recording. Output formats vary with individual needs

Systematics Association Special Volume No. 3. "Data Processing in Biology and Geology", edited by J. L. Cutbill, 1970, pp. 105–113.

for special displays of data and also depend on the input formats of analysis programs. To this must be added the complexity and frequent arbitrary alterations to computer operating systems. But by far the most serious difficulty is that one cannot specify many of the requirements in advance.

DATA FORMATS

Since the data files will dominate most application, the first step is to choose a standard format which all files can follow. This must be able to accept complex data but without high overheads when the data are simple.

The first restriction is imposed by the machine hardware. We use an ICL System 4 which is a character addressable machine with 8 binary bits per character. Thus integers can be stored as 16- or 32-bits binary numbers, non-integers in floating point form (32 or 64 bits) and text as sequences of 8-bit characters in EBCDIC code. Data consisting of binary patterns not equivalent to printable symbols can also be stored as sequences of 8-bit characters. In our files each entry or record consists of a sequence of elements such as these organized to show the relations between them.

Most files need their own pattern of organization. Thus a file of collecting localities might contain a section on location and a section for published references to the locality. The reference section might have subsections for author, date, title, and so on. A record might contain more than one reference, and each reference more than one author. The elements must be organized to preserve these relations.

We do this by a simple device. Each record consists of a variable number of characters and forms one field. This field is divided into three subfields: identifier, length, and contents.

The identifier indicates that the field contains a record. The length enables the program to move the record as a single unit. The contents consists of a sequence of fields with exactly the same structure—identifier, length, and contents—as the record itself. In our trivial example the record contents would consist of fields for location and reference. The type of field is indicated by the identifier. The reference field would contain fields for author, data, etc. These are not further divided but contain the characters of the author's name, the date, and so on. Two authors of one paper would be represented by two fields each with the author identifier. No confusion with authors of other papers can arise because these would be found in a different field with a reference identifier. But if we want a list of all authors each reference field can be examined in turn and the author fields extracted.

In order to process records organized in this way a program needs additional data besides that contained in the record. It must know which fields are composite, which fields contain actual data, the order in which fields may occur, and for the basic fields the type of data (e.g. integer, floating point number, characters). This is supplied as a table constructed from a description prepared by the designer of the file (for example, see below).

FIELD	LEVEL	TYPE	NAME	GROUP	NEXT
0	0	GROUP	LOCALITY		
1	1	GROUP	LOCATION	0	5
2	2	ALPHA	COUNTRY	1	3
3	2	ALPHA	TOWN	1	4
4	2	ALPHA	NOTES	1	
5	1	GROUP	REFERENCE	0	
6	2	ALPHA	AUTHOR	5	7
7	2	ALPHA	DATE	5	8
8	2	ALPHA	TITLE	5	9
9	2	GROUP	SOURCE	5	13
10	3	ALPHA	JOURNAL	9	10
11	3	ALPHA	VOLUME	9	12
12	3	ALPHA	PAGE	9	
13	2	ALPHA	NOTES	5	

The order of this table is that in which the field identifiers would be found if each field was present once and once only in a record. The level indicates the nesting of fields and the type the nature of the contents. The name is that by which the user will refer to a field. The actual identifiers are the numbers in the column headed "field". "Group" and "next" columns contain pointers within the table which are needed often by the program and so are worth calculating in advance.

This structure can be displayed better as below.

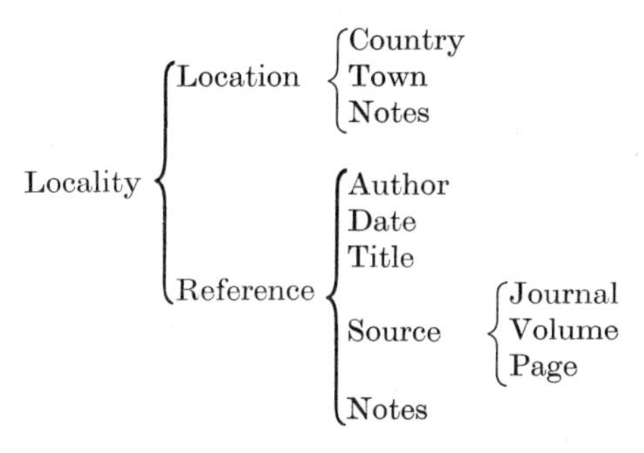

The same name may be used for two or more fields provided a unique reference is possible by combining names (location notes, reference notes).

This technique has three main advantages. Complex information can be stored as can be seen from the papers of Piper *et al.* (p. 16), Forbes *et al.* (p. 311) and Cutbill *et al.* (p. 255) in this volume. Secondly missing data take up no space. This is important as our structures often have 100–200 fields but an average record contains data in only 20 or 30. Finally it allows one to write general purpose programs. A program that can obey the instruction "examine the locality file and list all the references to localities whose country is Spitsbergen" can equally well obey the instruction "examine the portrait file and list the names of subjects whose clothing is hat and sex is female". It merely has to load the table appropriate to the file.

PROCESSING OF RECORDS

Some aspects of processing such as the construction and checking of new records can be standardized. But research needs are unpredictable and it is necessary to have a programming language in which to write sequences of instructions. No normal computer language is suitable as their basic operations are not those we require. Besides arithmetic, *go to* and conditional operations, we need to say things like—is there a reference field, does it contain an author field, is the author Jones, is there another author field, set p equal to reference, get next record, insert p into it, and so on. The types of data elements are numerous—numbers, fields, files, records, pieces of text, parts of records, dictionaries, lists, and trees. New types may have to be added for particular jobs. In addition all the operations needed are not yet known and it must be easy to add new ones as required. We are experimenting with a simple language with these basic facilities. There is only space for an outline here.

There is no compiler in the conventional sense. At run time the instructions are read and translated into a decision table which controls the action of the program through a driving routine. The instructions consist of sentences composed of a sequence of phrases. Each phrase specifies an operation and the data items on which the operation is to be performed, thus

operation (argument, argument, etc.);
operation (········); operation (········):

When the end of a sentence is reached, the resulting decision table is obeyed, deleted and the reading of instructions continues.

Basic operations are coded in normal languages, in our case an assembly language, but Fortran and similar languages could be used. At the start of the run only the code for a few essential operations are in core. The others exist as precompiled libraries and are brought into core by the operation "load". Thus mathematic operations are made available by *load* (math).

Operations are either routines which alter one or more of the arguments, or functions which leave an answer on a list. Functions can occur as arguments. Thus we can write

set (A,plus(B,div(mult(C,E),F)));

which is explicit but rather clumsy if compared with $A = B+(C \times E)/F$. When we have passed the experimental stage the interpreter which constructs the decision tables will be improved to allow more readable formats.

So that the order in which the phrases are obeyed can be controlled, each operation returns a result—true or false. On true the next phrase is obeyed. On false a jump is made. The full format of a phrase is thus

label,operation(argument),label.

The operation "*goto*" always returns the result false. The operation *if* tests the result of a function argument.

Subroutines are set up by the operation *define*(x). When obeyed this preserves the rest of the decision table, enters x in the dictionary and ends the execution of the sentence. x can now be referred to as an operation in later sentences. All such subroutines can be called recursively. Working space and data items are set up by the operation *type* which enters the names in the dictionary and reserves, and if necessary, formats storage.

The operations which have so far been defined include those for opening and closing files, reading and writing records, for locating, modifying, and inserting fields into records, and for translating the contents of records into text form for output. In addition arithmetic operations and logical comparisons are available. Together these give a basic capability for the building, searching, and outputting files.

The programs are rather inefficient because of the dictionary searching and checking of data types which must be done while the program is running. But this is not a serious problem. Many basic operations are long and have been coded efficiently. Only the mathematical operations are really inefficient. Most programs are used only a few times and the extra cost involved in writing and testing an efficient program would be greater than the eventual saving. If a program in frequent use is working and if the user is quite sure he will not alter it again, the most frequently used routines and sequences of phrases can be rewritten as basic operations. As an aid the system keeps a count of the number of times every routine is used during a run. When suitable timing routines become available it will be possible to determine the proportion of time spent in any part of the program.

DATA INPUT

Although cheap on-line methods of data input such as described by Suszynski (this volume, p. 57) will become more widely used, most data preparation is still done using card punches and paper tape typewriters. The first step in processing is to read the card decks or paper tapes and store the text on magnetic tape or disc. Few medium-sized computers have satisfactory systems for input of paper tape. We have written a program with some essential facilities. It accepts 5, 7 or 8 track tape in any code with or without shifts, except those containing non-escaping characters. It images each line, resolving tabulation, erases and composite characters such as underlining. Simple editing facilities are available. Thus |K causes all previous characters in the line to be ignored and |W causes the previous word to be ignored.

Data on specimens is often repetitive, so we allow the typist to declare abbreviations while typing a document. Thus the text

|ABBR {a} *anderssoni*|
|ABBR {b} *arcticus*|
|ABBR {c} *Schwagerina* {a} (Schellwien)|
|ABBR {d} *Triticites* {b} (Schellwien)|
{c} and {d}

would read as: '*Schwagerina anderssoni* (Schellwien) and *Triticites arcticus* (Schellwien)" with obvious applications.

Once the text is on disc it must be printed and proof read. Line-printers seldom possess a proper character set. Our print program has an option that prints a line of marks under each line of text to indicate unavailable characters. Thus

```
input — "Schwagerina anderssoni (Schellwien)"
output — "SCHWAGERINA ANDERSSONI (SCHELLWIEN)"
          +** * ******* **********  |
```

The program numbers lines and can print short lines in two columns to improve the printing speed.

After proof reading the disc copy of the text usually needs editing to correct mistakes. Text editors will be familiar to anyone who has used multi-access facilities so there is no need to go into detail. The edit program reads instructions and makes a new copy of the text. It is possible to say things like

L/*Schwigerina*/ (copy to next line containing "*Schwigerina*")
E/*ig*/*ag*/ (replace *ig* by *ag*)

and so on. It is also possible to delete and insert blocks of lines and to join up pieces of text.

DATA CONVERSION

In order that the text can be converted to a structured record, it must contain additional data to help the translation program. There are four ways in frequent use—fixed field, fixed sequence, tagging, and free format. In the fixed field method the position of the characters in each line determine the position the data will occupy in the structured record. This is the usual method used on punch cards and is the least useful. In the fixed sequence method the fields are variable length, terminated by separator characters and their order determines their position in the record. This is only useful for simple structures. In a tagging system each field has a label to indicate its position in the record. Thus: ⟨locality⟩ ⟨location⟩ ⟨country⟩ Spitsbergen ⟨town⟩ Brucebyen ⟨reference⟩ ⟨author⟩ Smith, A. B. ⟨author⟩ Brown, P. Q. ⟨title⟩ . . . ⟨reference⟩ ⟨author⟩ Jones, L. V. . . . and so on. This method is very flexible and can be used for any data structure that can be accommodated in our records. Its disadvantage is the extra typing needed for the tags even if these are shorter than in the example.

Free format input allows data to be written in conventional untagged form and the conversion program uses signals like punctuation and initial capital letters to work out which element is which. Unfortunately there are seldom enough signals in existing data and in many cases it is necessary to know the meaning of a word before it can be assigned to the correct field. We have written routines to analyse types of structured data such as taxonomic names, dates, grid references and geographical co-ordinates.

Our programs assume that input will be tagged, but users may include directives indicating that certain parts of the text are free format, fixed sequence, or fixed field. This allows users to experiment with different formats without additional programming.

The conversion program can check the tags and that the structure they imply is valid. It also checks that numeric fields contain valid numbers and can be asked to edit character fields to remove redundant spaces and newline characters. These are only a few of the checks that may be needed. To simplify further tests the conversion program creates a holding file with the data in a partially converted form. Records are retained in this file until they have passed all tests. The conversion is then completed and the record added to the main files.

PROGRAM AND FILE CONTROL

Even a simple archive project may have 200 text documents and files in existence at one time and it is all too easy to overwrite a magnetic tape, or delete a document that is still wanted. A batch of work might consist of reading in several paper tapes, running the editing program on several documents, running the conversion program and so on. Often a batch contains 30 or more jobs, some of which are to be run only if earlier ones are successful. This places a considerable responsibility on the operator. Add 20 control cards per job and a 10 s overhead in loading jobs and one has a serious barrier to the use of the system. For this reason we have written our own control program for running batches of jobs.

Essentially this consists of a short piece of coding which reads from a list of jobs, loads the appropriate code for each job and passes it information about the documents and files which it is to process. It is possible to jump from list to list and a jump always occurs if a job signals an error when it ends. Jobs which are dependent on each other may be safely included in one batch. The overhead between jobs is 2 s instead of 10 s.

Documents are held in a random access file and indexed by name and number. The routines accessing documents from this file, if given only a name, automatically select the highest numbered version when reading and create the next in sequence when writing. A random access file is also used as a file index. It contains tables describing record structures, a list of magnetic tapes available to each user and a list of files. These are also indexed by name and number. The routines for creating new files automatically select a magnetic tape and prevent two versions of one file being put on one tape. These routines also maintain file histories which contain for each version the creation date, the name of the program writing the file, the identifiers of documents and files input to the program, as well as the tape identifier and length of the new file. It is possible to override the selection of a tape and specify that the file should be on disc. This saves a lot of tape loading while testing programs.

The control system is not comprehensive. There are no automatic facilities for duplicating important files or recovering after a file has been lost. However we have tried to accumulate in an accessible form all the information necessary to efficient running of a project.

COMPATIBILITY AND AVAILABILITY

The program package is about 90% written and tested and is being used experimentally on several projects at Cambridge. Completion date is autumn 1970 and we expect to run it for a further year before it is sufficiently clear of faults to be released generally. It is not easy to pass from the stage when the package runs well provided an expert is around to look after it to the stage when it can be used by relatively inexperienced people. This is not entirely a matter of good design and some of the problems are worth mentioning in more detail.

The package is written in IBM 360 assembly language and implemented on an ICL System 4-50 with two replaceable disc units, three magnetic tapes, paper tape punch and reader. It requires about 40 K bytes of core and runs under the J Disc Operating System.

Unfortunately, each new version of this operating system has contained alterations which forced us to recode part of the package. There is no sign of these alterations ceasing. An attempt to transfer the package to a System 4–70 with an identical configuration and operating system failed because of undocumented incompatibility in the hardware of the two machines.

Because the package is written in IBM 360 assembly language and the Disc Operating Systems of the two ranges have a lot in common, an implementation on an IBM 360 should not be too difficult. We estimate that it would need about 6 man-months' work. However the language prevents easy implementation on ranges such as the ICL 1900 series. Normally compatibility is made easier by use of a higher level language. Unfortunately Fortran, Algol, Cobol, and PL1 are quite unsuited to this kind of programming and when we started work three years ago nothing else was available. There are now available a number of languages designed for system and compiler writing and which contain most of the essentials that we need. If our package proves to be as useful as we hope then it will be worth rewriting (and at the same time improving) it in one of these languages.

ACKNOWLEDGEMENTS

This work has been financed by a grant from the Office for Scientific and Technical Information and we are particularly grateful to C. C. Leamy for his very constructive attitude to our ideas from the earliest planning stages. C. J. Powell has been involved with much of the programming and Mrs B. Falkner has typed all the programs and documentation as well as finding time to prepare for processing about 12 000 entries from the catalogue of the Sedgwick Museum.

9. The British Biological Recording Network

FRANKLYN PERRING

Biological Records Centre, Monks Wood Experimental Station, Huntingdon, England

INTRODUCTION

The Biological Records Centre is responsible for collecting data on the distribution of most of the British flora and fauna in the country as a whole and in nature reserves in particular. It must store these data, so that they may be retrieved and presented in cartographic or tabular form to meet the needs of biogeographers, ecologists, and the conservation organizations. It operates in two ways: where there is a strong national group or society able to undertake the organization of a survey itself, the Centre acts mainly as an adviser and data processor; where for some reason there is no suitable national group or society the Centre itself undertakes the organization.

The first major national distribution mapping scheme in Britain was that run by the Botanical Society of the British Isles 1954–64. As I was involved with that scheme from its beginning, and because the Records Centre evolved from it, the description of the Biological Recording network in Britain will be based on the botanical example. However the ornithologists have a similar, though more complex network, and, through the Lepidoptera Mapping Scheme which is being run by the Centre, a network of entomological recorders is being developed which we hope will largely follow the botanical pattern.

THE IMPORTANCE OF THE GRID

Satisfactory maps can only be produced when records are collected evenly throughout the area to be surveyed. In the British Isles we have been using 10 km grid squares as our basic unit of recording since 1954. For a primary survey of any group we are concerned only to collect

Systematics Association Special Volume No. 3. "Data Processing in Biology and Geology", edited by J. L. Cutbill, 1970, pp. 115–121.

the answers to the simple question, "Which species are present in each of the 3600 squares which cover the Country?" It is important, especially with a large volunteer force of amateur naturalists available, not to be tempted to elaborate at the outset. It is easy to devise more complex schemes, and the answers will often be provided for some parts of the country—but never for the whole. The advantages of using 10 km squares as recording units at the country level are as follows:

1. The problem is finite. There are a known number of squares, and experience soon indicates for each group the approximate maximum of records which are likely to be made. When a square has been "completed" the volunteer can be asked to work in another under-recorded area.
2. The progress of a scheme can be reported by distributing maps showing numbers of species recorded from each square, which acts as a stimulus to the volunteer force.
3. The volume of data which will have to be handled can be calculated in advance from samples made in a small number of squares. Thus the data handling costs can be accurately estimated in advance.
4. Data are received in blocks for particular squares: all the plant or animal records have the same date, grid reference, county, etc. so that data input is very economical. In the past our records have been punched into cards, but in future all data from primary surveys will be punched into paper tape. The basic unit of information being species number, date, grid square, county and status. Programmes will be developed which will make it possible to check the accuracy of the data, eliminating any records which fall in the sea, and showing where grid squares are not compatible with county numbers.
5. Map-making can be carried out on tabulators which produce, economically, a high quality product suitable immediately for printing. For 15 years botanical maps have been produced from 40-column cards on a small tabulator with 25 consecutive print units. The maps can be sent from machine to printer after only a minimum amount of touching up. In seeking a replacement suitable for reading 80-column cards or tape we have looked at data-plotters of two kinds, those with a drawing pen and those with a printing head. They are both slow compared with tabulators or line-printers and do not yet produce the symbols we require, in particular a large solid dot, of the cartographic quality needed for publication. We have thus decided to replace our 40- with an 80-column tabulator fitted with 100 consecutive print units. This will be built for us and give a selection of five special symbols made to our order. In addition the line spacing will be modified to give the same vertical and horizontal spacing.

A pack of 80-column cards will be prepared by the computer, one card for each horizontal line of the map arranged from north to south. Each column of the card will correspond to a vertical line of the map, excluding those used to indicate species number and card order. The position of the hole in the column will determine which symbol appears in the particular 10 km grid square. For making maps of the British Isles two runs of 100 cards will be necessary, one for Great Britain and the other for Ireland which is on a different grid. However, even using a tabulator of modest speed (300 lines per minute), one or two minutes will be sufficient for the printing of each map. By working with an off-line tabulator in this way we can make as many copies as we wish without using expensive computers or peripherals. The tabulator will also provide the Centre with a fast card-reader, which we need to meet increasing demands for print-out of new data. The system is also very economical. Data-plotters can cost £10 000, whereas the new tabulator will cost only £4500 and will list cards as well as print maps.

6. Surveys can be carried out at regular intervals, and because the method is repeatable, conclusions about changes in species distribution can be validly drawn. This is of importance to conservation. It is intended that a second edition of the "Atlas of the British Flora" (Perring and Walters, 1962) should appear in 1985 when the date line will be advanced from 1930 to 1950, and that a completely new survey should be organized about 2000. If surveys are made every 50 years this should ensure that no once widespread species declines to a point of danger before the decline is noticed. For rare species this is too long a time interval and more frequent surveys are necessary. Maps of the rare species of vascular plants in Britain were prepared between 1954 and 1958. During the last two years we have carried out a further survey of the rarest 300 species, which included all those which occurred in 15 or fewer 10 km squares, to discover the extant localities. We hope in the next two years to collect information on the population size for each locality, so that we shall have a basis for making exact comparisons of the total population of each of these species in the country every decade.
7. The 10 km square is part of a network of grids which make use of the same data at various levels: county, country or continent. At a continental level our data are now being incorporated into maps being prepared by the Secretariat of the European Flora Mapping Scheme in Helsinki, Finland. The unit chosen for mapping is the 50 km square of the Universal Transverse Mercator grid. This unit is small enough to produce meaningful maps, but not so small that there are too many for the task to be completed. There are about the

same number of 50 km squares in Europe as there are 10 km squares (3600) covering an area the size of the British Isles. The grid is square within 6° bands of longitude but these regular areas are joined by "zones of compensation". At these points, by allowing "50 × 50 km squares" to vary in width between 40 and 60 km, it is possible to remove squares in a regular manner proceeding to higher latitudes.

There are two important reasons for using the U.T.M. grid in an area as large as Europe.

(i) The units are squares of approximately equal area. Latitude/longitude gives a grid which changes from squares to rectangles and increases the number of units proceeding from south to north, which could distort a distribution map.

(ii) Maps showing the grid are available for almost the whole of Europe at 1/500 000 and, in many countries like Great Britain, maps showing the grid on a larger scale are coming into general use.

Map publication, which is in no way automated, is following the systematic order of "Flora Europaea" (Tutin *et al.*, 1964). Each country completes its own portion of the map by hand. The pieces are then built up into maps for the whole of Europe in Helsinki. Each country must maintain its own documentation of the source of each record.

Entomologists are now following the methods developed by the botanists. Initiative taken by Professor Leclerq of Gembloux, Belgium and J. Heath of the Biological Records Centre has led to the setting up of an Invertebrate Survey of Europe, to which workers in half-a-dozen countries have already agreed to contribute. They will use the 50 km squares and the same base map, but in addition an 80-column record card has been prepared which is suitable for all invertebrates. For the next three years the headquarters of the Secretariat will be at Monks Wood.

In many counties in the British Isles, botanical recording is based on subdivisions of the 10 km square. Where labour and time permit, the unit chosen is the 2 × 2 km square, the tetrad, of which there are 25 per 10 km square, and an average of 500 per county. Elsewhere 5 × 5 km squares have been chosen, giving about 100 units per county.

The Botanical Society of the British Isles has a Recorder appointed in each county. Their function is to answer questions about the plant life of their area. Recently the Society has suggested that a card index, the property of the Society, should be kept by each Recorder. Many indexes already existed, but many more have now been started. In future, details of distribution, at least for the common species, will be stored at county level, the data being summarized in the form of lists for each 10 km square and sent at intervals to the B.R.C. for storage at the national level.

SELECTIVE STORAGE

It is the aim of the B.R.C. to hold in its index the most important and most frequently asked for data on the distribution of plants and animals in the British Isles. For the commonest species all we need to know is whether or not it has been recorded in a 10 km square during the most recent survey; such data has solely a cartographic function and can be stored on magnetic tape or disc until required for that purpose. For rare species or those which need expert determination, it is much more likely that details will be asked for by outside enquirers, or will be required for compiling commentaries on maps produced. For this reason we are acquiring documentation as full as possible for the following categories of records.

(i) National Rarities: Species which occur in 15 or fewer 10 km squares.
(ii) Local Rarities: Any records for a species which are separated by more than 100 km from any other record for that species, even though the species as a whole is not a national rarity.
(iii) Species in critical genera or groups, or taxa below the species level where a specimen has been determined by an acknowledged expert.

In practise this selection of records for which details are required can be operated through the County Recorder system. Guide lines on the type of records, and the methods of submitting them to the B.R.C. have been prepared. As in the past, there will in the future be two kinds of records: lists of species additional to a 10 km square; and Individual Record Cards for specimens which fall into categories (i)–(iii) above. These latter, which form perhaps 10% of our data bank, may still be submitted on 40-column cards. They are big enough, cheap, travel well and are familiar to thousands of amateur naturalists. They can be sorted and stored by species/county as original documents. From these, duplicated packs of interpreted 80-column cards are being prepared which will contain species name, locality, county, date, collector's initials, and source in a machine and user readable form. One will be stored in county order, the other in species order by grid squares within counties. From these questions can be answered by visual inspection or by a print out on the tabulator. These records, before reaching their final storage location, will have been passed through the computer so that data of cartographic importance will be added to the appropriate section of the magnetic tape or disc store for each species.

THE ROLE OF THE MUSEUM

For a group of organisms like the vascular plants which are catered for by a well organized society like the B.S.B.I., a data network based on County Recorders, keeping card indexes in their own homes, is adequate: but it is not ideal. Important records ought to be supported by voucher specimens. If these are collected a herbarium must be started and maintained. When the recordership changes, will the successor have room, in cupboards of the same size, in which to house them?

Ideally, such information about the plant and animal life of an area ought to be centred on a museum—natural history ought to be a natural part of local history, but too often it plays second fiddle to archaeology. Wherever possible a museum should serve a county and be the biological records centre for that county. No doubt the work will be done by local amateurs, but local authorities ought to provide facilities, space to work and to store card indexes and specimens. Only if the facilities are provided can we achieve the complete network which we require.

If the Maud proposals for the reform of local government in this country are accepted we should have fewer units, each strong enough financially to support an adequate museum. This could mean a break away from the vice-county system on which biological recording in this country has so long been based, but once the necessity for change and its advantages were accepted, all the biological societies in this country might sit down together to see whether a new system based on regional museums could be set up. We should aim at an organization and a museum for each block of 10 km squares.

CONCLUSION

I make no apologies for devoting most of my paper to the problems of data collection rather than to data processing. We must remember our primary objectives and not allow fascination with the sophistication of the machinery to blind us. In general the use which will be made of each record in our data bank is low. Perhaps 75% of it may only be used once in 25 years. The commoner a species the fewer extra records which can be added to a map, so it may be quicker to do this by hand. Therefore we must keep down our input costs and structure our output so that the majority of questions can be answered cheaply. I am sure there are some who would suggest that we use optical scanning of marked sheets. This is certainly a method we shall be looking into again on an experimental basis as soon as our own systems man is appointed within the next few months, but there are problems with disciplining several thousand unknown volunteers in the correct marking of fragile forms in all sorts and conditions of weather. Moreover our recording cards are used for a wide range of non-data processing functions, so that the economics of long production and standardization have to be considered. We should obviously start with a small scheme under our full control.

Experience over 15 years has shown that nearly all the questions asked of our data bank are either species or locality orientated. We feel that, on the grounds of economy, we should only be ready to answer these questions easily. Even if a few questions are expensive *per se* as long as they *can* be answered, the *overall* cost may be less than if a very complex and comprehensive search routine had been devised.

This is a small country, Monks Wood is very central; people are glad to call and do a little searching themselves. Often a sight of the primary data is valuable in solving problems which arise, so that a visit is desirable anyway. In the absence of inexhaustible funds I believe it is more important to aim at as complete a collection of data as possible which is reasonably accessible, than to have incomplete data at the touch of a button.

REFERENCES

Perring, F. H. and Walters, S. M. (1962). "Atlas of the British Flora". Nelson, London.

Tutin, T. G. *et al.* (1964). "Flora Europaea". Cambridge University Press, London.

10. Machine Languages for Representation of Geological Information

C. J. Dixon*

École des Mines de Paris, France

ABSTRACT

There are no theoretical reasons why information should be stored in machines in the same form as the languages of human communication. Investigations into the vocabulary of geology and the syntax of geological description show the desirability of using a special language for the machine into which information is translated on input and from which text is reconstructed on output. The languages of human communication consist essentially of a vocabulary and a grammar which, for reasons of history and generality, do not necessarily correspond to the symbolism and structure of the information in any particular field of study. Furthermore, the human language used to convey information in a field of study forms a subset of that language. A project is described that aims at the definition of a machine language for geological information, investigations into the problems of input text processing and text construction for output purposes. The language consists in principle of a series of identifiers and a data structure. Identifiers are of four types: labels, element descriptions, relationship descriptions and qualifiers, and the data structure is based on the notions of list structures used in dynamic programming. Methods are suggested for representation of the identifiers as symbols containing their essential semantic content. Such symbols may correspond to words, phrases and complex word constructs in human language. During the experimental stages, it has become necessary to use a metalanguage corresponding to the machine language, but in a suitable form for human comprehension.

During the few years that binary machines have been used to handle geological information, some progress has been made in handling files of numerical data, and in the use of indexing or documentation systems.

The latter have as their main aim, a reduction in the amount of

* Present address: Imperial College, London, England.

Systematics Association Special Volume No. 3. "Data Processing in Biology and Geology", edited by J. L. Cutbill, 1970, pp. 123–134.

reading, or examination of collections of objects, that must be done to obtain the information relevant to a particular enquiry or study. However in some fields the time is coming where even a machine documentation system provides such a large volume of information, that the use of machines to assist the geologist in synthetizing and rationizing the information, becomes justifiable. Provided the information consists of homogeneous sets of numerical data, and provided that mathematical notions exist that permit the use of mathematical analysis, this presents no serious problem.

Unfortunately, much observational geological information does not readily lend itself to such methods, and it becomes necessary to try and place the information in a machine information system allowing it to be handled in any chosen way.

However, the representation of geological information in a machine information system requires a deep understanding of the nature of the information. To this aim, a group of workers at the École des Mines de Paris and the Royal School of Mines, London, began a project in 1968 under the general title of "Project Geosemantica 70". The initial aims of this project were set out by Laffitte (1968) and some of the preliminary conclusions are described in this paper, and in a number of others (Dixon, 1969; Capitant *et al.*, 1969).

Not unnaturally in a project involving two "Schools of Mines", it is the information concerning mineral deposits that has been our main preoccupation. Mineral deposits pose some interesting information problems. First of all, their study touches all aspects of the geological sciences, although none in their most detailed and refined form. Mineral deposits are not simple objects; there exists no notion of a "mineral deposit" other than on a commercial basis where economic considerations play a role as important as geology. There is no concept of a "species" of mineral deposit. They range from isolated lenses of sulphide minerals to stratigraphic horizons underlying thousands of square kilometers. The only basis on which they can be grouped and understood, is on the basis of the geological environment in which they are situated, which in itself is not a clear concept.

What we can do is observe and describe various bodies of rock and their environments at various scales, and show how each is related to the others in larger and larger groupings. The first of our conclusions is therefore, that an information system must have a structure that reflects these features.

Any particular "observable metallogenic unit" (see Dixon, 1967) consists of certain bodies of rock containing mineral of potential economic interest as an associated environment of rocks, structures, etc. Geologists study these geological objects, they observe, they measure, and perforce they must communicate the results of these observations and measurements. They do so, of course, in the language of the subject, which is a subset of a language of human communication, augmented with numerical data in units of measurement acceptable to other scientists. We can say that geological information is created at the time of observation and measurement, and is then communicated by coding it in the geological language. What I intend to describe here is methods of representing this information in a binary machine in a way that corresponds to the geologists' observations and conclusions, using the language of the geologist as the communication link between the actual geology and its machine representation.

The aim is to do so with the minimum of disruption of traditional geological language. Clearly the problem can be separated into two parts—the design of the method of representation, and the interface problem with the language of geologists. Concerning this interface problem, there is one basic fact worth pointing out at the outset. Geologists communicate partly by linear character languages and partly in pictures. Technically, pictures (or patterns) are difficult to input into a binary machine, and we must take careful note of this fact because the prior transformation of graphical information into a human language risks the introduction of ambiguity.

STRUCTURES IN GEOLOGICAL INFORMATION

In spite of our attempt to keep separate the representation problem and the interface, we must look to descriptive text for some ideas on the structure of the information. Descriptive geological texts are usually broken up by headings and subheadings. Let us look at an example of such a pattern of headings.

Geographical Setting

General Geology

- Stratigraphy and Lithology
  - "The Upper Volcanic Groups"
  - "The Red Shales"
  - Basic Rocks
  - Stratigraphic Relationships
- Structure
  - "The Eldorado Anticline"
  - Late Faulting
- Mineralization
  - Mineralogy
  - "The San Juan Orebody"

Other Orebodies
Alteration
Zoning

Metallogenic Conclusions

At a first glance this seems to be quite a normal breakdown of the subject matter. Many of the headings are what may be termed "subject rubrics" which are identifiers, that tell us in general what subject matter is to be found in the text that follows. However, mixed up with the subject rubrics are proper names (placed in quotation marks) that are not subject rubrics but imply the existence of natural groupings of the geological objects being described. We can therefore identify in a descriptive text firstly a structure of subject matter and secondly a natural geological structure.

We can find the same thing in the body of a text. For example:

"plunging at 45° west"

is a phrase consisting of a subject identifier and a numerical descriptor. By contrast:

"alternation of red slates and quartzites"

is of the other type, that is to say it tells us something about the actual geological configuration of rocks that make up the environment being described.

At least one other type of elementary proposition may be found in text, for instance:

"gentle folding followed by normal faulting"

which states explicitly a chronological relationship.

One feature of geological description is the implication of information. The first phrase above contains a subject identifier but in the second and third there are descriptors that are identified implicitly. It is necessary to state that "45° west" is a plunge, because it could also be several other things, but it is not necessary for a geologist to identify the descriptor "red slates" as a rock, it is however necessary to do so if it is a machine that carries out the interpretation.

MACHINE REPRESENTATION OF STRUCTURE

Let us consider in detail the representation in a binary machine of what I have called natural geological structure. Between each pair of geological objects in an environment, there are two possibilities—either they touch or they do not. If they do touch, then there exists a relationship between them which is basically geometrical but has a

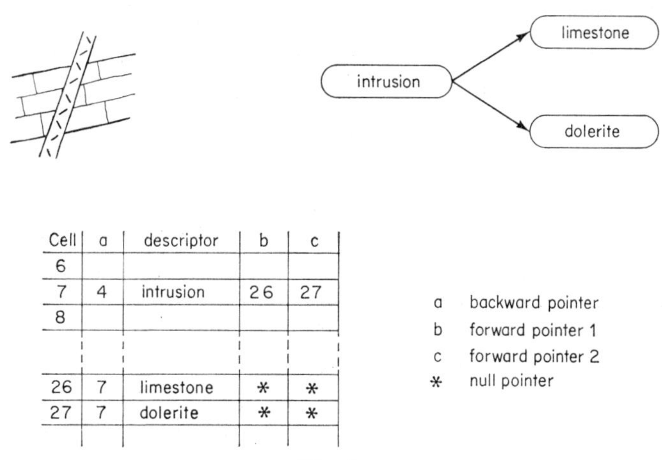

Cell	a	descriptor	b	c
6				
7	4	intrusion	26	27
8				
26	7	limestone	*	*
27	7	dolerite	*	*

Fig. 1.

geological cause. For each such relationship that exists, we can state a relationship descriptor which contains the information on the geometrical and geological aspects of the relationship. In the simplest case we can place this descriptor in a cell associated with two pointers to two other cells that contain the descriptors of the two objects between which the relationship exists (Fig. 1). Clearly we can extend this idea to include the possibility of another relationship descriptor in one or both of the two cells (Fig. 2). There are occasions where the same

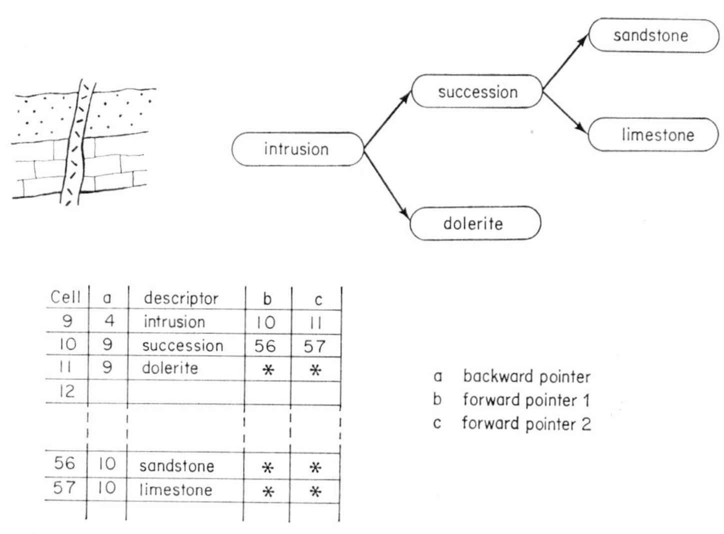

Cell	a	descriptor	b	c
9	4	intrusion	10	11
10	9	succession	56	57
11	9	dolerite	*	*
12				
56	10	sandstone	*	*
57	10	limestone	*	*

Fig. 2.

relationship holds between more than two objects. This can be solved by introducing a dummy relationship just as we use "and" in text. For instance:

"alternation of limestone, sandstone, and shale".

With these few simple rules we can build up large binary trees which represent any spatial configuration of geological objects.

By choosing suitable machine structures we can represent any other "structure" that exists in the data. Chronological structure, for instance, can be thought of as a network and represented in more or less the same fashion as in network analysis. What I called "subject matter" structure earlier, can be represented as a standard multiple branched tree structure like an overlay structure in programming.

A further refinement we can add is to represent in a machine, several structures referring to the same set of descriptors. Quite simply, we can replace the descriptors in our structures by pointers to a table of descriptors (Fig. 3). The table of descriptors can be held in the machine's

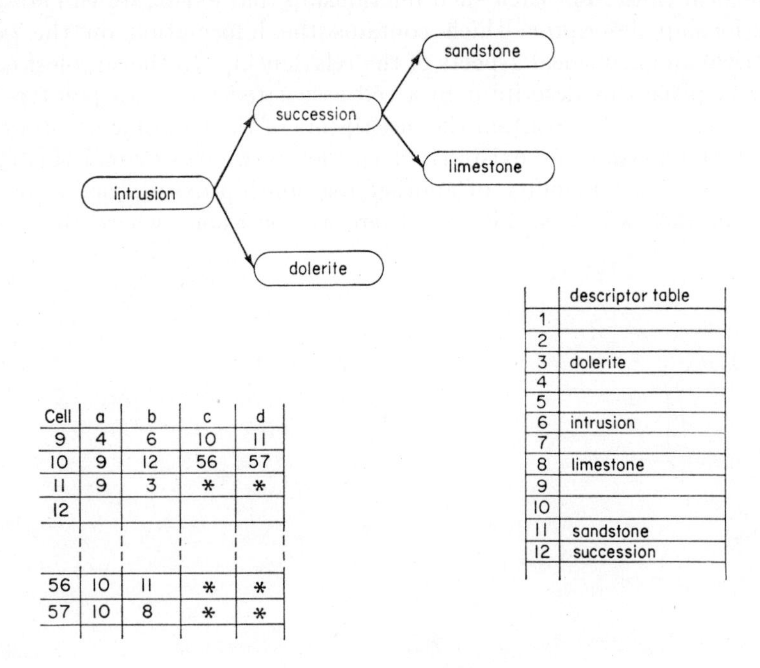

Cell	a	b	c	d
9	4	6	10	11
10	9	12	56	57
11	9	3	*	*
12				
56	10	11	*	*
57	10	8	*	*

	descriptor table
1	
2	
3	dolerite
4	
5	
6	intrusion
7	
8	limestone
9	
10	
11	sandstone
12	succession

Fig. 3.

collating sequence, so permitting its use as a "key word" list for more conventional treatment.

LABELS

We have already seen that it is very difficult to define a mineral deposit and therefore the idea of a "file" consisting of a certain number of "records" is difficult to implement. What we can do is to define certain groups of objects that can be thought of as belonging together and give them a name. In fact we do this all the time. The single word "Bingham" conveys to a knowledgeable geologist a certain assemblage of rocks and bodies of mineralization in the state of Utah in the U.S.A. Because it is a place name, one can say that it conveys a certain amount of locational information in itself, but certainly not enough. It is simply a label which can be used as a means of access to information, whilst conveying none in itself. Descriptive geological texts are rich in all sorts of labels often used as a means of implying information that is complex. The label "The Chalk" for instance like all other formation names, conveys implicitly a whole gamut of lithological, stratigraphical and geological information. In many languages, labels are important in text interpretation, because they are proper names beginning with an upper case letter and can readily be identified.

We can use the same idea in a machine. Each grouping of objects that has been thought of by a geologist as one entity can be labelled. At any point in the natural structure we can place labels rather than object descriptors which are in fact pointers to other labelled groups. Thus if a certain area contains a series of mineral deposits, each consisting of a series of orebodies; then each of these orebodies can be represented as a labelled group which fits into the descriptor of each of the larger groupings and so on. Although the treatment of a data set like this is more complex than a record by record data-file, it does solve the problem of what to call a "record".

INTERFACE CONSIDERATIONS

Although we have referred several times to descriptive texts, we have done so in order to gain an understanding of the structures existing in descriptive geological information. Our machine representation scheme is quite independent of the interface. Less progress has so far been made on the interface problem. In broad terms we can say how we can go about constructing the interface.

A number of studies in other fields shows how the problem of interpretation can be solved. For example, Bell (1968) has shown that provided text is free from certain types of syntactical ambiguity, a relatively simple syntactical analysis permits the isolation of a series of elementary propositions from the text. We can build up our structures if we can isolate from a geological description, phrases that:

(i) state one simple subject identifier and one descriptor;
(ii) one simple relationship between two objects or groupings;
(iii) one simple chronological relationship.

So far many examples of geological descriptive texts seem rather intractable but we can at least hope to make some progress with texts subject to certain restrictions.

METALANGUAGES

The data structures discussed earlier are rather difficult to handle during the experimental phase, and in consequence it became desirable to use a metalanguage which described the structure in a more convenient form. Taking again the case of natural geological structure we can construct such a metalanguage by using the direct analogy that exists between binary trees and algebraic expressions.

Suppose we let

O_n be any object descriptor
L_n be any label
R_n be any relationship descriptor

then the relationship:

"The Ambridge group consists of oölitic limestone overlain by marl"

can be represented as the following:

$$L_1 = R_1(O_1, O_2)$$

where L_1 = "Ambridge group"
O_1 = oölitic limestone
O_2 = marl
R_1 = normal sedimentary succession.

Again we can generalize, so that in any expression of the form:

$$L_n = R_m(x, y)$$

x and y can be either object descriptors or labels or further expressions of the same type. A series of expressions of this sort can be transferred into our machine data structure without difficulty.

What appeared later in our investigation was that this metalanguage could be used as a routine input language. The reason is connected with the already mentioned feature of graphics accompanying texts.

GRAPHICAL INFORMATION

We must digress here to discuss a very important point about geological information. The configuration of objects in space that constitutes a piece of geology is three-dimensional. If we add to this the chronology of events that have brought this configuration into being, we have a four-dimensional field. Every time we draw a diagram, section, plan, stratigraphic table or paragenetic sequence chart, we are recognizing that descriptive text is not the best way of communicating such information. Text is written in a linear language. Up to the present, the input of anything other than a linear character language into a computer, poses certain problems, but at the same time we can perhaps elaborate an input language for the information in our four-dimensional field that is less ambiguous than normal human language. Perhaps in a modified form our metalanguage is an example of a linear language, suitable for computer input that can be used to transform diagramatic information.

DESCRIPTORS

So far we have only examined the representation of structure. Just as a human language has a vocabulary and a syntax, so our machine representational scheme must have a "vocabulary" in addition to structures. This will consist of a series of descriptors capable of describing any object or any relationship. To use the words of our own language is perhaps simplest, but has several important disadvantages. In brief these problems are as follows.

1. The total meaning of the words in the human geological vocabulary is not evident in the words themselves. One can note that "Biotite granite" could be identified by a computer as a type of granite but the word "granite" tells the machine nothing except a statistical probability that it is either a rock or mineral because it ends in "ite".
2. The words in the geological vocabulary are not used in the same sense by all geologists. One can note here that there are, however, quite extensive "schools of thought" within which considerable conformity exists.
3. The vocabulary of geology is rich in synonyms and words of

equivalent meaning, plus the fact that essentially the same words in different languages with minor but important differences in spelling.

4. The words and word-groups in the geological vocabulary are very variable in length, and to use them as descriptors in a machine means carrying a large overhead of program or the introduction of large redundancy.

The proposed solution to these problems is the use of so-called semantic symbols. I do not propose to discuss this subject in great detail because it has, and is to be described in greater detail elsewhere (see Dixon, 1968; Laffitte, 1968; Dumort, 1968; Capitant *et al.*, 1969).

The construction of a semantic symbol vocabulary is based at the outset on a classification of objects and relationships. This is relatively simple in particular, it is simpler than trying to produce a classification of terms in the human geological vocabulary.

Many excellent classifications of rocks, minerals, etc., already exist and others are the subject of intensive study.

The result of any classification is a list of basic notions corresponding to observable facts and measurements, which constitute a set of "atoms" of information. To put it another way, we can elaborate a series of attributes, which each object either possesses or not. In a binary machine we can therefore represent any object as a string of bits showing the possession or non-possession of each attribute. At any chosen level of precision, the set of attributes is chosen such that this pattern of bits is different for each object we chose to regard as different. Such a string of bits is what we call a "semantic symbol". It is a symbol in that it represents something else, and semantic in the sense that it conveys automatically the meaning in terms of our set of classification attributes. An important feature of such symbolism is that the set of attributes need not be hierarchical. It is worth noting that many classifications of geological objects fall down because their authors have tried to force them into a hierarchical scheme.

INTERFACE CONSIDERATIONS

I have been at pains to point out that our symbolism is based on the classification of objects, not of terms. This again enables us to treat the interface with human language as a separate problem.

The interface for object and relationship descriptors consists of a structured dictionary and a fast look-up algorithm. Such a dictionary and access method is at present being developed as part of Project Geosemantica 70. One important feature is that we can structure the dictionary according to the main human languages and "schools of thought" amongst geologists, so solving the problem of differences in meaning and usage.

BASIC DESCRIPTORS AND QUALIFIERS

In practice it is not convenient to place the whole description of an object in a single symbol so we can introduce a notion of basic descriptor symbols and qualifier symbols, parallel with nouns and adjectives in human languages. A basic description of an object is a description of its substance or essential nature, of a relationship it is its essential geometrical nature. All other information is stored in the form of descriptors. For instance a body of rock would have as its basic descriptor a symbol representing the lithology or petrography of the rock. Information such as its age would be stored as qualifier.

Qualifiers can be classified and represented as semantic symbols except for the special case of numerical data.

A MACHINE DATA LANGUAGE

What has been described here is a design for a geological machine data language capable of being used to store descriptive geological information in great detail. It is economical, in that it contains virtually no character representation, consisting only of bit string symbols and pointers. Each bit in the storage device conveys an atom of real geological information. It does require support in the form of large interface dictionaries which are perfectly possible with modern direct access storage devices and data set organizations. It can be interfaced at present with normal geological vocabulary and with special syntax forms, and later should be capable of being interfaced with simple text.

Just how useful it will be, and how economic to use, will depend on the results of tests with actual data sets created in the language. This cannot be done in advance of the existence of the interfaces, which will take time. But it gives such possibilities of solving many of the hitherto intractable problems concerned with geological information that it should lead to a significant advance in our capability to handle geological information in machines.

REFERENCES

Bell, C. J. (1967/68). Implicit information retrieval. I.B.M. Research Report RC-2026. (See also International Conference on Mechanized Information Storage and Retrieval Systems, Cranfield, England.)

Capitant, B., Dixon, C. J. and Laffitte, P. (1969). "Project Geosemantica 70—Symbolisme Sémantique" (in preparation).

Dixon, C. J. (1967). The structure of metallogenic data files I.U.G.S. Meeting on Computerization of Mineral Deposit Information, New York, 1967. (See also I.U.G.S. Newsletter, June 1967.)

Dixon, C. J. (1969). The machine representation of geological information. A.I.M.E. Internationl Computer Applications Symposium, Salt Lake City, September 1969.

Dumort, J.-C. (1968). L'Analyse sémantique: outil dans le domaine de l'informatique géologique. *Bull. B.R.G.M.* Series 2, Section IV, No. 3.

Laffitte, P. (1968). Limites actuelles de l'informatique géologique. *Bull. B.R.G.M.* Series 2, Section IV, No. 3. (See also L'Informatique géologique et la terminologie. *Mineralium Deposita*, No. 3, June 1968).

11. Some Geological Data Structures: Arrays, Networks, Trees and Forests

T. V. LOUDON

Institute of Geological Sciences, London, England

ABSTRACT

Geologists frequently record and process data as two-dimensional arrays. Use of this structure can help to make data collection more systematic, can help groups of workers to standardize their methods, and can provide an effective framework for computer analysis. Even data which cannot conveniently be recorded as a table, because of columns which are largely blank, or because most of the information is contained in a few rows, can be stored and analysed in the computer as an array.

The array, however, is not always an appropriate data structure in geology. For example, network and tree structure can more effectively represent some type of geological data, such as sequence of events in geological time, or a description in which general and detailed observations are combined. Choice of an appropriate data structure is important for recording data and for handling the data in a computer.

ARRAYS

When geologists first responded to the availability of computers, the organization of data for punched cards brought a new formality to their approach to data collection and analysis. The field geologist, making a series of observations on one bed after another, found that observations for each bed could be made to fit on a punched card. Data from successive beds could be recorded on successive cards. He could organize his observations into categories, for example: thickness, facies type, main lithology, subordinate lithology, minor lithology, degree of exposure, weathering, colour, grain size, sorting, fossil content, sedimentary structures, other comments. Each category could be assigned a position on the card. The categories, or observed variables, could be chosen to be

Systematics Association Special Volume No. 3. "Data Processing in Biology and Geology", edited by J. L. Cutbill, 1970. pp. 135–145.

independent, and in many cases, the set of possible observations of a variable could be chosen to be mutually exclusive, so that only one value need be recorded for each variable.

The process of recording data could in some cases be formalised further. The set of possible values for a variable could be foreseen, and drawn up as a list. Thus, possible lithologies might be: conglomerate, sandstone, siltstone, shale. An abbreviation could be recorded in place of the full name, thus, CG, SD, ST, SH, or a numerical code could be used, where 1 represented conglomerate, 2 sandstone and so on. There are major drawbacks to this rigid framework for collecting and recording data, but before looking at these and the reasons for them, it is worth keeping in mind the advantages that are gained.

Firstly, by making explicit statements of what is to be observed, it is easier for a number of geologists to agree on vocabulary and definition and to build up files of descriptions which all can share. Secondly, by clarifying the aspects of the sediment about which observations are made, the absence of a comment acquires new significance. The fact that no record is made under the heading Sedimentary Structures would indicate that none was observed. In a less rigid descriptive framework it might merely mean that none was looked for.

In the field, recording of data may be simpler if a spread-sheet is used. The spread-sheet is divided into rows and columns, and each column has a heading indicating the type of information that the geologist should record in it. Each row in the sheet is used to record data on one item, such as a depositional unit. Examples of spread-sheets can be found in Brisbin and Ediger (1967). An alternative method is to record observations in a field note-book using a check-list as a guide. The check-list corresponds to the headings on the spread-sheet, and reminds the geologist of the attributes of the rock that he had decided to record. The check-list has perhaps greater flexibility than the spread-sheet since new categories can be easily added in the field. On the other hand, the results of using a spread-sheet are usually neater, and easier to transcribe on to punched cards. Agreed terminology and operational definitions can be listed on a sheet or in a report for reference purposes.

There are also advantages in using a rigid recording framework from the point of view of referring to and analysing the data. If a linked set of observations can be brought together in a rectangular table with each column representing observations of one variable or attribute and each row representing one item or point, then they are said to constitute a two-dimensional array. A single row or column of the table would form a one-dimensional array. It is also possible, though less common, for geologists to record data as a three-dimensional array. This would be printed as a set of tables, each with the same row and column headings, and might arise, for example, if complete sets of readings were repeated at weekly intervals.

The two-dimensional array, particularly if it consists of quantitative measurements, is also known as a matrix, and a special branch of algebra, matrix algebra, is available for discussing this data structure. Matrix algebra has proved to be a simple, powerful and elegant tool and straightforward accounts of it can be found and understood even by geologists with no mathematical training (see, for example, Krumbein and Graybill, 1965).

A wide range of statistical methods is available for analysis of data arrays. Intuitively, there are benefits in having data that are complete and consistent for every item. Any pair of items can be compared on the same basis, and any pair of variables can be compared for every item.

DIFFICULTIES WITH ARRAYS

When data are looked at in a formal mathematical framework, difficulties may arise that seem, at first sight, to be due to the rigidity of the framework. A second look may show that the problem is not that the framework is rigid but that it is inappropriate. The mathematical approach should not be discarded simply because the wrong structure was chosen. The array, however, is such a useful structure that geologists have adapted it for use with many types of data.

The geologist who attempts to fit his data into the structure of a two-dimensional array is likely to find sooner or later that he is obliged to adopt various ad hoc devices to force his observations into the structure. For example, he might choose the variable "lithology" as one of the columns in his data matrix. Among his chosen categories might be "sandstone" and "shale". His system of recording data might work admirably until he had to describe a unit in which sandstone and shale are finely interbedded. Time is limited and he might wish to treat the unit for descriptive purposes as one bed, but there is only space to record one lithology, since all categories must be mutually exclusive. The easiest solution might be to erect a new category "interbedded sandstone and shale". Probably that would serve well enough but it does not seem quite satisfactory. A completely new category has been added, and the procedure for recording data is obscuring the fact that this is not really a new lithology at all, but a mixture of two that both occur frequently elsewhere in the section. Moreover, other beds of mixed lithology are likely to be found, and the number of new categories could quickly become inconveniently large.

In these circumstances, the geologist could resort to another method of modifying the matrix. The concept of "lithology" could be represented by two or more variables. Instead of thinking of each bed as being of one lithological type, he could record "dominant lithology" as one variable, "subordinate lithology" as another and "minor lithological components" as a third. A new danger now appears. It may be that the vast majority of beds consist of a single lithology. Two new variables would then have been created and most of the time would not be used. The field geologist would make field records on large sheets of paper, most of which would be left blank. Empty spaces would be left on the punched cards and read in by the computer to fill locations in its store with the symbol "blank". Nor is there any guarantee that the revised system would be adequate. A system might be devised in which lithology occupies three columns and in the course of using this system, a unit might be encountered which had four lithological components.

The basic difficulty seems to be that the useful information in a measured section is not evenly shared between the beds. It may be the unusual item that deserves notice because it contains the significant information.

A section that contains a few fossiliferous beds or one in which a sample was taken every 100 ft for geochemical analysis provides a more extreme case of information being concentrated in a few items. A possibility in these circumstances is to record the palaeontological or the geochemical data in a separate subfile. A variable in the main data array could be used to indicate whether or not fossils were found in the bed. The actual species recognized could then be recorded in the palaeontology subfile. The same bed numbers could be used in both files to provide a means of cross-reference from one to the other.

If spread-sheets are used, a similar procedure can be followed. The geologist would have several types of spread-sheet available. One sheet might perhaps be used to record the overall characteristics of the bed. If the geologist found that a particular bed was fossiliferous he might note the fact on the main sheet, then select a separate sheet designed for recording details of the palaeontology, and on this record the same bed number followed by the required information about the fossil content. If geochemical samples were collected at intervals, relevant information could be recorded on a third type of sheet designed for this purpose. The main data in this case would be obtained not in the field but in the geochemical laboratory. There is no reason why the geochemistry spread-sheet should not be partly completed in the field, and passed on with the set of specimens, to be completed in the laboratory.

The situation of an array which is largely blank can arise in a somewhat different form. The geologist might examine a large number of attributes in every one of a set of items, but only occasionally observe properties that should be recorded. For example, trace fossils, sedimentary structures, maximum grain size, shape of grains, slump structures, cleavage, joint orientation and composition of cement might be looked for in every bed in a measured section, but observed in only a few. Creation of subfiles would not help, because attributes are not necessarily found together in the same items, and to prepare a subfile for every attribute is to lose the advantages of structuring the data as an array.

One method of recording such data is to use a system of abbreviations for both the variable and the measurement, or, with qualitative data, for both the property and the attribute. In the example given above, the geologist might record *S.S.* C-bdg *M.G.S.* 1·0 to indicate that in a particular bed he had observed a sedimentary structure, namely cross-bedding, and had been able to obtain a measurement of the maximum grain size, which was 1·0 phi unit. Instead of writing the measurement in a particular column of a table to indicate its meaning, the measurement is preceded by a label which specifies its correct location in the data array. As two words must be written instead of one, this method has no advantage unless the variable or property was observed in considerably less than half the items. Since the first of each pair of words locates the other in the array, the order in which properties are listed need not necessarily be consistent.

In the example, abbreviations were used rather than writing the words in full. The purpose of this was to save effort in completing a field notebook and in transcribing the data on to cards or tape. The manner in which the abbreviations were constructed does not therefore have any special significance. However, there is ambiguity if the same abbreviation is used for different properties or variables, or for two or more attributes of the same property. If various abbreviations are used for the same word, the record is unnecessarily difficult for a geologist, or a computer, to understand. A list of abbreviations and their meanings is likely to be required at some point, and it is convenient to develop this as a check-list before and during the recording of the data.

The input statements of the usual scientific computer languages, such as Fortran and Algol, cannot write and store a data array directly in this form, and a special section of the programme must be written for this purpose. If it is written in Fortran it will be convenient to have all the abbreviations the same length. If the section of programme is in machine language, the end of one word and the beginning of the next can be indicated by a selected symbol, such as an oblique stroke, a comma, a blank, or two or more adjacent blanks. With this procedure, terms which are seldom used can be written in full to avoid memorising or looking up the abbreviations. Punched cards, which are easy to use for editing and up-dating data stored as a complete matrix, can still be used with the techniques described above for recording a sparse matrix, but the cards would be used as though they were a continuous sequential medium, like paper tape or magnetic tape.

NETWORKS

The elements of the two-dimensional array or matrix are connected to one another in two directions. One can visualize a rectangular network of threads linking an element horizontally to other measurements or attributes of the same item, and vertically to measurements of the same variable on other items. This structure is not of course limited to data about beds of rock in a measured section. It is equally applicable, for example, to percentages of various minerals measured in a set of thin sections, results of geochemical analyses on rock powders, or measurements of length, breadth, height and weight of a population of fossil brachiopods.

A special subject, known as network analysis, has been developed for the study of networks. The network is not usually a rectangular grid as just described but has a more general form. A number of points or "nodes" are connected in pairs (see Fig. 1). The connection between two points is considered separately from the link between any other

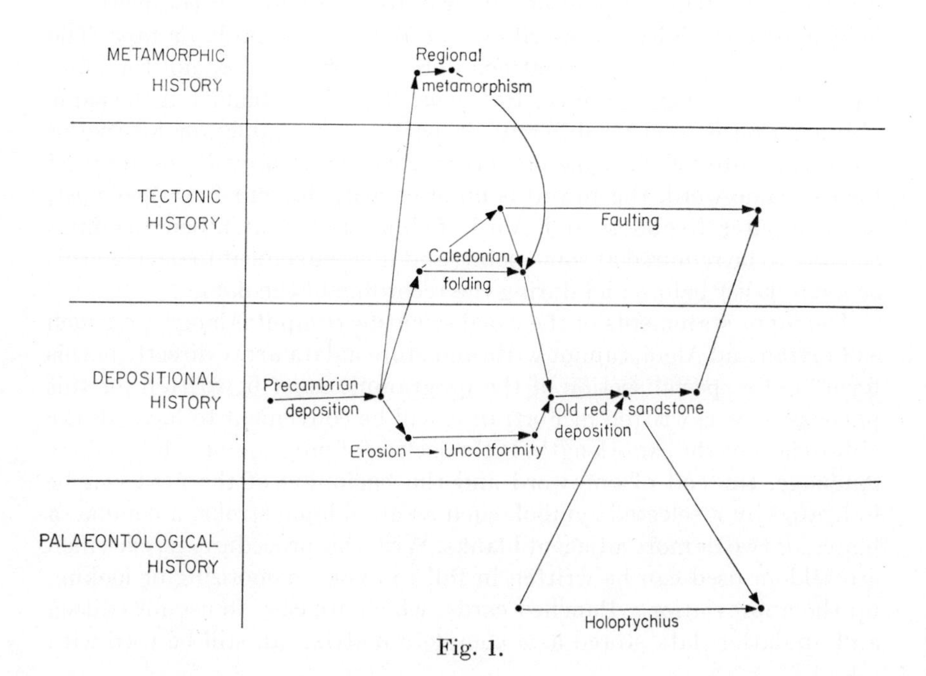

Fig. 1.

two, whereas in a matrix, the links run through the entire data structure. Network analysis is used in the study of electrical circuits, and recently has been used extensively by operational research workers in "critical-path" and similar techniques to study sequences of operations involved in the completion of some task. These methods appear to be relevant to stratigraphical problems, although they have not yet, to my knowledge, been used in that context. In operational research, nodes in the network are sometimes termed "events" and the links between them are termed "activities". Thus, deposition of Old Red Sandstone rocks in an area might be considered as an activity, and the beginning and end of Old Red Sandstone deposition could be termed "events". Activities thus have duration in time, whereas events are practically instantaneous, and can be used as markers to specify a recognisable point on a time scale.

A geological map generally contains information about a pattern of events and activities which could, if one wished, be depicted as a network. An example is shown in Fig. 1, in which activities are shown in the conventional manner, as lines with an arrowhead showing the end of the activity and the occurrence of an event. The lowest arrow in the diagram, for example, joins two events, the origin and extinction of the fossil fish Holoptychius. Since a specimen of this fish was found in the Old Red Sandstone of the area, the deposition of that bed presumably was later than the origin, and earlier than the extinction of Holoptychius. Arrows have therefore been added to show that relationship. An arrow merely indicates that the event at one end preceded the event at the other end. The length of the line, and the slope of the line, to left or right, is a matter of convenience in drafting the diagram, and there are no implications about duration of activities.

An examination of the diagram shows that it can represent time-relationships inferred from a map, which are generally of the form: "x preceded y". For instance, an igneous dyke cutting a bed of sandstone might indicate that deposition of the bed preceded intrusion of the dyke. Even in an area with a relatively simple geological history, it might be known that a preceded b, b preceded c, d preceded c, d preceded e, and yet the relationship of a or b to d or e might remain quite unknown. An attempt to represent information of this kind as a matrix is unlikely to be successful. A matrix brings together whole sets of elements and makes it possible to examine the relationship among all of them simultaneously. Links in a network, on the other hand, refer to only two items at a time, and deductions are drawn from the network by tracing paths from event to event.

Nevertheless, network analysis is a powerful tool. As shown in Fig. 1 it is possible, though not necessary, to separate the network into segments corresponding to different topics. The data can in this way be structured as linked subfiles within a network. At most, a few hundred events could be usefully drawn on a diagram. Many thousands of events have been handled with ease in a computer in critical-path studies. A maximum and minimum estimate of time involved in an activity can be included in the information presented to the computer, or, if a reasonable estimate can be made on the basis of rates of erosion, deposition or evolution, a best estimate and probable error can be presented. With this additional information, the computer can calculate an upper and lower estimate for the time of occurrence of a specified event. It can determine the critical-path, that is the sequence of activities on which the time-scale is most critically based. This information is useful to the geologist as it enables him to concentrate his efforts on refining the data that are most significant. The computer can readily accept new or revised data and indicate their effect on the overall picture.

As mentioned above, the use of network analysis with stratigraphic data does not seem to have been fully explored, but experience with formally identical problems in operational research suggests that networks are an appropriate structure for much stratigraphic data, and that network analysis has an essential role to play in developing large-scale computable models of geological history. The methods are fully described in many text-books, and a wide variety of computer programmes are available.

TREES AND FORESTS

It was suggested above that a network was an appropriate structure for describing time-relationships of a series of events inferred from a geological map. But a description of the geology, as it might appear in a report accompanying the map, would have a different structure. The geologist might choose to frame his description somewhat along the following lines. First, he might describe the most general features of the geology. Then, one by one, he might discuss important topics in more detail. He might feel that each topic could be broken down into component parts, each of which he might consider in turn. The descriptions would thus have a hierarchical structure in which a set of paths was followed, each leading from the general to the particular. Having gone into the desired detail on one topic, the description returns to a higher level, then follows another path into details on another subject. The structure can be shown as a diagram which starts as a single line, but which branches repeatedly. For obvious reasons the structure is known in mathematics as a tree. An evolutionary tree is a familiar geological example.

There seem to be two main reasons for using a hierarchical tree structure in a complex geological description. One is the need to specify clearly what is being described and to set it in its geological context. The other is the economy of thought that can be achieved by using the same terms for analogous situations in a variety of contexts. "Medium-grained", for instance, has different meanings when applied to a siltstone and to a conglomerate. Properties such as size, shape, orientation, proportion and arrangement can be observed, and can often be described by the same terminology, in entities of a wide range of size and character, such as continents, depositional basins, geological formations, facies, beds, sedimentary structures, fossils, grains, crystals or even molecules (see Griffiths, 1961). As can be seen, one of the above entities could be a component of another. Thus grains and fossils might be components of a bed, which in turn was part of a formation and so on. Clearly, in any description, the object under consideration must be clearly identified and its relation to its components and to the entity of which it is itself a component must be indicated. The statement that an object is rather large, reddish-purple, ellipsoidal in plan view, elongated in a north-south direction, with a hinge-line a short distance from one edge, is of little value unless it is known whether the object is a fossil or a basin of deposition. If the latter, it could be important to know which basin it is, and whether rock-types mentioned elsewhere are part of it, or unrelated to it.

The following description of a hypothetical rock unit provides an example of a written description showing hierarchical relationships that can be readily structured as a tree, but are difficult to express as part of an array.

"The basal six feet of strata comprise sandstone and shale. The sandstone is cross-bedded, medium-grained, somewhat coarser towards the base. It contains occasional, scattered, sub-angular grains of purplish quartz. The shale forms thin partings which contain flattened thin-shelled lamellibranchs." Having specified in the first sentence the set of objects under consideration (the basal six feet of strata), the geologist subsequently considers a class of objects drawn from the set (sandstone), then certain constituents of the sandstone (quartz grains). In the next sentence, another class of objects is considered (shale) drawn from the six feet of strata. Then one of its constituents is mentioned (lamellibranchs), and their attributes described (flattened, thin-shelled).

If a complete measured section is described in such a way that each bed is represented by a tree, the complete description comprises a set of trees, known again for obvious reasons, as a forest. The geologist would record such information in the field in written form, and it would not normally be possible to use a spread-sheet, although a check-list could be used. There would, of course, be complete freedom to add new terms to the check-list as required.

Computer analysis of data structured as a forest is a more difficult problem than analysis of arrays, and if the geologist's main concern is to collect data for a computer, tree structures should be avoided where possible. They may, nevertheless, prove useful in supplementing an array, to record parts of a description which would otherwise be difficult to present to the computer. List-processing techniques (see Berkeley and Bobrow, 1966) make it possible to store and manipulate such data within the computer.

Written English, which is a string of words, is itself a somewhat artificial medium for representing a tree structure. Various devices are needed to allow the reader to mentally organize the data, choosing an appropriate structure as the data are presented. Summaries, headings, subheadings, cross-references, recapitulation, sentence and paragraph construction, and connective words, phrases and sentences may all be required to help the reader in this task. Perhaps computer methods will eventually provide a more natural framework for recording and communicating observations about the natural environment.

CONCLUSIONS

When a systematic set of observations is made and the results are recorded as data, the pattern of relationships between individual data items constitutes the data structure. Data have no meaning unless the relationships between items are known. The structure plays an important part in determining how much information can be extracted from a data set.

The need to consider data structures explicitly is, of course, largely due to the development of computer methods, in which implicit assumptions, taken for granted by human beings, are not acceptable. Detailed studies of data structures and their representation in the computer have therefore been carried out by computer scientists (see, for example, Hopgood, 1969). The geologist who intends to submit his data for computer analysis may therefore wish to consider first, the type of analysis that may be used, and thus the type of relationship that he must record and the manner in which he can represent them. It may be that the data can be conveniently represented by one of the more common structures, such as an array, a network, or a tree.

REFERENCES

Berkeley, E. C. and Bobrow, D. G. (eds) (1966). "The Programming Language LISP". MIT Press, Cambridge, Massachusetts.

Brisbin, W. C. and Ediger, N. M. (eds) (1967). A national system for storage and retrieval of geological data in Canada. Report of the ad hoc Committee on Storage and Retrieval of Geological Data in Canada. (S. C. Robinson, Chairman).

Busacker, R. G. and Saaty, T. L. (1965). "Finite Graphs and Networks". McGraw-Hill Book Company, New York.

Griffiths, J. C. (1961). Measurements of the properties of sediments. *J. Geol.* **69**, 487–498.

Hopgood, F. R. A. (1969). Compiling techniques. *Macdonald Computer Monographs*, 8.

Krumbein, W. C. and Graybill, F. A. (1965). "An Introduction to Statistical Models in Geology". McGraw-Hill Book Company, New York.

12. Stratigraphic Modeling by Computer Simulation

GRAEME BONHAM-CARTER* AND JOHN W. HARBAUGH

Stanford University, California, U.S.A.

ABSTRACT

A very simple computer-simulation model of a sedimentary basin is developed which incorporates factors such as depth of water, rate of supply and composition of sediment, the effect of base-level control, subsidence of the crust, and eustatic fluctuation. Output from the model comprises a sequence of vertical stratigraphic sections, representing the configuration of water, lithofacies and basement rocks through time. The sections are displayed graphically using the computer's line printer and depict time lines as well as lithofacies boundaries. The model is illustrated with results from six experiments. A Fortran IV listing of the program is included.

INTRODUCTION

Although computer simulation is a technique relatively new to geology, the philosophy of simulation is already well established. In stratigraphy, for example, it is common practice to develop theoretical models to explain an observed distribution of rock types. Such theoretical models may be written down in words, or shown graphically as a series of diagrams. For example, the history of development of a sedimentary basin, as depicted by such models, can be regarded as a type of mental or graphic simulation. In effect, the geologist makes some simple assumptions concerning the process that he believes to be important in the development of a stratigraphic sequence, and then "thinks through" the outcomes of his assumptions. In other words, he develops a model, then uses the model to simulate the stratigraphic sequence.

One of the best-documented models of this type was published by Sloss (1962). The Sloss model is a graphic model of a hypothetical

* Present address: Department of Geology, University of Rochester, Rochester, New York 14627.

Systematics Association Special Volume No. 3, "Data Processing in Biology and Geology", edited by J. L. Cutbill, 1970, pp. 147–164.

sedimentary basin, and involves assumptions concerning the depositional environment, such as depth of water, wave and current action, distance from shore, types and sources of sediment, tectonic warping, and fluctuations of sea level. Below, we outline the basic components of the Sloss model, then describe a computer model which incorporates many of the same assumptions.

The computer model is superior to the graphic model in a number of ways. Above all, the assumptions are formalized and expressed in quantitative terms. Experiments with the model are, therefore, completely objective and repeatable. Furthermore, a wide variety of simulation experiments can be carried out in a very short time, each experiment thoroughly documented automatically with printed cross-sections through the sedimentary basin. Although it is definitely an advantage to express the assumptions as quantitative relationships, we must remember, however, that the underlying assumptions are no better than those expressed qualitatively in the graphic model.

SLOSS MODEL

The Sloss model deals principally with deposition of clastic sediments on continental shelves. Sediments are delivered to the edge of the sea by streams flowing across a coastal plain. When brought to the sea, each sediment particle is transported until it finds a position of rest and becomes available for incorporation in the sedimentary sequence. The position of rest is a function of kinetic energy (waves and currents), material (composition and particle size), and boundary conditions (bottom slope and roughness). The interaction of these factors produces an equilibrium surface, base level, above which a particle cannot come to rest and below which deposition and burial are possible. At any instant in time, given an adequate supply of sediment, the interface between water and sediment tends to coincide with base level. Successive interfaces, representing successive instants of time, can be interpreted as a record of the relative rate of subsidence of the depositional basin.

The gross geometry or shape S of a body of sedimentary rock varies as a function of the quantity Q of material supplied to the depositional site, the rate of subsidence R at the site, the rate of dispersal D, and the nature of the material supplied M:

$$S = f(Q, R, D, M)$$

In this expression, R is a measure of the rate of subsidence expressed as a receptor value, defined as the available volume below base level created per unit time by subsidence. The proportion of different particle sizes of sedimentary material M is assumed to be constant. This assumption implies that weathering and erosion in a heterogenous source area yield coarse, medium, and fine clastic particles to the depositional area in unchanging proportions.

With these assumptions, Sloss developed the process elements and the resulting stratigraphic responses in a series of diagrams. These diagrams depict the shape and position of lithofacies boundaries on a vertical section through a sedimentary basin. By making $Q > R$, for instance, a regressive sequence was produced, whereas by making $Q < R$ a transgressive sequence was formed.

COMPUTER MODEL

The computerized extension of the Sloss model incorporates the following factors: (1) quantity of material supplied, which may include from one to five different sediment size fractions; (2) initial geometry of the sedimentary basin, expressed as water depth; (3) tectonic warping (subsidence) through time and from place to place in the basin; (4) position of base level or equilibrium surface defined with respect to sea level for each particle size class.

The model treats only two spatial dimensions, representing a vertical section through a sedimentary basin. The other horizontal dimension is not represented, but could be incorporated if a more advanced version were to be developed. Time is divided into discrete steps, and space is represented by a sequence of columns, each of unit width, which represents water depth and thickness increments of various sediment types (Fig. 1). Subsidence (or uplift) is represented by sliding the columns up or down relative to sea level. Fortran arrays are used to store water depths and sediment thicknesses for each vertical column for each time increment.

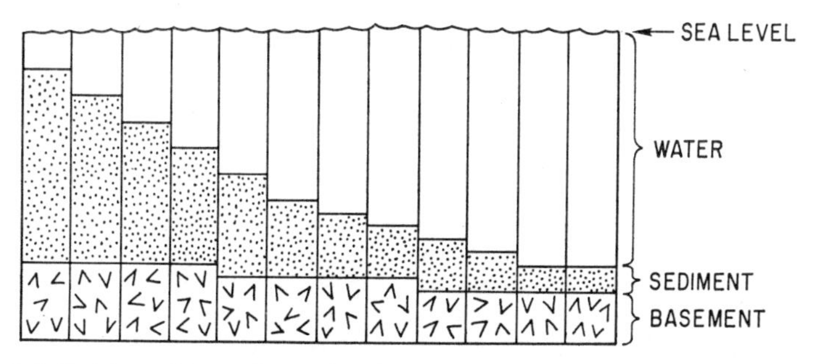

Fig. 1. Subdivision of two-dimensional sedimentary basin into series of discrete vertical columns representing water, sediment, and basement.

Each column of sediment and water is displayed as a row of symbols on the computer printer, and the resulting sequence of rows forms a cross-section through the sedimentary basin. (Fig. 3).

Sediment transport and deposition

Transport and deposition of sediment is treated heuristically in the computer model. It is assumed that during any time interval, a certain increment of sediment (the sediment "load") enters the basin from a source area on one side of the basin and is then transported from column to column. Deposition may take place in each column, the amount deposited being debited from the sediment load, and the remainder passed onto the next column, where the process is repeated. The sequence proceeds from the sediment-source side of the basin toward the seaward side. The amount of sediment deposited in a particular column depends on (a) the amount of sediment available for deposition and (b) the water depth in that column in relation to base

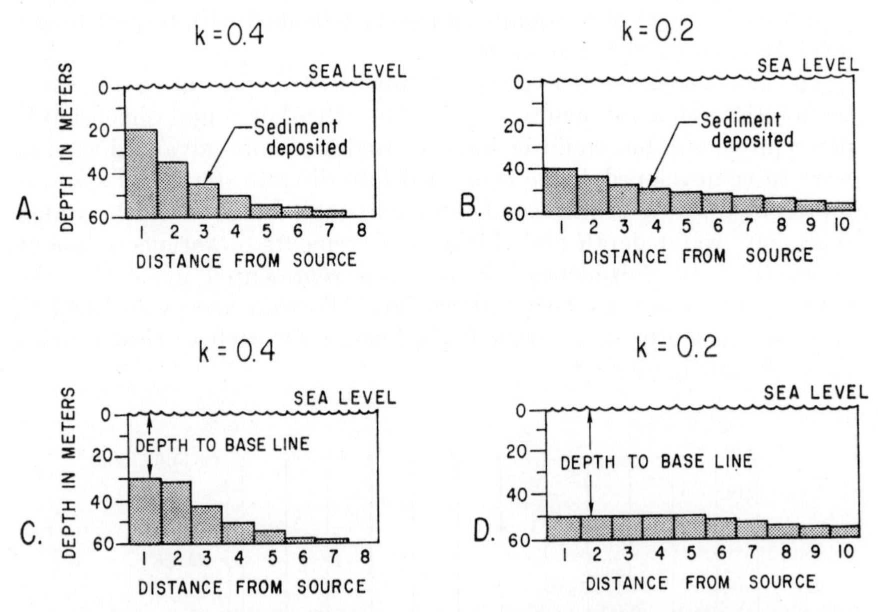

Fig. 2. Four diagrams representing vertical sections through sedimentary basin into which uniform volume of sediment is supplied from source at left. Initial water depth is uniform. Diagrams A and B assume that base level does not exert influence and illustrate effects of varying the "decay constant" k. Diagrams C and D illustrate effects of two different base levels, coupled with values of k equivalent to those of diagrams A and B, respectively. Volume of sediment introduced in all four diagrams is equivalent to column 100 m high.

level. Mass balance is observed by accounting for all sedimentary materials as they move through the system.

The rules governing transportation and deposition of sediment are extremely simple and are outlined below. Part of the sediment load reaching a particular column is deposited if the water depth is greater than depth to base level. In columns that contain water sufficiently deep so that base level exerts no control, the proportion of sediment deposited for each particle-size class is represented by a curve that declines exponentially toward the seaward side of the basin (Figs. 2A and 2B). The accounting system arithmetic involved in this process can be envisioned as follows.

Let the sediment load entering the basin be L, then the proportion of this load that is deposited in the first column is k. Thus, the amount deposited is kL, and the remaining load that is shunted on to the next column in the sequence is $L - kL$ or $L(1 - k)$. The quantities deposited in successive columns are illustrated in the following tabulation:

Column	Sediment load	Sediment deposited
1	L	kL
2	$L(1 - k)$	$kL(1 - k)$
3	$L(1 - k)^2$	$kL(1 - k)^2$
4	$L(1 - k)^3$	$kL(1 - k)^3$
5	$L(1 - k)^4$	$kL(1 - k)^4$
⋮	⋮	⋮
n	$L(1 - k)^{n-1}$	$kL(1 - k)^{n-1}$

The amount deposited in the second column is found by multiplying the remaining load by k, giving $kL(1 - k)$. In column 3, the remaining load is $L(1 - k)^2$ and the amount deposited is $kL(1 - k)^2$. Subsequent columns are treated similarly. Thus we may generalize: for the nth column, the sediment load remaining is $L(1 - k)^{n-1}$ and the amount deposited is $kL(1 - k)^{n-1}$. This relationship is an adaptation of the familiar law of growth and decay in which k is the decay constant.

The significance of different values of k is open to interpretation. It is clear from Figs. 2A and 2B that k pertains to slope of deposits. If slope were solely a function of grain size, larger values of k might correspond to coarse sediment capable of reposing on steeper slopes and being less mobile than fine sediment. Maximum slope angle and mobility, however, do not bear a simple relationship to grain size. Thus, it is difficult to interpret k in terms of a simple physical relationship, and instead, it should be regarded as a parameter used in conjunction with a relationship that is largely heuristic.

If the model dealt only with the transport of sediment according to the growth and decay law, the calculations would be simple and could be readily carried out by hand. If we introduce constraints, however, the solutions are not simple. If we introduce base level, above which sediment of a particular particle size cannot come to rest, we must continually check the elevation of the sediment-water interface in each column to insure that it is not above base level. Figures 2C and 2D illustrate the effect of base level as a constraint, employing specified values of k and holding other factors constant.

The model incorporates various decision rules that govern sediment deposition. We can distinguish three situations: (1) where the sediment-water interface is above, or equal to, base level (as specified for a particular particle-size class), deposition is not possible, and all the load is shunted on to the next column; (2) where the sediment-water interface is slightly below base level, only part of the quantity of sediment that would otherwise be deposited is accomodated. Sufficient sediment is deposited to bring the column to base level, and the remainder is passed on to the next column; (3) where the sediment-water interface is sufficiently far below base level, all of the sediment available for deposition is accomodated.

These relationships can be expressed algebraically. If water depth in a column is D, and depth to base level is B, then the amount deposited S is given by one of the following relationships. If water depth is less than, or equal to depth to base level, no sediment is deposited:

$$S = 0 \qquad D \leqslant B$$

If water depth minus the quantity kL is less than, or equal to depth to base level, the amount deposited is equal to depth minus base level:

$$S = D - B \qquad (D - kL) \leqslant B$$

Otherwise, the amount deposited equals the quantity kL, and base level has no influence on sedimentation, thus

$$S = kL$$

Figures 2C and 2D illustrate these algebraic base level control relationships. In Fig. 2C, depth to base level is 30 m. Without base level control of the total of 100 m of sediment supplied, 40 m would have been deposited in the first column, reducing its water depth to 20 m. But 20 m is shallower than base level. Thus deposition of $D - B = 60 - 30 = 30$ m of sediment takes place in column 1, and the remaining load of 70 m of sediment passes on to column 2. The sediment load reaching column 2 is equivalent to $100 - 30 = 70$ m of sediment. The new value of kL is thus $0{\cdot}4 \times 70 = 28$. This time, if all 20 meters of sediment are deposited, water depth is given by $D - kL = 60 - 28$

= 32, which is greater than depth to base level. Thus 28 m are deposited in column 2. Similar calculations are carried out for columns 3–8. In Fig. 2D, depth to base level is 50 m and $k = 0 \cdot 2$. Base level imposes a depth limit for sediment in the first five columns. In columns 6 to 10, the amount deposited in each column dies away exponentially.

The treatment of more than one grain size further complicates the calculations. Each grain size is assigned: (1) a specific value of k; (2) an initial sediment load to be released from the source for each time increment; (3) a depth to base level. In an experimental run, many columns receive sediment of more than one grain size during a time increment. Mixtures of sediment of varying size in a particular column are graphically represented by using a symbol that represents the particle-size class that forms the largest proportion of the volume deposited during the particular time increment.

Crustal subsidence

The effect of subsidence, or conversely, eustatic changes in sea level, can be represented by adding, or substracting, values to the water depth in each column. Increasing water depth in a single column by some number of meters has the effect of depressing the entire column downwards by the same number of meters. In this model, there are two ways in which water depth values can be changed. First, provision is made in the program for adding a specific increment to water depth in each column during each time increment. The increment to water depth may vary from column to column, but is constant for each time interval. This simulates uniform subsidence with time and also may be used to represent changes in sea level. Alternatively, subsidence can be related to deposition by a simple proportionality constant. Each column can be depressed by an amount equal to thickness of sediment deposited multiplied by a proportionality constant F. At one extreme, if F equals one, the amount of subsidence equals the amount of deposition. At the other extreme, subsidence does not occur if F equals zero.

Subsidence need not be instantaneous with deposition. As an alternative, subsidence may lag behind deposition by some whole number of time increments. For example, if the lag length is three time increments, subsidence will occur at the end of every third time increment, and the amount of subsidence in a particular column is obtained by multiplying F by the total quantity of sediment deposited in that cell during the previous three time increments. The lag cannot be made shorter than one time increment because of the division of time into discrete steps in the model. Unit lag causes virtually instant response. It could be argued that there may be an appreciable lag between the loading of the actual crust and its subsequent subsidence.

The model makes it convenient to experiment with the effects of different lag lengths.

EXPERIMENTS WITH MODEL

The computer program representing the model is listed in Table 1 and an example of input is shown in Table 2. The response of the model

TABLE 1. Fortran program for two-dimensional sedimentary-basin model. About two-thirds of total program consists of subroutine CROSEC, which is used for plotting stratigraphic cross-sections with line printer. Program was run on IBM 360, model 67, but with minor modifications could probably be run on almost any computer having a Fortran compiler.

```
C.....SIMPLE TWO-DIMENSIONAL SEDIMENTARY BASIN MODEL
C.....  NTIM          NO. OF TIME INCREMENTS
C.....  NCOLS         NO. OF COLUMNS OF SEDIMENT
C.....  NFRACT        NO. OF SEDIMENT SIZE FRACTIONS
C.....  SED(NT,I,L)   NO. OF SEDIMENT UNITS DEPOSITED IN N-TH TIME INCREM.
C                         IN I-TH COLUMN, AND L-TH SIZE FRACTION
C.....  SEDINP(L)     NO. OF SED UNITS IN INITIAL SED LOAD (L-TH SIZE FR)
C.....  SEDIN(L)      TEMPORARY ACCOUNT FOR L-TH SIZE FRACTION
C.....  DEPTH(I)      WATER DEPTH IN I-TH COLUMN
C.....  SUBSID(I)     AMOUNT OF EXOGENOUS SUBSIDENCE (OR COULD BE USED TO
C                         REPRESENT SEA LEVEL CHANGES) ADDED TO WATER DEPTH
C                         IN I-TH COLUMN EVERY LAG-TH TIME INCREMENT
C.....  SUBFAC        SUBSIDENCE FACTOR RELATING SUBSIDENCE TO DEPOSITION
C.....  LAG           TIME LAG IN TIME INCREMENTS BETWEEN DEPOSITION AND
C                         RESULTING SUBSIDENCE (LAG=1 IMPLIES INSTANT RESP)
C.....  EQUIL(L)      EQUILIBRIUM DEPTH BELOW WHICH L-TH SIZE FRACTION
C                     NOT BE DEPOSITED
C.....  CON(L)        CONSTANT DETERMINING RATE OF DEPOSITION FOR L-TH SIZ
C.....  KPRINT        PRINTED OUTPUT EVERY KPRINT-TH TIME INCREMENT
C.....  NPLOT         X-SECTION PLOTTED EVERY NPLOT-TH TIME INC
C.....  KPLOT         TIME LINE PLOTTED EVERY KPLOT-TH TIME INC ON X-SECT
      DIMENSION SUBSID(30),SEDIN(5),SEDINP(5),EQUIL(5),CON(5),A(5),D(5)
      COMMON DEPTH(30), SED(30,30,5), TITLE(18), NCOLS, NFRACT
    1 FORMAT(18A4)
    2 FORMAT(7I5, F5.0)
    3 FORMAT(16F5.0)
    4 FORMAT(1H1, 18A4/ 5X, 'TIME INCREMENT ', I5/ ' COLUMN', 3X,
     1 'DEPTH', 3X, 5('SIZE', I2, ' DEPOS  LEFT    '))
    5 FORMAT(1H , I4, F10.1, 3X, 5(6X,2F6.1,4X))
C.....READ INPUT PARAMETERS
  999 READ(5,1) TITLE
      READ(5,2) NTIM,NCOLS,NFRACT,KPRINT,NPLOT,KPLOT,LAG,SUBFAC
      IF (LAG.LT.1) LAG=1
      READ(5,3) (SEDINP(L), L=1,NFRACT)
      READ(5,3) (EQUIL(L), L=1,NFRACT)
      READ(5,3) (CON(L), L=1,NFRACT)
      READ(5,3) (DEPTH(I), I=1,NCOLS)
      READ(5,3) (SUBSID(I), I=1,NCOLS)
C.....PLOT INITIAL CROSS SECTION BEFORE SEDIMENTATION
      CALL CROSEC(0,KPLOT)
C.....BEGIN MAJOR DO-LOOP, ONCE THRU PER TIME INCREMENT
      DO 110 NT=1,NTIM
      IF (MOD(NT,KPRINT).EQ.0) WRITE(6,4) TITLE, NT, (L, L=1,NFRACT)
C.....ALLOCATE SEDIMENT LOAD TO BE DEPOSITED THIS TIME INC
      DO 40 L=1,NFRACT
   40 SEDIN(L)=SEDINP(L)
C.....BEGIN LOOP, ONCE THRU PER SED COLUMN
      DO 80 I=1,NCOLS
```

```
C.....CALCULATE AMOUNT OF SED OF EACH SIZE READY FOR DEPOSITION (A(LLL))
      DO 50 L=1,NFRACT
      SED(NT,I,L)=0.0
   50 A(L)=SEDIN(L)*CON(L)
C.....ENTER LOOP ONCE THRU PER EQUILIB DEPTH STARTING WITH DEEPEST
      DO 70 LL=1,NFRACT
      L=NFRACT-LL+1
      ATOT=0.0
      DO 55 LLL=1,L
   55 ATOT=ATOT+A(LLL)
      IF (ATOT.LE.0.01) GO TO 80
C.....CALCULATE TOTAL AVAILABLE SPACE FOR DEPOSITION. IF 0 GO TO 70
      B=DEPTH(I)-EQUIL(L)
      IF (B.LE.0.0) GO TO 70
C.....DETERMINE ACTUAL AMOUNT OF DEPOSITION. DEPOSIT SIZE FRACTIONS
C       IN PROPORTION TO THEIR TOTALS IN THE LOAD
      DEPOS=AMIN1(ATOT,B)
      Z=DEPOS/ATOT
      DO 60 LLL=1,L
C......AMOUNT OF SED OF SIZE LLL DEPOSITED IS D(LLL), ADDED TO SED
C       ARRAY, SUBTRACTED FROM A(LLL)
      D(LLL)=A(LLL)*Z
      SED(NT,I,LLL)=SED(NT,I,LLL)+D(LLL)
      SEDIN(LLL)=SEDIN(LLL)-D(LLL)
   60 A(LLL)=A(LLL)-D(LLL)
      ATOT=ATOT-DEPOS
      DEPTH(I)=DEPTH(I)-DEPOS
   70 CONTINUE
   80 IF (MOD(NT,KPRINT).EQ.0) WRITE(6,5) I,DEPTH(I), (SED(NT,I,L),
     1  SEDIN(L), L=1,NFRACT)
C.....IF NT(MODULO NPLOT) EQUALS ZERO PLOT CROSS SECTION
      IF (MOD(NT,NPLOT).LT.1) CALL CROSEC(NT,KPLOT)
C.....ADD SUBSIDENCE TO DEPTH IF NT(MODULO LAG) EQUALS ZERO
      IF (MOD(NT,LAG).GT.0) GO TO 110
      DO 100 I=1,NCOLS
      SUM=0.
      IF (SUBFAC.LT.0.00001) GO TO 100
C.....DETERMINE AMOUNT OF SEDIMENT LOADED ONTO COLUMN SINCE PREVIOUS
C       SUBSIDENCE ADJUSTMENT
      DO 90 LG=1,LAG
      INDEX=NT-LG+1
      DO 90 L=1,NFRACT
   90 SUM=SUM+SED(INDEX,I,L)
      SUM=SUM*SUBFAC
  100 DEPTH(I)=DEPTH(I)+SUBSID(I)+SUM
  110 CONTINUE
      GO TO 999
      END
      SUBROUTINE CROSEC(NT,KPLOT)
C.....SUBROUTINE FOR DRAWING GRAPHIC SECTIONS
      DIMENSION PLOT(120), PLOT1(120), SYMBOL(5)
      COMMON DEPTH(30), SED(30,30,5), TITLE(18), NCOLS, NFRACT
      DATA SYMBOL,DOT,EYE,BLANK,RLT/'0','$','A','B','C','.','I',' ','<'/
    1 FORMAT(1H1, 18A4/ 5X, 'TIME INCREMENT ', I5/)
    2 FORMAT(1H , 120A1)
    3 FORMAT(1H0)
      WRITE(6,1) TITLE, NT
C.....FOR EVERY COLUMN IN VERTICAL SECTION, DO DOWN TO 80
      DO 80 II=1,NCOLS
      I=NCOLS-II+1
C.....SET BLANKS INTO ALL POSITIONS OF PLOT ARRAYS
      DO 10 K=1,120
      PLOT1(K)=BLANK
```

```
   10 PLOT(K)=BLANK
C.....INDEX DENOTES POSITION IN PLOTTING ARRAY BELOW SEA LEVEL - ONLY
C      POSITIVE VALUES ARE PRINTED. USE I'S FOR WATER
      NDEP=ABS(DEPTH(I))+0.5
      INDEX=0
      IF (DEPTH(I).LT.0) INDEX=-NDEP
      DEV=ABS(DEPTH(I))-NDEP
      IF (NDEP.LT.1) GO TO 30
      DO 20 K=1,NDEP
      INDEX=INDEX+1
   20 IF (INDEX.GE.1) PLOT(INDEX)=EYE
C.....FILL SEDIMENT POSITIONS WITH APPROPRIATE SYMBOLS
   30 IF (NT.EQ.0) GO TO 55
C.....FOR EACH TIME INCREMENT DO DOWN TO 50
      DO 50 NN=1,NT
      N=NT-NN+1
C.....SUM SED FRACTIONS. FIND DOMINANT SED FRACTION AND ALLOCATE SYMBOL
      SUM=0.0
      LBIG=0
      BIG=0.0
      DO 35 L=1,NFRACT
      IF (BIG.GT.SED(N,I,L)) GO TO 35
      BIG=SED(N,I,L)
      LBIG=L
   35 SUM=SUM+SED(N,I,L)
      NSUM=SUM+0.5+DEV
      DEV=SUM+DEV-NSUM
      IF (NSUM.LT.1) GO TO 50
      DO 40 K=1,NSUM
      INDEX=INDEX+1
      IF (INDEX.LT.1) GO TO 40
      PLOT(INDEX)=SYMBOL(LBIG)
      IF (K.EQ.NSUM.AND.MOD(N,KPLOT).EQ.0) PLOT1(INDEX)=DOT
   40 CONTINUE
   50 CONTINUE
C.....INSERT 'LESS THAN' SIGNS FOR BASEMENT ROCKS
   55 DO 60 K=1,2
      INDEX=INDEX+1
   60 IF (INDEX.GE.1) PLOT(INDEX)=RLT
      IF (INDEX.LT.1) GO TO 70
      WRITE(6,2) (PLOT(K), K=1,INDEX)
      WRITE(6,2) (PLOT1(K), K=1,INDEX)
      GO TO 80
   70 WRITE(6,3)
   80 CONTINUE
      RETURN
      END
```

TABLE 2. Listing of input for program in Table 1 to produce output shown in Fig. 3.

```
TEST RUN 1, ONE GRAIN SIZE, NO SUBSIDENCE
  10   20    1    1    1    3    1    0
  10
   3
   5
   1    2    3    4    5    6    7    8    9   10   11   12   13   14   15   16
  17   18   19   20
   0    0    0    0    0    0    0    0    0    0    0    0    0    0    0    0
   0    0    0    0
```

consisting of a sequence of stratigraphic cross-sections produced is shown in Fig. 3. Other output is listed in Table 3. As the cross-sections reveal, without subsidence, and only a single grain size, the model produces a "deltaic" deposit that grows progressively out into deep water, forming a series of "foreset" beds that dip progressively more steeply as the delta builds outward.

A similar experiment, in which crustal subsidence occurs, is shown in Fig. 4. The sediment-water interface remains at a constant elevation after the first time increment, each column subsiding by an amount equal to the thickness of sediment deposited immediately before. Under these conditions, the response of the crust is to subside the most where

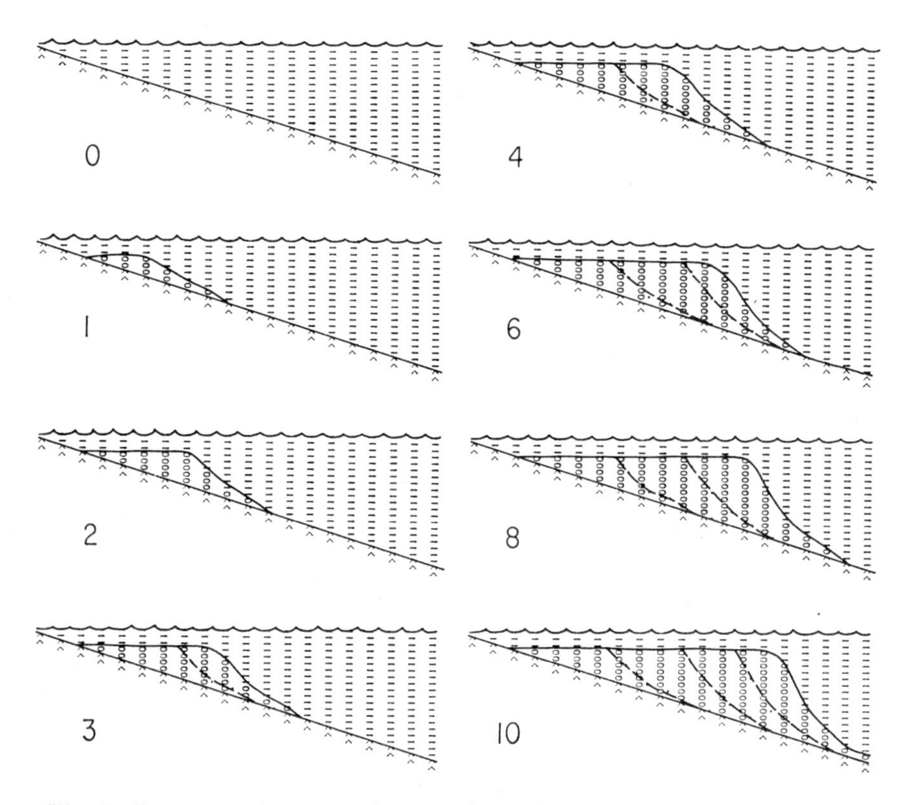

Fig. 3. Sequence of cross-sections produced by sedimentary-basin model in experimental run in which crustal subsidence does not occur and which involves sediment of a single size class. Time increments 0, 1, 2, 3, 4, 5, 6, 7, 8, 9, and 10 are shown. I = water, O = sand, $ = silt, < = basement. Model responds by producing regressive sequence of "deltaic deposits". Slope of deposits is affected by initial slope of sea floor and by "decay constant" k, which has been set at 0·5. Sediment-water interface is plotted every third time increment, thus forming series of stratigraphic time lines.

the maximum amount of sediment has been deposited. The overall form of the deposits is that of a lens, reminiscent of the lens-shaped mass of Cenozoic sediments of the Gulf Coast.

TABLE 3. Example of tabular output from run of sedimentary-basin model. Output is based on input data listed in Table 2, and pertains to time increment one. Each row pertains to an individual column in cross section. Column labeled DEPTH refers to depth after deposition in arbitrary units, column labeled DEPOS refers to amount of sediment deposited in equivalent arbitrary units, and column labeled LEFT pertains to load of sediment that remains to be transported to next column. Column DEPOS thus accounts for all sediment supplied (10·0 units) during time increment (that which is deposited is debited from that which remains, and what is left is shunted to column LEFT).

TEST RUN 1, ONE GRAIN SIZE, NO SUBSIDENCE
TIME INCREMENT 1

COLUMN	DEPTH	SIZE 1 DEPOS	LEFT
1	1.0	0.0	10.0
2	2.0	0.0	10.0
3	3.0	0.0	10.0
4	3.0	1.0	9.0
5	3.0	2.0	7.0
6	3.0	3.0	4.0
7	5.0	2.0	2.0
8	7.0	1.0	1.0
9	8.5	0.5	0.5
10	9.7	0.3	0.3
11	10.9	0.1	0.1
12	11.9	0.1	0.1
13	13.0	0.0	0.0
14	14.0	0.0	0.0
15	15.0	0.0	0.0
16	16.0	0.0	0.0
17	17.0	0.0	0.0
18	18.0	0.0	0.0
19	19.0	0.0	0.0
20	20.0	0.0	0.0

In Fig. 5 a regressive sequence has been produced by introducing two grain sizes and making subsidence equal to zero. At time 1, sand is deposited close to shore, and silt is deposited farther from shore. Beginning at time 2, and continuing during subsequent time increments, sand begins to build out over silt, the effect of base level (set to three depth units for sand) is to cause the zone of maximum deposition to migrate progressively seaward. Note that time lines intersect the facies boundary. Irregularities in the facies boundary are produced by numerical rounding associated with the use of whole numbers of graphic symbols on the computer printout.

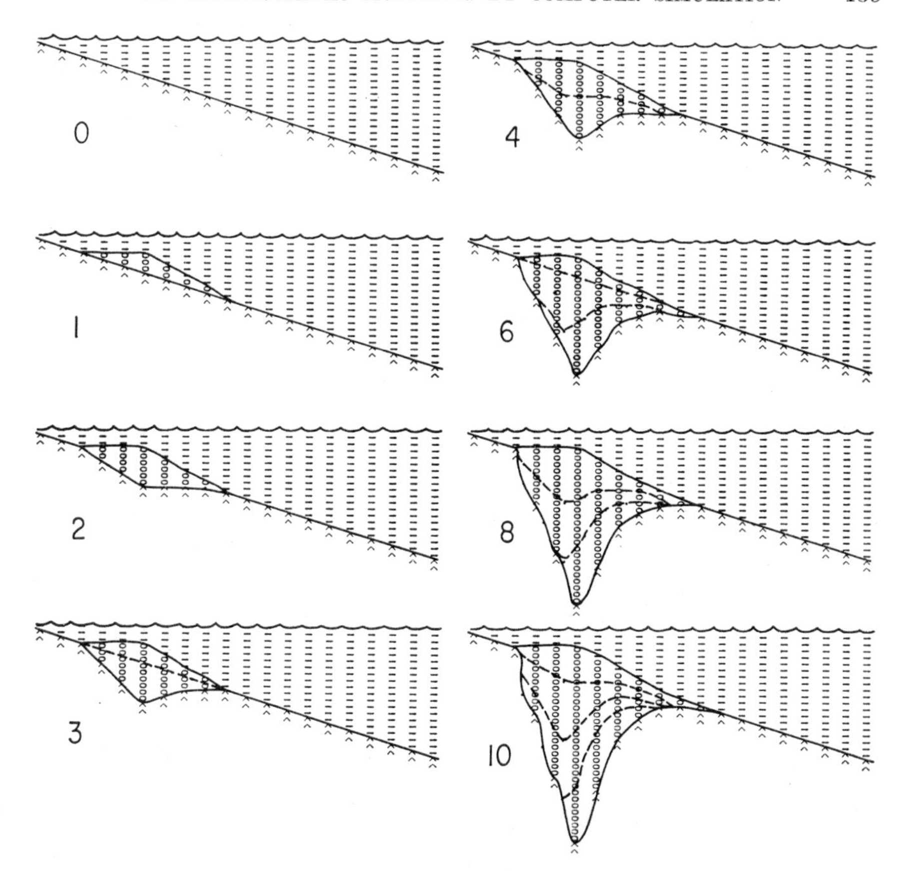

Fig. 4. Response of model when controlling parameters are similar to those in experiment shown in Fig. 3, except that subsidence takes place in each column in proportion to volume of sediment deposited in that column. Each increment of subsidence lags behind deposition by one time increment. Note that configuration of sediment-water interface is unchanged after first time increment. Greatest amount of subsidence occurs where maximum quantity of sediment is deposited.

In the experiment shown in Fig. 6, the same controlling parameters are employed as in Fig. 5, except that rate of subsidence is made equal to rate of sedimentation, with a lag of one time increment. As in Fig. 4, the configuration of the sediment-water interface remains constant through time, and the site of maximum deposition remains in one place. Because there are two facies, however, we note that vertical boundaries separate the sand and silt facies.

The effect of lag is introduced in the experiment shown in Fig. 7. A time lag of three time increments results in interfingering of the sand

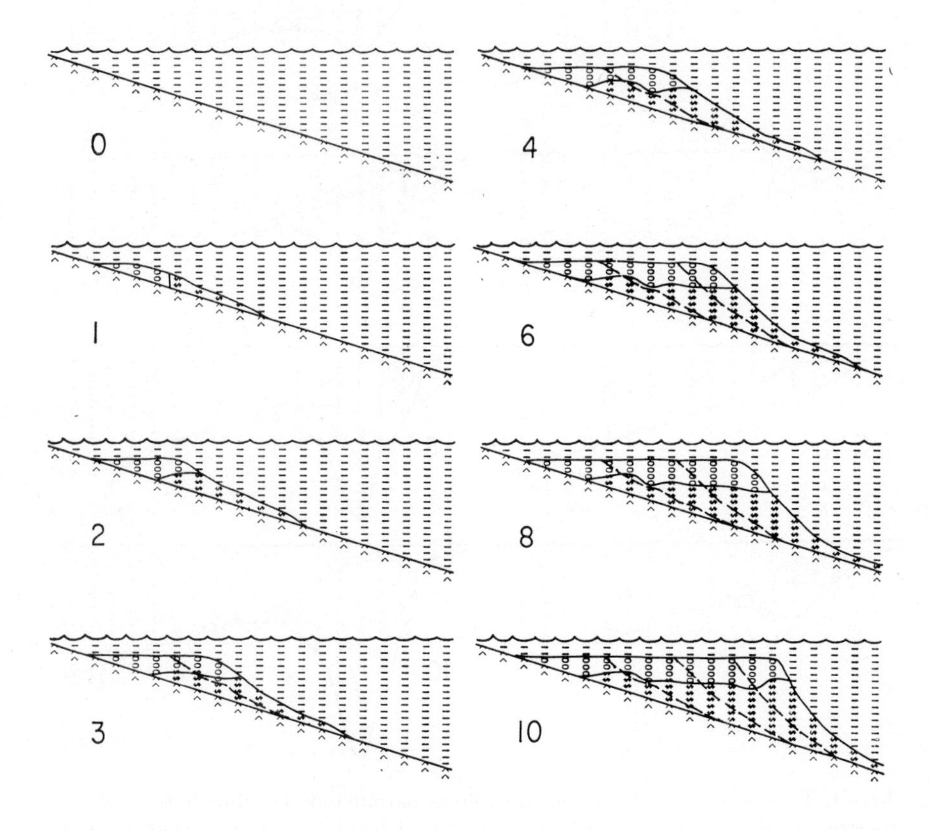

Fig. 5. Regressive sequence that was obtained by assuming that sediment load comprises two grain sizes, sand and silt, and that there is no subsidence of crust. Symbol O denotes that sand is dominant, $ that silt is dominant. At time 1, sand is deposited close to "shore", with silt further offshore. At time 2 and subsequent time increments, however, base level prevents sand from depositing close to shore, and the site of deposition moves progressively seaward, causing sand to build out over silt. Time lines transect facies boundaries.

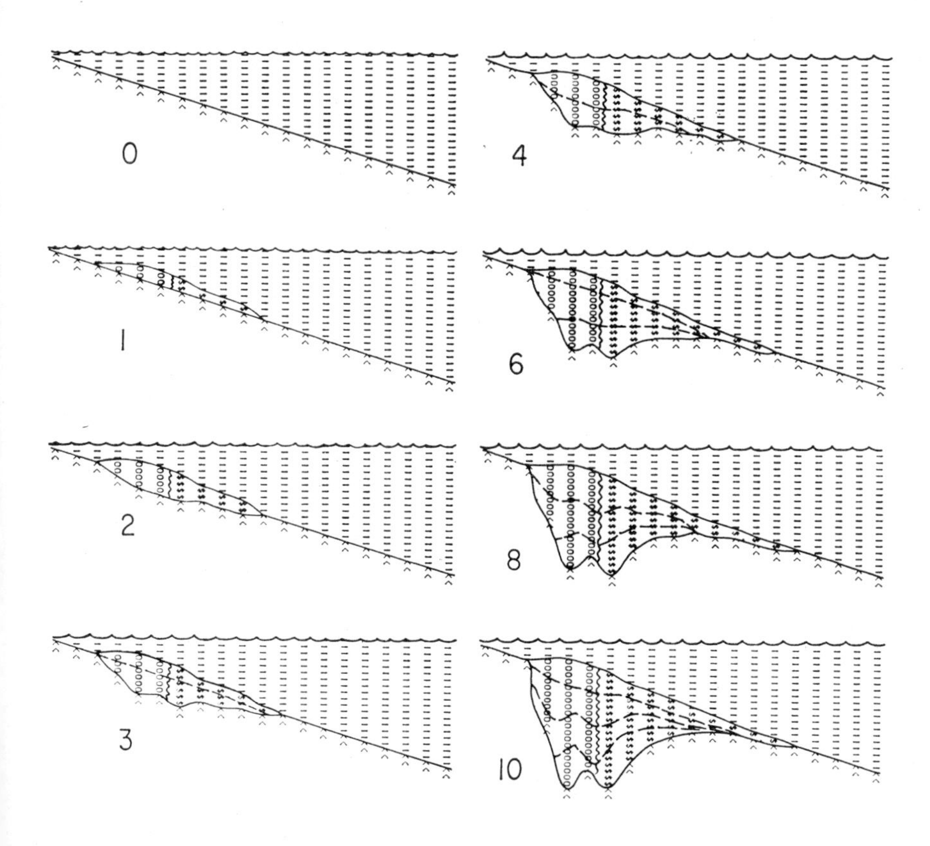

Fig. 6. Input controls are identical to those which produce output shown in Fig. 5, except that subsidence has been made a function of deposition. At each time step, crust subsides by amount exactly equal to amount of deposition during previous time step. Vertical facies boundaries in a lens-shaped body of sand and silt are produced.

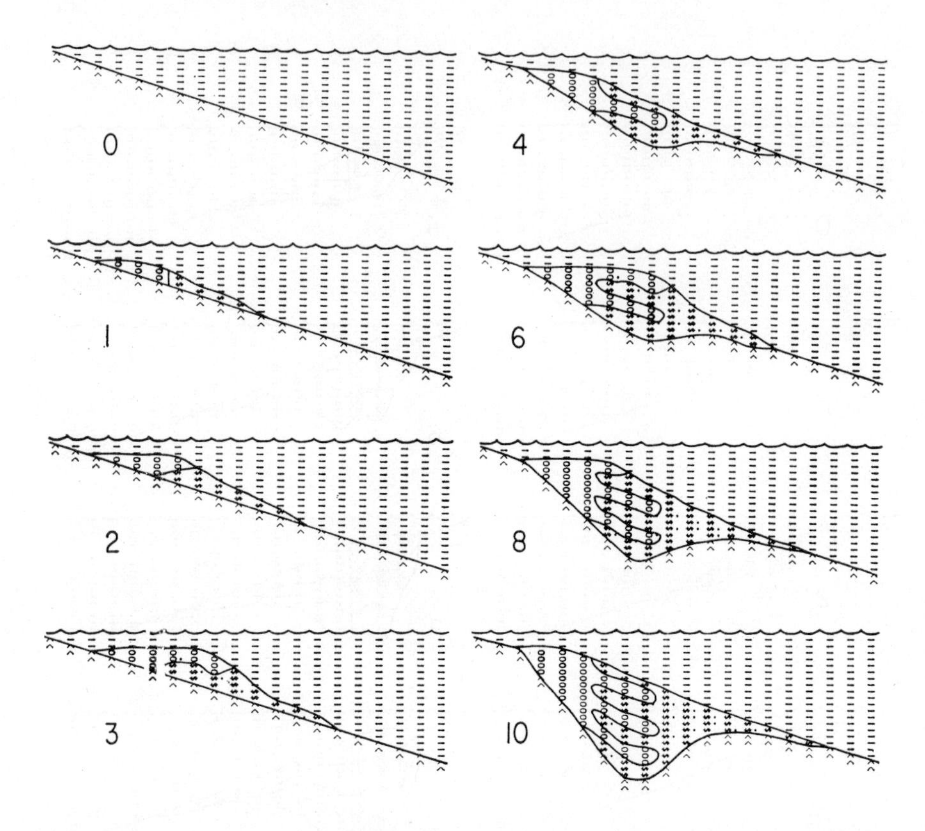

Fig. 7. Input controls are identical to those which produced output in Fig. 6, except subsidence lags behind deposition by three time increments. Time increments 0, 1, 2 and 3, are thus identical to those of Fig. 5, with no subsidence. At time 4, subsidence produces transgression of silt over sand, which is followed by further regression of sand over silt during 5 and 6. This sequence is repeated producing interfingered deposit.

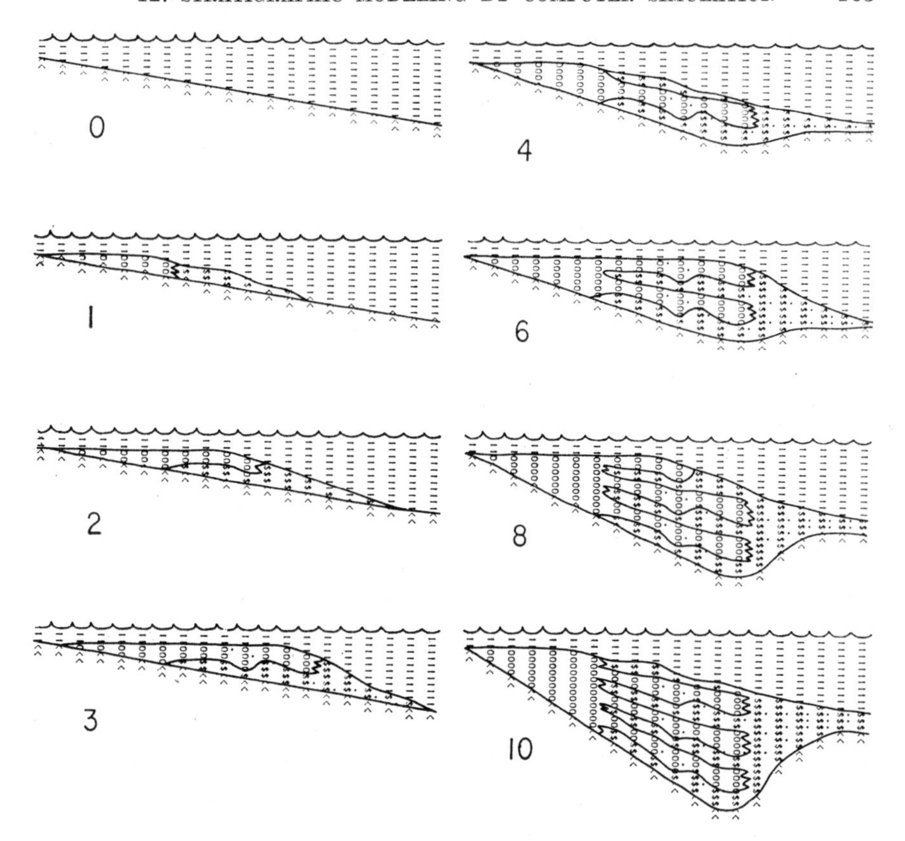

Fig. 8. Response of model similar to that shown in Fig. 7, with two grain sizes and with subsidence lagging behind deposition by three time increments and occurring every third time increment. Experiment differs from that shown in Fig. 7, having a more gentle initial slope of basin floor, and larger sedimentary load supplied during each time increment.

and silt layers. At times 1, 2 and 3, a regressive sequence is formed similar to that shown in Fig. 5. At the end of the third time increment, however, each column subsides by an amount equal to the total thickness of sediment deposited in the column during the preceding three time increments. During increments 4, 5 and 6, a new regressive sequence builds out over the previous one, causing the facies to interfinger. In the experiment shown in Fig. 8, the basin slope has been reduced, but otherwise the controlling parameters are similar to those used to produce the output in Fig. 7. As a result the region of interfingering is more pronounced, the effects of transgression and regression being more widespread.

The sedimentary basin model can be used for making a wide variety of simple sedimentation simulation experiments in its present form. The program can be readily employed to simulate eustatic changes in sea level, and could also be modified to incorporate additional factors that affect the deposition of sediment. The usefulness of such a model lies in its ability to test the effects of simple assumptions, and to formalize concepts that are qualitative and geometric.

ACKNOWLEDGEMENT

This work was supported by the U.S. Office of Naval Research, Geography Branch, Contract N00014-67-A-0112-004, Task Number NR 388-081.

REFERENCE

Sloss, L. L. (1962). Stratigraphic models in exploration. *J. sedim. Petrol.* **32**, 415–422.

13. A General Purpose Computer Program to Produce Geological Stereo Net Diagrams

DAVID BONYUN AND GEORGE STEVENS

Department of Computer Science and Department of Geology, Acadia University, Wolfville, Nova Scotia, Canada

ABSTRACT

This paper describes a program which produces on the line printer a circular stereo net density diagram as used by geologists. Several different cases are handled, depending on the type of data available (micro-planar, micro-linear, field planar or field linear), the type of diagram required (standard or beta), and the type of computation required (step function influence, or normal decay).

Included as appendices are the listing of the programs, sample data, sample requests and sample diagrams.

One of the major features of the program is its ability to show that if poles are assumed to exert influence according to an exponential decay function, the resulting diagram is not significantly different from that produced according to standard technique (a step decay function).

THE PROBLEM

The problem of producing stereographic diagrams from linear and planar data and under conditions of rotation and plane intersection ("β diagrams") has been considered with reference to computers, by several others for the past few years. The output required is a circular diagram indicating the density distribution of the point projections of hundreds of lines and poles of planes on an arbitrary reference plane, usually the horizontal. The projection used is the Lambert–Schmidt Equal-Area projection. The densities are measured as percentages of the total number of poles per 1% area of the whole diagram.

Four different types of basic data are considered: field planar, field

Systematics Association Special Volume No. 3. "Data Processing in Biology and Geology", edited by J. L. Cutbill, 1970. pp. 165–188.

linear, microscopic planar and microscopic linear. Each of these types of data is corrected to sets of three-dimensional coordinates which are the poles on a reference hemisphere. These three-dimensional coordinates are projected to two (note that only *one* hemisphere is used and hence there is only *one* projected pole per element) on the plane from the point on the hemisphere opposite the point of tangency of the projection plane. Each small neighbourhood on the plane is considered in turn to see how many poles influence it, and this number is altered to a percentage of all poles.

NORMAL DECAY OF POLE INFLUENCE

Standard manual methods long in use by geologists allow a pole to exert influence over its immediate neighbourhood according to a step function. If an arbitrary point is within a distance, d, of the pole, it is taken that the pole is influencing the arbitrary point. Otherwise, there is no influence.

We consider the case that each projection exerts an influence on every point in the projection circle according to a normal function with standard deviation $\sigma = d$, the usual step distance. It was found that the resulting density diagram was not significantly different from that resulting for the conventional step function diagram, but required much more time to produce.

PLANE INTERSECTION (BETA DIAGRAMS)

A feature of the program is its capacity to produce a density diagram of the poles of intersection lines of planes. Given n planes, the $n(n-1)/2$ intersections are computed by a subroutine, INPLD, and then projected; the density distribution diagram of these is then produced. (This is the beta-diagram).

ROTATION

The program also has the capacity to perform various rotations, so as to generate a new diagram referred to some other plane of projection instead of the usual horizontal. In addition, the data shown on *any* plane of projection can be rotated (i.e. experience torque) about any chosen axis, with the resulting re-oriented data still referred to the original plane of projection.

The method of specification for a new projection plane is to identify a point on the sphere as the pole of the new plane. This point is given as an azimuth plus a vertical angle which is the complement of the new plane's dip angle. Once the new projection plane is determined, all poles on the sphere are projected to it, and the density diagram is produced.

Torque about any axis is achieved by the additional specification of a rotation angle about the chosen axis, with the sense ("right up" or "left up"). The final projection is referred back to the original plane.

PROGRAM DESCRIPTION

The program package consists of five subroutines and four mainline programs. Our first attempts at producing these diagrams resulted in ellipses rather than circles and this historical anecdote is recorded in the titles of the main line programs. These programs are EGG, EGG1, EGG2 and EGG3. The subroutines are LIMIC, PLFIE, PLMIC, INPLD. Another subroutine, DGS, was used with the version of EGG1 which considered the normal decay problem.

EGG is the main program. It accepts data, determines what type it is and calls the appropriate subroutine to convert the data as given to usable form. In the case of a request for a rotation, it links to the rotation program EGG2. After the data has been made usable (three-dimensional cartesian coordinates of the poles on the hemisphere with negative z-coordinate), it is stored on disc in case the next request is for rotation; then projection equations are handled. Scaling is introduced to make the projection circle have the same radius (10 units) as the sphere rather than $10\sqrt{2}$ as would be expected from the Lambert projection.

The rotation program, EGG2, is a routine which reads from a disc the previously stored set of three-dimensional coordinates of the poles. It accepts the three coordinates of another point on the sphere which is taken to be the directed end of an axis of rotation and the first two transformations make this point the top point of the new projection plane.

The amount of rotation, also on the data card, together with the indicator L or R (up on left or up on right), is then used to complete the transformation. The new coordinates of the poles are not stored on the disc for subsequent use. A subsequent request for rotation from the same disc results in a rotation from the *original* data.

EGG3, linked from EGG and from EGG2, sorts the points on the projection into descending y-coordinate order.

EGG1, linked from EGG3, is the routine which computes influences, and prints the diagram. An optional deck of punched cards may be obtained if further copies of the diagram are required. The technique is to consider small sub-areas of the circle. A sub-area is taken to be centered on a rectangle in which a printed character may be put. It is

$\frac{1}{10}$ in wide and $\frac{1}{8}$ in high. The circle is 11 in in diameter when drawn. The reference point is the centre of such a rectangle.

For each sub-area (taken row by row for printing purposes) only those poles which have proximate y-coordinate values are considered as possibly exerting influence. The number which are within d (1 unit) of the centre of the rectangle are counted and converted to a percentage of the total number of projections—i.e. d is the radius of a circle whose area is 1% of that of the diagram. A symbol representing this percentage is then printed in the appropriate position in the line.

Because of the nature of the diagram, a projection near the edge of the circle exerts influence on points near the other end of the diameter on which it lies. This makes it necessary to consider as possible candidates for influence not only those projections with y-coordinates less than d different from the y-coordinate of the centre of the sub-area, but also some points diametrically across the circle.

The rim of the circle is also of special interest. The points on the rim corresponding to each scan-line are handled separately and printed out as vertical columns on either side of the circular diagram.

A different version of EGG1 was employed to consider the effect of normal decay of pole influence. Here, all poles must be considered as having an effect on each rectangle centre. The amount of influence is determined by the distance from a pole to the point on the sphere corresponding to the centre of the rectangle, according to the formula:

$$\text{Influence} = e^{-l^2/2}$$

where l is the great circle distance. The antipode of a pole might be nearer to the antiprojection of the rectangle centre; consequently, the influence is considered to come from either pole or its antipode, whichever is nearer.

This routine has been adapted easily to compute diagrams indicating density based on a step function where a pole is assured to have a discrete circle of influence *on the sphere* rather than on the projection plane.

The subroutine DGS performs the counting of influence on point on a sphere by calculating, for each pole, the great circle distance from the pole to the point. If the distance is greater than one quarter of a circumference, the pole's coordinates are negated so that, in effect, the other end of the vector line is assumed to be the effective end in computing influence.

The subroutine, INPLD, handles the request for intersection. Whereas, in general, up to 600 readings may be given to EGG for processing, only 35 planes are acceptable to INPLD for the generation of the beta diagram. The readings for these data planes are destroyed during the generation of the intersection poles as up to 595 points may be generated (as when there are precisely 35 planes).

Standard analytic geometry techniques are used together with the fact that the coordinates of the poles are, in fact, a set of direction numbers for the vector from the centre of the sphere to the pole.

The other subroutines, LIMIC, PLFIE and PLMIC are simple routines for converting the data, as given, into a form which corresponds to field linear data.

DATA REQUIREMENTS

Four basic sources of data are possible. These may be catalogued as field linear, field planar, micro-planar and micro-linear. The program has different requirements for each of the four types as outlined below.

1. *Field linear*: a bearing (0°–360°) and a plunge (0°–90°; 0° on horizon; 90° straight down).
2. *Field planar*: a bearing (0°–360°) and a dip (0°–90°) and a letter, N or S, to indicate direction of dip.
3. *Micro-linear*: a bearing (0°–90° or 270°–360°) and a plunge (0°–90°), plus a letter, L or R, to indicate direction. These data are read from a Federov 5-axis universal stage.
4. *Micro-planar*: a bearing (0°–90° or 270°–360°) and a dip (0°–90°) and a letter, N or S, to indicate direction.

The data is given ten readings per card.

Besides the data cards with the reading, certain control cards are required by the program. The required cards are itemized below.

(1) *A divider card* bearing the symbols to be printed on the diagram according to percentage density (prepunched—see Appendix 3).

(2) *An identification card* with the information to be printed below the diagram (50 usable columns).

(3) *A parameter card*: the first five columns with number of readings; column 10 with the type of data or request:

1—Microscopic planar
2—Microscopic linear
3—Field linear
4—Field planar
5—Beta diagram
6—Rotation

(4) *A second parameter card ONLY* if previous one requested a β diagram. Column 5 should contain either a 1 or 4 to identify the type of planar data to come.

(5) The data cards—ten readings per card. OR (5) In the case of rotation, a rotation card with the bearing, dip, torque, cant, and R or L. The last non-rotated data are always used for the rotation.

After the diagram has been drawn, the three rotational angles (two to bring the end of the axis to the top of the new projection plane; and one which is the cant of this plane from the old one) are listed together with two sets of three-dimensional coordinates. These two sets are the new coordinates respectively for: (1) what was the top of the old projection plane (identified as N), and (2) the point which when projected to the old plane produced the centre point (designated +).

As has been said earlier, the rotation feature of the program really handles two separate (although clearly not unrelated) problems. The first is a new density diagram resulting from a new projection plane; the second is a new density diagram resulting from torque along some axis. Both are handled by three-dimensional coordinate transformations and both are "returned" to the original orientation. That is, in the first case two transformations take place (to bring the pole to the top centre of the diagram) and then the first is reversed so that the azimuth of the point is the same as it was to start; in the second, three transformations take place (to bring the pole to the top of the diagram and then apply the torque) and the first two are then reversed. This makes the axis of torque appear not to have moved.

APPENDIX 1

The Lambert Projection

Given a point on a sphere (x, y, z), $x^2 + y^2 \times z^2 = r^2$, the Lambert projection causes it to appear on a plane diagram in the following manner. Let (X, Y) be the coordinates of the point on the plane and let the plane be paralled to the x, y plane through $(0, 0, -r)$ (see diagram below). The point (x, y, z) is joined to $(0, 0, -r)$ and the resulting chord is rotated through $(0, 0, -r)$ to the projection plane. There are three implications to the system:

1. Only points with negative z-coordinates may be so projected.
2. Distance from $(0, 0, -r)$ to (x, y, z) = distance $(0, 0, -r)$ to $(X, Y, -r)$.
3. Areas on the sphere become equal areas on the plane although shapes may become distorted.

To find X, Y in terms of x, y, z, r, it is necessary to note that:

$$\frac{X}{x} = \frac{Y}{y} = \text{some constant } 1; \quad \text{i.e. } X = xl,\ Y = yl$$

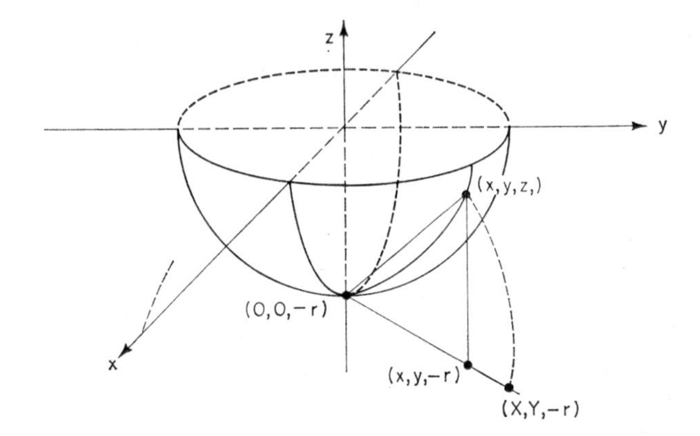

Now $X^2 + Y^2 = x^2 + y^2 + (z + r)^2$, from remark 2 above. Therefore:

$$x^2l^2 + y^2l^2 = x^2 + y^2 + (z + r)^2$$

$$l^2 = 1 + \frac{(z + r)^2}{x^2 + y^2} \quad \text{provided } z \leqslant 0$$

To keep the circle with the same radius as the original sphere, namely r, it is necessary to scale down X, Y so that each is only $1/\sqrt{2}$ of the above, i.e. l may be taken as

$$\sqrt{1 + \frac{(z + r)^2}{x^2 + y^2}} \cdot \frac{1}{\sqrt{2}} = \sqrt{\frac{1}{2}\left(1 + \frac{(z + r)^2}{x^2 + y^2}\right)}$$

To obtain the inverse transformation, i.e. given X, Y, r and the facts that $X = lx$, $Y = ly$

$$l = \sqrt{1 + \frac{(z + r)^2}{x^2 + y^2}} \cdot \frac{1}{\sqrt{2}}$$

We find x, y, z as follows:

$$X\sqrt{2} = x\sqrt{1 + \frac{(z + r)^2}{x^2 + y^2}}; \qquad Y\sqrt{2} = y\sqrt{1 + \frac{(z + r)^2}{x^2 + y^2}}$$

Therefore:

$$2(X^2 + Y^2) = \frac{x^2(x^2 + y^2 + z^2 + r^2 + 2zr) + y^2(x^2 + y^2 + z^2 + r^2 + 2zr)}{x^2 + y^2}$$

$$= 2r^2 + 2zr$$

$$= 2r(r + z)$$

Therefore: $z = \dfrac{X^2 + Y^2 - r2}{r}$

Since

$$x = \frac{X}{l}, y = \frac{Y}{l}, \quad z = \frac{X^2 + Y^2 - r^2}{r}, \quad \text{and} \quad x^2 + y^2 + z^2 = r^2$$

we have:

$$\frac{X^2}{l^2} + \frac{Y^2}{l^2} + \left(\frac{X^2 + Y^2 - r^2}{r^2}\right)^2 = r^2$$

from which

$$\frac{1}{l^2}(X^2 + Y^2) = r^2 - \left(\frac{X^2 + Y^2 - r^2}{r^2}\right)^2$$

$$l = \sqrt{\frac{r^2(X^2 + Y^2)}{r^2 - (X^2 + Y^2 - r^2)^2}} = \sqrt{\frac{r}{r - z}}$$

APPENDIX 2

Listings of the Programs and Subroutines

```
              SEPTEMBER 1969          LISTING OF EGG PROGRAMS AND SUBROUTINES.
// JOB
// FOR
*LIST SOURCE PROGRAM
*ONE WORD INTEGERS
      SUBROUTINE LIMIC (A,I)
      COMMON NRTH,NSTH,NL,NR,II,JJ
      IF(I-NL)1,2,1
    1 IF(I-NR)3,4,3
    3 WRITE (1,5) I,II
      GO TO 20
    2 IF (A-90)6,6,7
    6 A=90-A
      RETURN
    7 IF (A-270)8,9,9
    8 WRITE(1,10)A,I,II
   20 GO TO (21,22),JJ
   22 CALL EXIT
    9 A=450-A
      RETURN
    4 IF(A-90)11,11,12
   11 A=270-A
      RETURN
   12 IF(A-270)8,13,13
   13 A=630-A
   21 RETURN
    5 FORMAT ('INVALID  CHARACTER  ',A1,'  ITEM ',I3)
   10 FORMAT ('INVALID READING ',F5.0,2X,A1,2X,'ITEM ',I3)
      END
// DUP
*DELETE             LIMIC
*STORE      WS  UA  LIMIC
// FOR
*ONE WORD INTEGERS
*LIST SOURCE PROGRAM
      SUBROUTINE  PLFIE (A,I)
      COMMON  NRTH,NSTH,NL,NR,II,JJ
      IF(I-NRTH) 1,2,1
    1 IF(I-NSTH) 3,4,3
    3 WRITE (1,5) I,II
      GO TO (6,20),JJ
   20 CALL EXIT
    2 A=A+90
      RETURN
    4 A=A-90
      IF (A)15,6,6
   15 A=A+360
    6 RETURN
    5 FORMAT ('INVALID CHARACTER ',A1,' ITEM ',I3)
      END
// DUP
*DELETE             PLFIE
*STORE      WS  UA  PLFIE
// FOR
*LIST SOURCE PROGRAM
*ONE WORD INTEGERS
      SUBROUTINE PLMIC(A,I)
      COMMON NRTH,NSTH,NLEFT,NRITE,II  ,JJ
      IF(I-NRTH) 1,2,1
    1 IF(I-NSTH)3,4,3
    3 WRITE(1,5) I,II
      GO TO 20
    2 IF(A-90)6,6,7
    6 A=180-A
      RETURN
    7 IF(A-270)8,9,9
    8 WRITE(1,10)  A,I,II
   20 GO TO (21,22),JJ
   22 CALL EXIT
    9 A=540-A
      RETURN
    4 A=360-A
   21 RETURN
    5 FORMAT('INVALID CHARACTER ',A1,' ITEM ', I3)
   10 FORMAT('INVALID  READING ',F5.0,2X,A1,2X,'ITEM ',I3)
      END
```

```
// DUP
*DELETE             PLMIC
*STORE       WS  UA  PLMIC
// FOR
*LIST SOURCE PROGRAM
*ONE WORD INTEGERS
      SUBROUTINE INPLD (NDCT,NTD)
      DIMENSION A(600,2), B (3,30)
      COMMON N1,N2,N3,N4,I,JJ,R,N,A
      NTD = 7
      NDCT=2
                                                          5 6 6  748 K 11
    2 CALL EXIT
    1 READ (61'1) B
      K=1
      II=N-1
      DO 3  I1=1,II
      I2= I1+1
      DO 3  I3=I2,N
      X= B(2,I1)* B(3,I3)-B(2,I3)* B(3,I1)
      Y= B(3,I1)* B(1,I3) - B(3,I3)*B(1,I1)
      Z= B(1,I1) * B(2,I3) - B(1,I3)*B(2,I1)
      D = SQRT (X*X+Y*Y+Z*Z)
      XA= R*X/D
      XB= R*Y/D
      XC= R*Z/D
      IF(XC)101,102,102
  102 XA=-XA
      XB=-XB
      XC=-XC
  101 CONTINUE
      WRITE (61'K) XA,XB,XC
      K=K+1
    3 CONTINUE
      N=K-1
      RETURN
      END
// DUP
*DELETE             INPLD
*STORE       WS  UA  INPLD
// FOR
*LIST SOURCE PROGRAM
*IOCS(TYPEWRITER)
*ONE WORD INTEGERS
*IOCS(CARD,1132PRINTER)
*IOCS(DISK)
C     PROGRAM FOR DR GEORGE STEVENS - GEOLOGY  DEPARTMENT
C     SW 10 ON TO CHECK DATA
C     DIMENSION AND FORMAT                                              DUMP 002
      DIMENSIONA(600,2),NAME(25), L(37),LB(2),ID(600)
      COMMON NRTH,NSTH,NLEFT,NRITE,I,JJ10
      COMMON  R,N,A,L,D,LB,NAME
      COMMON NTD
      DEFINE FILE 61(600,6,U,N61)
   50 FORMAT(2I5)                                                       DUMP 004
   51 FORMAT(10(F4.0,F3.0,A1))                                          DUMP 005
   52 FORMAT(41A1)                                                      DUMP 006
  305 FORMAT(25A2)                                                      DUMP 014
C     DATA                                                              DUMP 016
C     N IS NUMBER OF POINTS                                             DUMP 017
C     NTD= 1-PLANAR MICRO, 2-LINEAR MICRO 3-LINEAR FIELD 4-PLANAR FIELD DUMP 018
C     NTD=    5-PLANE INTERSECTION         6-ROTATION
C     NDCT IS A MARKER FOR USE WITH INTERSECTIONS
C     A(I,1) IS AZIMUTH, A(I,2) IS DIP                                  DUMP 019
C     R IS RADIUS OF CIRCLE                                             DUMP 020
C     D IS DISTANCE USED BETWEEN POINTS FOR DENSITY                     DUMP 021
 3655 CONTINUE
      READ(2,52)(L(I),I=1,37),NRTH,NSTH,NLEFT,NRITE
      CALL DATSW(10,JJ10)
      AK=3.1415927/180.                                                 DUMP 025
      R=10                                                              DUMP 026
      D=1.                                                              DUMP 027
  300 READ (2,305)NAME                                                  DUMP 028
      READ(2,50)N,NTD                                                   DUMP 029
      IF(NTD-5) 5178,5179,4176
 4176 IF(NTD-6)5178,4177,5178
 4177 CALL LINK (EGG2)
 5179 READ(2,50)MTD
 5178 CONTINUE
      READ(2,51)(A(I,1),A(I,2),ID(I),I=1,N)                             DUMP 030
      NDCT=1
      N61=1
 6172 CONTINUE
```

```
      DO 1 I=1,N
      GO TO (6159,6158), NDCT
 6159 CONTINUE                                                          DUMP 031
      AX=A(I,1)
      IX=ID(I)
      IF(NTD)557,557,558
  558 IF(NTD-5)559,559,557
  557 NTD=6
  559 CONTINUE
      GO TO (502,303,304,302,6179,556),NTD
 6179 CONTINUE
      IF(MTD-1)6178,6177,6178
 6178 IF(MTD-4)556,6176,556
 6177 CALL PLMIC(AX,IX)
      GO TO 3041
 6176 CALL PLFIE(AX,IX)
      GO TO 3042
  556 CALL EXIT
  502 CALL PLMIC (AX,IX)
      GO TO 3041
  303 CALL LIMIC (AX,IX)
      GO TO 3041
  302 CALL PLFIE (AX,IX)
 3042 A(I,2)=90.-A(I,2)
 3041 A(I,1)=AX
      ID(I)=IX
C     POLAR CO-ORDINATES FIRST                                          DUMP 050
  304 AA=(90.-A(I,1))*AK                                                DUMP 051
      AB=R*COS(A(I,2)*AK)
C     NOW CARTESIAN                                                     DUMP 053
      XC=AB*COS(AA)
      YC=AB*SIN(AA)
      XD=XC*XC+YC*YC
      ZC=SQRT(R*R-XD)
      ZD=-ZC
      WRITE(61'N61)XC,YC,ZD
      GO TO 6133
 6158 CONTINUE
      READ (61'I) XC,YC,ZD
      ZC=-ZD
      XD=XC*XC+YC*YC
 6133 CONTINUE
      XL=SQRT(1+(ZC-R)*(ZC-R)/XD)/1.4142135
      A(I,1)=XL*XC
      A(I,2)=XL*YC
    1 CONTINUE                                                          DUMP 056
      GO TO (3655,552),JJ10
  552 CONTINUE
      IF(NTD-5) 6170,6171,6170
 6171 CALL INPLD (NDCT,NTD)
      GO TO 6172
 6170 CONTINUE
      CALL LINK (EGG3)
      END
// DUP
*DELETE               EGG
*STORECI    WS  UA   EGG       2
*LOCAL,PLMIC,PLFIE,LIMIC,INPLD
*FILES(61,GSVGL)
// FOR
*ONE WORD INTEGERS
*LIST SOURCE PROGRAM
*ONE WORD INTEGERS
*IOCS(CARD,1132PRINTER)
*NAME EGG3
C     SORT ROUTINE                                                      DUMP 057
      DIMENSION ID(600)
      DIMENSION A(600,2),          L(37),LB(2),NAME(25)
      DIMENSION SCO(3,2)
      COMMON NRTH,NSTH,NLEFT,NRITE,I,JJ10,R,N,A,L,D,LB,NAME
      COMMON NTD
      COMMON  P,Q,S,SCO                                                 DUMP 058
      ID(1)=1                                                           DUMP 059
      DO 72  I=2,N                                                      DUMP 060
      ID(I)=1                                                           DUMP 061
      M=I-1                                                             DUMP 062
      DO 72  J=1,M                                                      DUMP 063
      IF(A(I,2)-A(J,2)) 73,73,74                                        DUMP 064
   73 ID(I)=ID(I)+1                                                     DUMP 065
      GO TO 72                                                          DUMP 066
   74 ID(J)=ID(J)+1                                                     DUMP 067
   72 CONTINUE                                                          DUMP 068
      DO 75 K=1,N                                                       DUMP 069
```

```
      IF (ID(K)) 76,75,76                                               DUMP 070
   76 H1=A(K,1)                                                         DUMP 071
      H2=A (K,2)                                                        DUMP 072
      J=K                                                               DUMP 073
   77 I=ID(J)                                                           DUMP 074
      ID(J)=0                                                           DUMP 075
      H3=A(I,1)                                                         DUMP 076
      H4=A(I,2)                                                         DUMP 077
      A(I,1)=H1                                                         DUMP 078
      A(I,2)=H2                                                         DUMP 079
      IF(I-K)78,75,78                                                   DUMP 080
   78 H1=H3                                                             DUMP 081
      H2=H4                                                             DUMP 082
      J=I                                                               DUMP 083
      GO TO 77                                                          DUMP 084
   75 CONTINUE
      WRITE (3,71)
   71 FORMAT ( 1H1)                                                     DUMP 085
      WRITE(3,70)(A(I,1),A(I,2),I=1,N)                                  DUMP 086
   70 FORMAT( 5(2X,'(',F8.4,',',F8.4,')  '))
      CALL  LINK (EGG1)
      END
// DUP
*DELETE               EGG3
*STORECI    WS  UA    EGG3
// FOR
*LIST SOURCE PROGRAM
*ONE WORD INTEGERS
*IOCS(CARD,1132PRINTER)
*IOCS(TYPEWRITER)
*NAME EGG1
C     PREPARE LINE BY LINE
C     SW 0 ON TO PUNCH LINES AS WELL AS PRINT
      DIMENSIONA(600,2),LINE(111),L(37),LB(2), NAME(25)
      DIMENSION SCO(3,2)
      COMMON  NRTH,NSTH,NLEFT,NRITE,I ,JJ10
      COMMON  R,N,A,L,D,LB,NAME
      COMMON NTD
      COMMON  P,Q,S,SCO
  202 FORMAT(1H ,A1,2X,55A1,1H+,55A1,2X,A1,1H(,A1,1H))
   53 FORMAT(1H ,    A1,2X,111A1,2X,A1)
   54 FORMAT (1H0 ,'TOTAL    ',I5,2X,' POINTS' )
  203 FORMAT(A1,2X,55A1,'+'/55A1,2X,A1,'('A1,')'10X,I3)
  204 FORMAT(A1,2X,56A1/55A1,2X,A1,13X,I3)
   60 FORMAT(1H1)
  306 FORMAT (1H0, 10X,25A2)
      CALL DATSW (0,NO)
      JJ=0
      WRITE(3,60)
      DO 2 J=1,89
      JJ=JJ+1
      Y=R-(J-1)*R/44.
      YTP=Y+D+.114
      YBT=Y-D-.114
      XB1=-SQRT(R*R-Y*Y)
      XB2=-XB1
      DO 3 I=1,111
    3 LINE(I)=0
      LB(1) = 0
      LB(2) = 0
C     TO FIND BOUNDS FOR LOOP
      DO 90  NN=1,N
      IF(A(NN,2)-(YTP))91,91,90
   90 CONTINUE
      GO TO 187
   91 M1=NN
      DO 95 NN=M1,N
      IF (A(NN,2)-(YBT))96,95,95
   95 CONTINUE
      M2=N
      GO TO 81
   96 M2=NN -1
      IF(M2-M1)187,81,81
   81 CONTINUE
      DO 160 IP=M1,M2
      DIFF=A(IP,2)-Y
      DIFF=D*D-DIFF*DIFF
      IF(DIFF)160,160,1611
 1611 DIFF=SQRT(DIFF)
      LOW=(A(IP,1)-DIFF)*5.5+56
      IHI=(A(IP,1)+DIFF)*5.5+56
      LOW=LOW+100/(LOW+100)*(1-LOW)
      IHI=IHI/111*(111-IHI) +IHI
      DO 161 IEX=LOW,IHI
```

```
  161 LINE(IEX)=LINE(IEX)+1
      IF((XB1-A(IP,1))**2+(Y-A(IP,2))**2-D*D)171,171,172
  171 LB(2)=LB(2)+1
  172 IF((XB2-A(IP,1))**2 +(Y-A(IP,2))**2-D*D) 173,173,160
  173 LB(1)=LB(1)+1
  160 CONTINUE
  187 DO 151 NN=1,N
      IF(A(NN,2)-(2*D-Y  ))152,152,151
  151 CONTINUE
      GO TO 188
  152 M1=NN
      DO 154 NN=M1,N
      IF(A(NN,2)+(Y+2*D  ))155,155,154
  154 CONTINUE
      M2=N
      GO TO 156
  155 M2=NN
  156 DO 162 IP=M1,M2
      T = SQRT(A(IP,1)**2 + A(IP,2)**2)
      IF(T+D-R)162,162,163
  163 G=A(IP,1)*(1.-2.*R/T)
      H=A(IP,2)*(1.-2.*R/T)
      DIFF=D*D-(H-Y)*(H-Y)
      IF(DIFF)162,162,186
  186 DIFF=SQRT(DIFF)
      LOW=(G-DIFF)*5.5+56
      IHI=(G+DIFF)*5.5+56
      LOW=LOW+100/(LOW+100)*(1-LOW)
      IHI=IHI/111*(111-IHI) +IHI
      DO 164 IEX=LOW,IHI
  164 LINE(IEX)=LINE(IEX)+1
      IF((XB1-G)**2+(Y-H)**2-D*D)181,181,182
  181 LB(2)=LB(2)+1
  182 CONTINUE
      IF((XB2-G)**2+(Y-H)**2-D*D)183,183,184
  183 LB(1)=LB(1)+1
  184 CONTINUE
  162 CONTINUE
  188 DO 165  I=1,111
      X=-R+(I-1)*R/55
      IF(X*X+Y*Y-R*R)4,4,5
    5 LINE(I)=L(37)
      GO TO 165
    4 IF(LINE(I))36,36,37
   36 K=36
      GO TO 39
   37 DO 38 K=1,34
      IF(LINE(I)*100/N+1-K)39,39,38
   38 CONTINUE
      K=35
   39 LINE(I)=L(K)
  165 CONTINUE
      DO 166 I=1,2
      IF(LB(I))167,167,168
  167 LB(I)=L(36)
      GO TO 166
  168 DO 169 K=1,34
      IF(LB(I)*100/N+1-K)170,170,169
  169 CONTINUE
      K=35
  170 LB(I)=L(K)
  166 CONTINUE
      IF(Y)200,201,200
  201 WRITE(3,202)LB(2),(LINE(LGS),LGS=1,55),(LINE(LGF),LGF=57,111),LB(1
     1),LINE(56)
      GO TO (206,207),NO
  206 CONTINUE
      WRITE (2,203) LB(2),(LINE(LGS),LGS=1,55),(LINE(LGF),LGF=57,111)
     1,LB(1),LINE(56),JJ
  207 CONTINUE
      GO TO 2
  200 WRITE(3,53)LB(2),LINE,LB(1)
      GO TO (208,2),NO
  208 CONTINUE
      WRITE (2,204) LB(2),LINE,LB(1),JJ
  209 CONTINUE
    2 CONTINUE
      WRITE(3,54)N
      WRITE (3,306)NAME
      IF(NTD-7) 2559,3557,3556
 3557 WRITE(3,2557)
 2557 FORMAT(1H0,'BETA DIAGRAM')
      GO TO 2559
```

```
 3556 WRITE( 3,2556) P,Q,S,SCO
 2556 FORMAT(1H0,'ROTATION ANGLES  '3F6.0,/1X,'N BECOMES  '3F10.4/
     11X,'+ BECOMES  '3F10.4)
 2559 CONTINUE
      CALL  LINK (EGG)
      END
// DUP
*DELETE              EGG1
*STORECI    WS  UA  EGG1
*DUMP       UA  CD  EGG1
// FOR
*ONE WORD INTEGERS
*IOCS(CARD,1132PRINTER)
*LIST SOURCE PROGRAM
*IOCS(TYPEWRITER)
*IOCS(DISK)
*NAME EGG2
C     SW 6 ON TO RETURN ORIETATION TO PENULTIMATE POSITION
      DIMENSION A(600,2),          L(37),LB(2),NAME(25)
      DIMENSION SCO(3,2)
      COMMON NRTH,NSTH,NLEFT,NRITE,I,JJ10,R,N,A,L,D,LB,NAME
      COMMON NTD
      COMMON  P,Q,S,SCO
      DEFINE FILE 61 (600,6,U,N6)
      CALL DATSW(6,NN6)
      NMS=  1
      READ (2,100) P,Q,S,M
  100 FORMAT (3F5.0,4X,A1)
      NTD=8
      TH1=P*3.14159/180.
      TH2=Q*3.14159/180.
      IF  ( S )  404,405,404
  405 NMS=2
      TH3=0
      GO TO 406
  404 CONTINUE
      IF (M-NLEFT)201,202,201
  201 IF (M-NRITE)204,205,204
  204 WRITE(1,300)M
  300 FORMAT('ERROR  'A1)
      CALL EXIT
  202 MULT=-1
      GO TO 206
  205 MULT= 1
  206 TH3= MULT*S*3.14159 /180
      S=S*MULT
  406 CONTINUE
      COS1= COS(TH1)
      SIN1= SIN(TH1)
      COS2= COS(TH2)
      SIN2= SIN(TH2)
      COS3= COS(TH3)
      SIN3= SIN(TH3)
      DO 200 II=1,N
      READ(61'II)X1,Y1,Z1
      X2=X1*COS1-Y1*SIN1
      Y2=X1*SIN1+Y1*COS1
      X1=X2
      Y1=Y2
      Y2=Y1*COS2-Z1*SIN2
      Z2=Y1*SIN2+Z1*COS2
      Z1=Z2
      X2=X1*COS3-Z1*SIN3
      Z2=X1*SIN3+Z1*COS3
      GO TO    ( 400,401),NN6
  400 CONTINUE
      GO TO (407,408),NMS
  407 Y3 = Y2*COS2+Z2*SIN 2
      Z3=-Y2*SIN2+Z2*COS2
      Y2=Y3
      Z2=Z3
  408 X3=X2*COS1+Y2*SIN1
      Y3=-X2*SIN1+Y2*COS1
      Y2=Y3
      X2=X3
  401 CONTINUE
      IF (Z2) 207,208,208
  208 Z2=-Z2
      X2=-X2
      Y2=-Y2
  207 CONTINUE
      Z= -Z2
      XD= X2*X2+ Y2*Y2
      XL= SQRT (1+ (Z-R)*(Z-R)/XD) / 1.4142135
```

```
      A(II,1)= XL* X2
      A(II,2) = XL* Y2
  200 CONTINUE
      SCO(1,1)=-R*SIN1*COS3-R*COS1*SIN2*SIN3
      SCO(2,1)=R*COS1*COS3
      SCO(3,1)=-R*SIN1*SIN3+R*COS1*SIN2*COS3
      SCO(1,2)=R*COS2*SIN3
      SCO(2,2)=R*SIN2
      SCO(3,2)=-R*COS2*COS3
      GO TO   (402,403),NN6
  402 CONTINUE
      GO TO (409,410),NMS
  409 Y2=SCO(2,1)*COS2 + SCO(3,1)*SIN2
      Y3=SCO(2,2)*COS2 + SCO(3,2)*SIN2
      Z2=-SCO(2,1)*SIN2 + SCO(3,1)*COS2
      Z3=-SCO(2,2)*SIN2 + SCO(3,2)*COS2
      SCO(2,1)=Y2
      SCO(2,2)=Y3
      SCO(3,1)=Z2
      SCO(3,2)=Z3
  410 X2=SCO(1,1)*COS1 + SCO(2,1)*SIN1
      X3=SCO(1,2)*COS1 + SCO(2,2)*SIN1
      Y2=-SCO(1,1)*SIN1 + SCO(2,1)*COS1
      Y3=-SCO(1,2)*SIN1 + SCO(2,2)*COS1
      SCO(1,1)=X2
      SCO(2,1)=Y2
      SCO(1,2)=X3
      SCO(2,2)=Y3
  403 CONTINUE
      CALL LINK(EGG3)
      END
// DUP
*DELETE             EGG2
*STORECI    WS  UA  EGG2     1
*FILES(61,GSVGL)
// JOB
// FOR
*ONE WORD INTEGERS
*TRANSFER TRACE
*ARITHMETIC TRACE
      FUNCTION DGS(X,Y)
C     SW 7 ON FOR STEP DECAY ON SURFACE OF SPHERE
      DIMENSION A(600,2),L(37),LB(2),NAME(25)
      COMMON  NRTH,NSTH,NLEFT,NRITE,I,JJ10
      COMMON  R,N,A,L,D,LB,NAME
      CALL DATSW (7,N7)
      Z2=(X*X+Y*Y-R*R)/R
      XL=SQRT(R/(R-Z2))
      X2=X/XL
      Y2=Y/XL
      DGS=0
      DO 1 II=1,N
      X1=A(II,1)
      Y1=A(II,2)
      Z3=(X1*X1+Y1*Y1-R*R)/R
      XL=SQRT(R/(R-Z3))
      X3=X1/XL
      Y3=Y1/XL
    4 DS= (X2-X3)*(X2-X3)+(Y2-Y3)*(Y2-Y3)+(Z2-Z3)*(Z2-Z3)
      TH=2.*ATAN(SQRT(DS/(4.*R*R-DS)))
      DSS=R*TH
      IF(TH-1.57081)2,2,3
    3 X3=-X3
      Y3=-Y3
      Z3=-Z3
      GO TO 4
    2 GO TO (10,11),N7
   11 DGS=DGS+EXP(-DSS*DSS)
      GO TO 1
   10 IF(DSS-1.4142)13,13,1
   13 DGS=DGS+1.
    1 CONTINUE
       RETURN
      END
// DUP
*DELETE             DGS
*STORE      WS  UA  DGS
// FOR
*ONE WORD INTEGERS
*IOCS(CARD,1132PRINTER)
*IOCS(TYPEWRITER)
*NAME EGG1
C     PREPARE LINE BY LINE
C     SW 0 ON TO PUNCH LINES AS WELL AS PRINT
       DIMENSION XLB(2)
      DIMENSIONA(600,2),LINE(111),L(37),LB(2),        NAME(25)
```

```
      DIMENSION SCO(3,2)
      COMMON  NRTH,NSTH,NLEFT,NRITE,I,JJ10
      COMMON  R,N,A,L,D,LB,NAME
      COMMON NTD
      COMMON  P,Q,S,SCO
  202 FORMAT(1H ,A1,2X,55A1,1H+,55A1,2X,A1,1H(,A1,1H))
   53 FORMAT(1H ,   A1,2X,111A1,2X,A1)
   54 FORMAT (1H0 ,'TOTAL    ',I5,2X,' POINTS' )
  203 FORMAT(A1,2X,55A1,'+'/55A1,2X,A1,'('A1,')'10X,I3)
  204 FORMAT(A1,2X,56A1/55A1,2X,A1,13X,I3)
   60 FORMAT(1H1)
  306 FORMAT (1H0, 10X,25A2)
      CALL DATSW (0,NO)
      JJ=0
      WRITE(3,60)
      DO 2 J=1,89
      JJ=JJ+1
      Y=R-(J-1)*R/44.
      XB1=-SQRT(R*R-Y*Y)
      XB2=-XB1
      DO 3 I=1,111
    3 LINE(I)=0
      LB(1) = 0
      LB(2) = 0
      DO 90  IEX=1,111
      X=-R+(IEX-1)*R/55
      IF(X-XB1)95,96,96
   96 IF(X-XB2)171,171,95
   95 LINE (IEX)= L(37)
      GO TO 90
  171 DGSS=DGS (X,Y)
      IF (DGSS - .1) 36,37,37
   36 K=36
      GO TO 39
   37 DO 38 K=1,34
      IF(DGSS*100./N-K)39,39,38
   38 CONTINUE
      K=35
   39 LINE (IEX)=L(K)
   90 CONTINUE
      XLB(2)=DGS(XB1,Y)
      XLB(1)=DGS(XB2,Y)
      DO 166 I=1,2
      IF(XLB(I)-.1)167,167,168
  167 LB(I)=L(36)
      GO TO 166
  168 DO 169 K=1,34
      IF(XLB(I)*100./N-K)170,170,169
  169 CONTINUE
      K=35
  170 LB(I)=L(K)
  166 CONTINUE
      IF(Y)200,201,200
  201 WRITE(3,202)LB(2),(LINE(LGS),LGS=1,55),(LINE(LGF),LGF=57,111),LB(1
     1),LINE(56)
      GO TO (206,207),NO
  206 CONTINUE
      WRITE (2,203) LB(2),(LINE(LGS),LGS=1,55),(LINE(LGF),LGF=57,111)
     1,LB(1),LINE(56),JJ
  207 CONTINUE
      GO TO 2
  200 WRITE(3,53)LB(2),LINE,LB(1)
      GO TO (208,2),NO
  208 CONTINUE
      WRITE (2,204) LB(2),LINE,LB(1),JJ
    2 CONTINUE
      WRITE(3,54)N
      WRITE (3,306)NAME
      IF(NTD-7) 2559,3557,3556
 3557 WRITE(3,2557)
 2557 FORMAT(1H0,'BETA DIAGRAM')
      GO TO 2559
 3556 WRITE (3,2556) P,Q,S,SCO
 2556 FORMAT(1H0,'ROTATION ANGLES   '3F6.0,/1X,'N BECOMES   '3F10.4/
     11X,'+ BECOMES   '3F10.4)
 2559 CONTINUE
      CALL  LINK (EGG)
      END
// DUP
*DELETE              EGG1
*STORECI     WS  UA  EGG1
*DUMP        UA  CD  EGG1
// DUP
*DEFINE FIXED AREA           5
*STOREDATA  WS  FX  GSVGL   32
// JOB
```

APPENDIX 3

Sample Data

```
              SEPTEMBER 1969            LISTING OF SAMPLE DATA FOR EGG PROGRAMS

// XEQ EGG
0123456789ABCDEFGHIJKLMNOPQRSTUVWXY. NSLR
 DAVID BENT APR9 /69 ROCK NO 11 INCL.S
  100     1
 315 08S 282 14S 306 12N 325 23N 066 37N 078 37N 304 49S 079 12N 321 40S 038 17N
 082 13S 303 09N 301 17S 314 14S 314 17N 060 03N 304 21S 040 16S 329 23N 085 20N
 274 17S 287 11N 314 02S 312 44S 296 10S 271 01N 331 15S 342 16N 270 01N 088 04N
 357 03N 282 22S 330 25S 044 37N 039 32N 337 40S 347 04S 077 06N 001 03N 064 26N
 081 11N 043 41N 310 20S 039 35N 090 09N 274 20N 270 17N 298 01S 035 01S 061 03N
 058 19N 050 23N 270 12N 082 12N 286 12N 042 30N 045 14N 056 18N 060 10S 084 18N
 295 06N 046 09N 065 13N 276 25N 273 21N 088 12N 304 14N 048 30N 036 12S 290 08S
 088 04N 313 22N 056 14N 064 16N 054 12N 333 08S 060 12N 052 18N 048 04N 275 18S
 044 16N 279 11N 330 10S 082 11S 355 13S 334 08N 062 12N 083 14S 282 03N 288 12N
 050 16N 058 12N 320 36S 313 14N 302 12N 340 35S 056 08N 044 18N 343 02S 358 05N
0123456789ABCDEFGHIJKLMNOPQRSTUVWXY. NSLR
 DAVID BENT APR9 /69 ROCK NO 4 C AXIS
  400     2
 088 30L 034 11R 085 08R 040 28L 013 12L 055 01R 349 36R 040 13L 080 28R 014 32R
 039 12L 029 23R 271 13L 040 11R 296 73L 339 52R 090 54R 299 62L 025 58L 054 24R
 020 34R 354 24L 060 26R 272 36R 046 12L 001 32L 290 29L 282 31L 341 01L 319 01R
 291 34R 340 34R 355 18L 070 52R 002 02R 294 35R 075 01L 044 75R 315 50L 299 51R
 005 66L 345 36L 015 10R 019 47L 020 31R 012 41R 043 19L 089 25L 045 01L 071 01R
 355 26R 299 36L 286 10R 316 19L 290 15R 344 37R 068 16L 080 24L 041 65L 031 32R
 277 61L 042 58L 284 08L 280 22R 049 20R 344 30R 053 09L 010 23R 005 40R 303 19L
 003 30L 001 39L 010 01L 305 03R 316 06L 001 31L 074 14L 025 08L 354 56R 025 10L
 009 09L 066 01R 007 27L 072 28L 044 09R 082 20R 038 30L 039 11R 087 24R 060 01L
 296 21L 032 32L 283 09L 025 32R 280 27R 305 46L 086 24R 019 27R 021 53L 046 17R
 277 26R 042 26L 325 16L 305 59L 290 21L 278 67L 049 10R 279 09L 067 29L 070 29L
 294 04L 296 26R 048 26R 272 28R 054 26L 353 56L 022 29L 035 34L 288 01R 002 18R
 040 09R 328 01L 285 01L 050 69L 323 18R 061 31L 061 26L 085 26L 050 19R 284 32R
 329 24R 071 53R 335 22L 328 40L 014 20R 035 44R 039 18L 070 67R 061 16L 070 26R
 348 35R 047 26L 048 30L 021 60L 320 09L 079 48L 283 60R 039 26R 277 10R 051 18L
 294 38R 350 22L 029 51L 335 26L 293 38R 021 32R 324 10R 280 84R 333 29L 001 12R
 288 20L 064 36R 087 35L 342 02L 085 26R 302 22R 080 30L 311 37R 339 32L 301 63L
 052 33L 064 34L 299 46R 051 26R 017 16L 340 13L 306 38R 329 45L 035 45R 077 40R
 040 16L 340 44L 062 01L 020 52L 088 51R 285 30L 340 19L 027 08L 336 42R 300 06R
 008 11L 277 07L 027 26L 355 37R 061 30L 337 02R 088 01R 002 47R 074 08L 340 66R
 069 59L 343 19L 270 76R 038 26R 333 27R 328 20L 039 25L 024 24R 015 08R 021 35L
 289 33L 301 26R 270 53R 325 32L 052 42L 084 52R 305 40R 285 37R 333 19L 044 01L
 304 54L 335 16R 047 39R 346 13L 052 27R 351 36R 314 21R 297 28R 014 20R 282 24L
 028 30R 067 07R 067 20R 273 24R 320 16R 043 05R 046 47R 293 59R 332 55L 065 34R
 333 08L 330 69L 081 85R 016 63L 353 08R 063 33L 280 01R 316 51R 042 18R 281 17L
 005 12R 283 53R 030 42R 311 23R 082 25L 081 35L 045 06R 350 27R 028 52L 275 28L
 357 37L 026 41L 013 50R 038 75R 013 30R 052 69L 076 40L 346 05R 312 15R 333 01L
 027 38R 322 18R 282 45L 011 01R 312 13R 030 36L 355 23R 327 53R 043 20L 054 72L
 322 10R 291 27R 072 50L 057 32R 063 80L 326 33R 077 26L 080 35L 056 15R 065 10L
 278 28R 323 45R 028 39L 046 19R 346 01L 081 25L 046 10R 057 30L 322 18R 322 21R
 047 36L 280 46L 354 20L 293 18R 081 27R 332 07L 077 01R 082 29R 085 32L 019 02L
 019 62L 080 59L 295 25R 320 75R 324 55L 285 05R 271 78R 022 61R 038 25L 304 14L
 068 27L 022 53R 017 08R 315 35R 329 09R 304 01R 278 67R 010 27L 288 67R 083 17L
 006 16L 276 14R 333 31L 032 18R 077 38R 292 20L 020 52L 003 31L 005 30L 278 01L
 285 89L 042 07L 081 31L 023 28L 020 27R 052 22R 003 12L 333 51R 022 01R 310 61L
 055 37L 284 18L 270 26R 278 41L 061 40R 050 23L 048 30L 052 28L 305 36R 018 32L
 066 17L 332 32L 020 08L 305 21L 337 40R 040 31L 322 28L 272 20R 340 32L 022 29R
 038 30R 304 32L 356 18L 317 67L 065 22L 315 17R 079 56L 352 38L 010 05R 005 12L
 083 34R 076 46L 294 23L 319 34R 320 34R 082 19R 033 30R 050 34L 027 15L 355 20R
 016 05R 345 28L 348 34L 333 30L 313 16L 333 33L 018 28R 305 50R 330 38L 032 17R

0123456789ABCDEFGHIJKLMNOPQRSTUVWXY. NSLR
 PROJECT 20-67 SITE 22 LUNENBURG,LUNB.CO.
  100     3
 080 30   045 00   210 30   275 05   235 10   260 20   225 00   280 10   125 10   020 10
 275 30   260 40   080 05   015 15   045 00   240 10   020 20   235 35   130 15   105 05
 115 15   350 05   110 15   095 10   095 10   060 05   275 20   095 00   235 15   060 05
 090 15   155 15   295 15   230 25   170 15   100 05   190 35   055 10   350 10   050 10
 015 10   185 05   040 40   205 30   030 10   080 05   185 25   200 15   040 10   015 30
 240 00   285 10   060 05   000 10   050 25   025 15   235 25   270 40   330 35   220 10
 045 20   230 40   175 10   030 20   075 20   060 20   090 20   055 25   030 10   085 10
 050 15   180 10   150 25   085 00   240 15   035 10   020 15   135 00   085 05   070 10
 040 45   320 10   040 25   155 60   025 05   050 30   060 20   020 10   090 10   070 05
 025 50   030 15   060 15   345 20   025 05   330 20   230 10   080 05   025 05   245 25
0123456789ABCDEFGHIJKLMNOPQRSTUVWXY. NSLR
EXPERIMENTAL DATA FOR INTERSECTION - FIRST
   35     4
  64 70S   50 85S   90 40S   87 44S   78 33S   77 50S   74 85S   70 70S   94 35S   97 40S
  54 88N   48 90N   45 87N   40 85N   42 80N   39 80N   38 70N   50 80S   61 80S   64 75S
  98 50S  106 52S  108 40S    5 50N   10 55N   15 45N   30 45N   34 50N  172 50S  176 55S
  64 55S   68 35S   72 48S   78 63S   25 60N
```

```
0123456789ABCDEFGHIJKLMNOPQRSTUVWXY. NSLR
EXPERIMENTAL DATA FOR INTERSECTION - INTERSECTED
  30    5
   4
  64 70S  50 85S  90 40S  87 44S  78 33S  77 50S  74 85S  70 70S  94 35S  97 40S
  54 88N  48 90N  45 87N  40 85N  42 80N  39 80N  38 70N  50 80S  61 80S  64 75S
  98 50S 106 52S 108 40S   5 50N  10 55N  15 45N  30 45N  34 50N 172 50S 176 55S
0123456789ABCDEFGHIJKLMNOPQRSTUVWXY. NSLR
EXPERIMENTAL DATA FOR ROTATION - FIRST
 400    4
  58 74S  68 25S  57 90N 342 77N  88 76S 300 28N 334 43S  59 73S   5 56N 290 84N
 306 50S 279 74N 314 40N 358 39S 356 16N  59 28N  25 86N 333 24S  73 12N   1 28N
 280 81S  70 63N 312 77S 348 84N 273 51S  29 45N   8  5S 304 80N 280 82S  27 32S
   9 79N 271 77N   4 28N 337 34S  67 56N  27 69N 282 16N 332 57N  58 74S 272 89N
  34 23N  13 71S 281 21S  54 20S 342 48N  47 83S  69 12S  56 57S  57  8N 306 57S
 337 21S 327  6N 320 11N 340 32N 348 84N 285 14N  47 54S  12 82N   1 11S  58 64S
 351 72S  89  4S 345 13N  87 74S 330 16S 309 21S 277 42N  84 28S 282 42N 332 60N
  82 58S 302  2N 274  1S  39 24S  68 89S  35 89N 351 37N   5 67S   5 84S   2 17S
 289 27N  48 32S  12 21S 323 55N 299 20S  32 32S 345 71S 357 16S  38 88N 310 48N
 294 49N 355 36S  89 56N 321 20N 280 41S  30 62N  39 16S  46 16N 326 27S 276 50N
 280 57S  38 83N 318 78S  54 22S  79 77S  37 58S  90 89S  34 86S  46 23S  18 21S
 313 70S  53 85S  54 32N 321 76S 288 62S  34 33S 298 39N 337  3N  35 48N 313 19N
 348 59N 334 87N 350  8N  55 38N  79 36N 344 79S 343 37N 277 83N 293 78N  13 38S
  54 42S 331 79N  57 67S 320 22S 288 79N 274 59S  11 40S 359 38N  36  6S 278  7N
  86 50S 293  7S   3 42N 282 55S  24 84N  30 11S 307 52S  15  3S 316 13N 298 78N
 308 57N  71 13N  82 13N 316 66N  15 25S  17 33S  38 43N 280 47N 307 29S   3 74N
 313 31N  64 81N   3 76S  18  9S  13 28S  33  2N 324 47S  88 55S 284 83S  56 77S
   9 42S  48  3N 305 69N 337 79N  63 36S  61 43S 340 86N   4  1S  71  5N  26  8N
  37 57N  25 76S  65 13S 346 55S  72 50N 325 14S 311 14N 353 72N 273 32S  34  8N
  49 43N  31 12N 293 81N 332 81S 295 41S 330 83S 286 39S  31 14S 283 41N 332 88N
 271 55S 320 25N 301 63S 273 58N  65  4N  15 90S  74 82S 273 86N 356 24S   2 30N
 313 20N 334 25S 339 30N   5  4S  45 77N 275 85N 286 82S 271 17S 346 49N   1 32N
  42 76S 321 62N   8 36S 297 26S 306 31S 308 37S  79 39N 295 72S 276 77N  32  2N
 305 71N  65 33S 356 24N  16 39S 331 21S  56 73S 284 69S 284 17N  33 20S 296 70N
 271 17N 283 60N   9 54S  28 36N  82 11S  47 71N 331 22N 355 15N 312 17N  60 17N
  65 63S  63 30S  87 79N 299 81N  60 10N 282 21N 328 83S 350 28S  36 21S  33 40S
 314 65S 293 86S  25 38S  47 80N 336  9N  21 63N  61 15S 290 23N 330 83S 329 48N
  44 33S 359 66S  19 42N 336 87S  74 78S  40 72N  35 49N  78 49N 282 69N  46 33S
  17 64S 307 10S 345  7N 276 46N  26 46N  11 18S  79  1N   3 19S  86 66S 302 69N
 347  1N  55 27S 319 18N 312 52N  37 49N  30 47N 323 29S  51  3N  34 67N  73 43N
  63 58S  49  3N  78 18S 287 88S 336 56N 298 62N 345 68N  23 49S  78  3S 354  9S
  76 24S 290 87N 295 48S  34 34N 340 18S  31 23S  39 90N 350 90S 334  4N  55  8N
  43 62S 346 20S 303 63N 281 73N 340 75N 328 38S  79 90N 295 65S 298 20N  54 83N
 320 86S  90 17S  88 34S  69 41S  61 65S 318 43S 311 19N 293 53S 360 61N  44  8S
  56 61S 328 36S  83 49N 320 34S 348 42N 276 11N  66 83N 294 39S 276 49S  12 48N
 328 88N  21 15S 354 87S 277 70S 292 34S 339 51N 312 22N  15 59S 342 51N 304 35N
  43 56N 351 60N 330 52N 326 10N 345 85S  73 74N 276 52S 320 42S  35 22N  70 23N
 307 78S 282 56S 326 80S  77 70S  81 73N 282 23N  47 33N 290 37S  41 53S 299 54S
  23 45S 358 36N  21 58S   7 24N 342 50S 338 42N  21 39S 303 36N  11 36S 338 51N
 313 77N 316 77S 286 90S 316 61S 282 39N 303 79S  37  6S 295 11S  67 46S 316 37S
0123456789ABCDEFGHIJKLMNOPQRSTUVWXY. NSLR
EXPERIMENTAL DATA FOR ROTATION - ROTATED
  111    6
  225   35   85    L
0123456789ABCDEFGHIJKLMNOPQRSTUVWXY. NSLR
ROTATION TO VERTICAL    ONE
  111    6
   45   55    0    R
0123456789ABCDEFGHIJKLMNOPQRSTUVWXY. NSLR
ROTATION TO VERTICAL    TWO
  111    6
  225  -55    0    R
0123456789ABCDEFGHIJKLMNOPQRSTUVWXY. NSLR
TRIAL DATA TO COMPARE NORMAL VS STEP. STEP ON SPHERE
  111    4
  54 88N  52 88N  50 90N  51 90N  48 90N  46 85N  46 90N  45 87N  40 80N  40 85N
  48 82N  42 80N  41 78N  41 85N  40 80N  39 85N  39 80N  38 77N  38 70N  36 70N
  54 88S  52 80S  50 80S  50 90S  51 85S  48 88S  47 80S  54 75S  56 80S  57 75S
  58 85S  59 80S  61 80S  60 70S  64 75S  66 80S  64 70S  64 75S  50 85S  88 35S
  90 40S  91 38S  92 40S  90 44S  87 44S  86 40S  85 45S  85 37S  84 40S  80 50S
  80 45S  78 33S  77 50S  76 60S  75 50S  74 43S  74 35S  72 50S  70 40S  91 78S
  94 65S  94 35S  93 40S  95 30S  96 55S  97 40S  98 50S  98 35S  99 55S 102 40S
 102 60S 102 35S 106 52S 108 40S 115 40S 112 38S 120 38S 125 40S 130 40S   5 50N
  10 55N  15 45N  20 48N  25 60N  25 40N  30 45N  30 50N  32 45N  34 50N 170 40S
 172 50S 174 50S 176 55S 178 45S 180 40S  60 50S  62 40S  63 45S  64 55S  64 40S
  66 50S  66 45S  68 35S  71 58S  72 48S  74 32S  74 60S  76 55S  78 63S  80 58S
  80 38S
```

APPENDIX 4

1. Examples of simple diagrams

```
TOTAL       100 POINTS
            DAVID BENT APR 9/69 ROCK NO 11 INCL S (PLANER INCLUSIONS IN QTZ)
            DATA TYPE-MICRO-PLANAR-STUDENT EXAMPLE
CONTOUR INTERVAL 0-1,1-2 & 3, 2 & 3-4, 4-5 & 6, 6-7, 7-8, 8-9, 9-10 & 11
```

```
TOTAL       400 POINTS
            DAVID BENT APR 9/69 ROCK NO 4 C AXIS
            MICRO-LINEAR STUDENT EXAMPLE
CONTOURS    1, 2, 3, 4, %
```

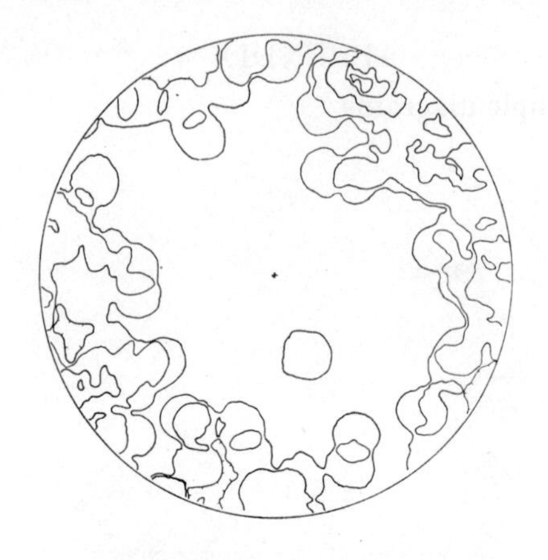

STUDENT'S EXAMPLE

TOTAL 100 POINTS

// ASM Project 20-67 SITE 22 LUNENBURG.LUNB.CO.

FIELD LINEAR

CONTOURS 1, 2, 5, 8, 10 %

2. Intersections

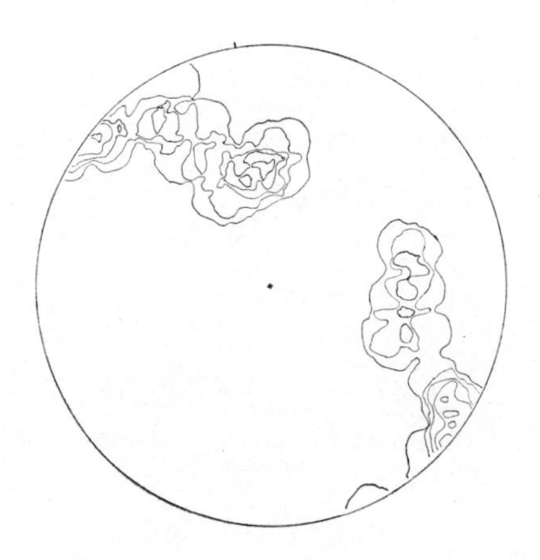

TOTAL 35 POINTS

EXPERIMENTAL DATA FOR INTERSECTION FIRST

BETA PLOT BEFORE

CONTOURS 2, 5, 8, 11, 14 %

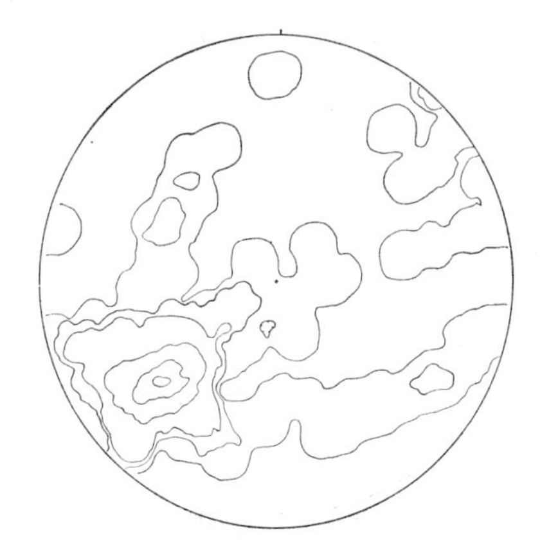

TOTAL 453 POINTS
EXPERIMENTAL DATA FOR INTERSECTION INTERSECTED
BETA DIAGRAM
BETA PLOT AFTER
CONTOURS 0, 1, 2, 10, 25, 35 %

3. Rotation

TOTAL 111 POINTS
EXPERIMENTAL DATA FOR ROTATION - FIRST
ROTATION BEFORE

CONTOUR INTERVAL 0, 1, 5, 10, 18 %

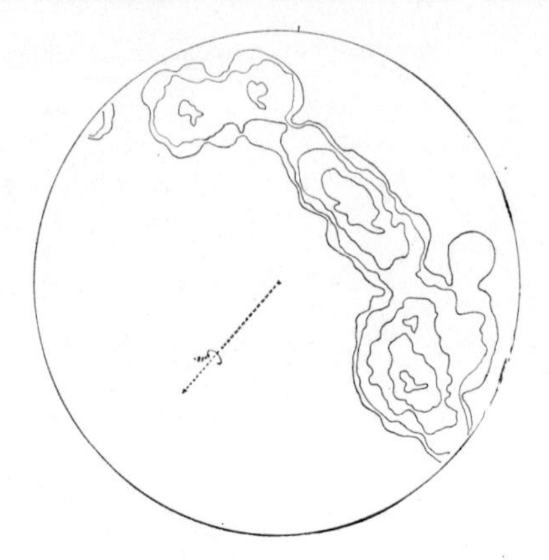

TOTAL 111 POINTS

EXPERIMENTAL DATA FOR ROTATION - ROTATED

ROTATION ANGLES	225	35	-85
	5.7784	0.9361	-5.7062
	2.7375	-8.8029	-3.8747

ROTATION ABOUT AXIS (225° -35°) by 85°

CONTOURS USED 0, 1, 5, 10, 18 %

TOTAL 111 POINTS

ROTATION TO VERTICAL ONE

ROTATION ANGLES	45	55	0
N BECOMES	0.0000	9.9999	5.7922
+ BECOMES	5.7922	5.7922	-5.7357

ROTATION TO PLACE GIRDLE AXIS (VIA PATH "A") (R#3)

CONTOURS 0, 2, 4, 6, 8, 10, 14 %

4. Step v. normal

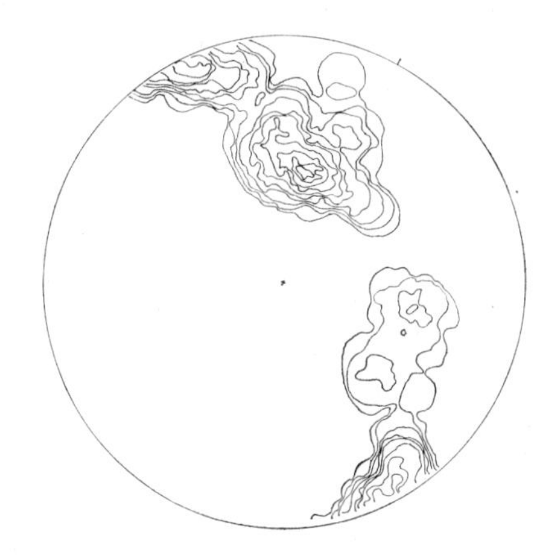

"Step Decay"

TOTAL 111 POINTS

TRIAL DATA TO COMPARE NORMAL VS STEP THIS IS STEP

CONTOUR INTERVAL 0, 1, 2, 4, 6, 8, 10, 13, 15, 17 %

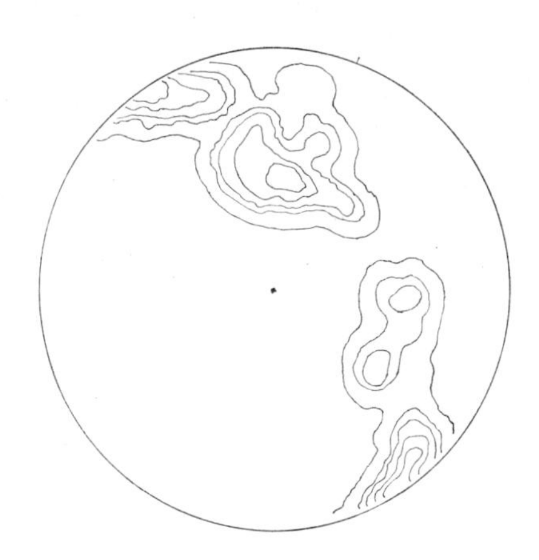

"Normal Decay"

TOTAL 111 POINT

TRIAL DATA TO COMPARE NORMAL VS STEP THIS IS NORMAL

CONTOUR INTERVAL 0, 1, 2, 4, 6. %

5. Computer printout from which Fig. 4a was traced to show normal form of result

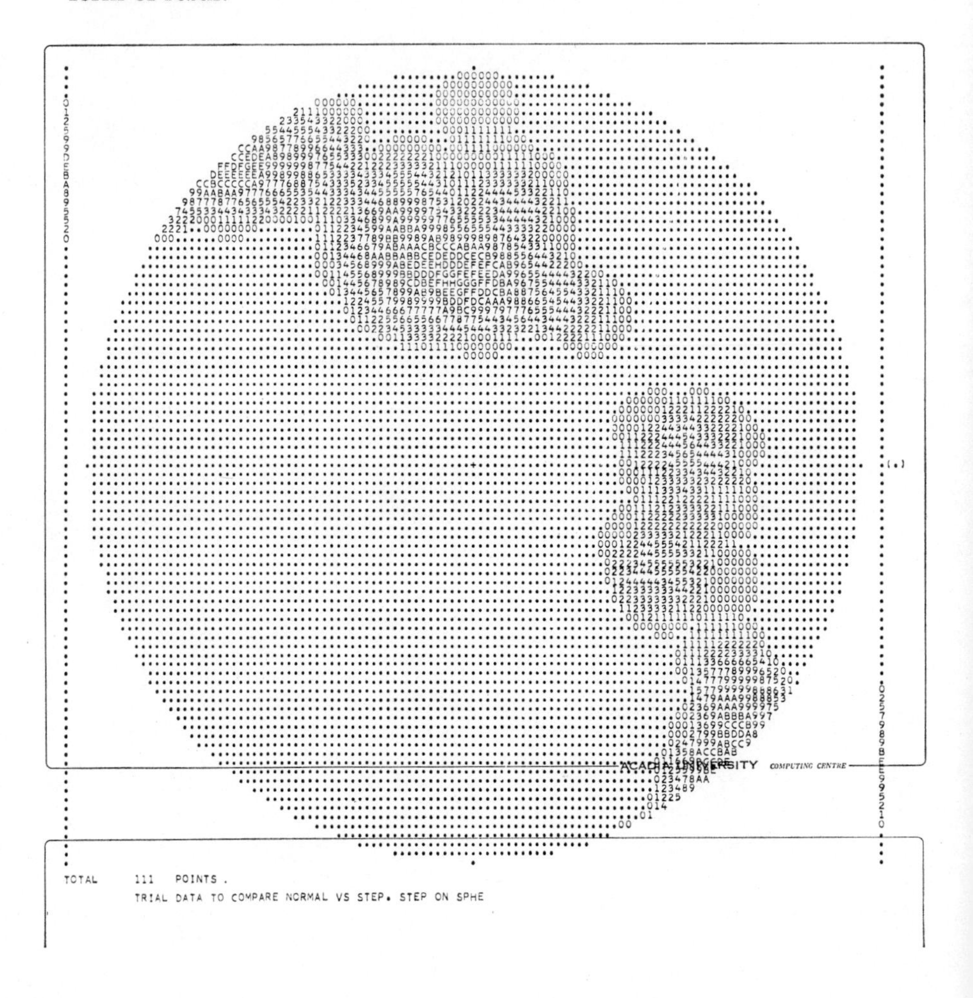

14. Generation of Keys by Computer

R. J. PANKHURST AND S. M. WALTERS

Department of Engineering, University of Cambridge, England, and Herbarium, Botany School, University of Cambridge, England

ABSTRACT

Keys are widely used by biologists and others as a simple but effective method of identifying specimens, but the construction or alteration of keys for anything more than a few species can be difficult and time-consuming. A computer program for generating printed keys from the primary data is presented. The resulting keys can be in either of the two popular formats. The input data consists mainly of a matrix of character values for specified objects (taxa) The characters may be binary or multi-valued, missing values are permitted, and characters can be given weights to cause the key to begin with what may be judged to be the more important characters. The program produces a dichotomous key wherever it can, and allows any number of characters to be included in each lead according to the user's choice.

INTRODUCTION

Biologists have used diagnostic keys as aids to identification since the eighteenth century (Voss, 1952). They are usually thought to be an essential part of a taxonomic monograph or a regional Flora, and are in most cases in dichotomous form, in which a series of alternatives is presented leading to the identification of the specimen in question. Such dichotomous keys may be "bracketed", when the alternative statements are adjacent, or "indented", when the alternatives are separated (Fig. 1). The advantages of the indented key lie mainly in the fact that it acts also to some extent as a "conspectus"—that is, the species which fall into a group according to some important character used in the key can be readily seen to be associated in this way. The disadvantages of the indented key become obvious if the key is at all extended; it is

Systematics Association Special Volume No. 3. "Data Processing in Biology and Geology", edited by J. L. Cutbill, 1970, pp. 189–203.

wasteful of space on the page and, even with careful indentation, it may be difficult to find the paired statement which contrasts with the one under consideration. Thus the indented convention is better for shorter keys, while the bracketed one is better for really long ones. Most Floras feel it necessary to be consistent, and treat all keys alike, so that the choice is a difficult one and editors choose differently. *Flora Europaea* and *Flora Malesiana*, for example, have indented keys, whilst *Flora URSS* has bracketed keys.

```
                              KEY TO BRITISH EPILOBIUM SPECIES

1   STIGMA FOUR-LOBED,                                                                            2
 2   FLOWER DIAMETER OVER 10 MM,                                                         E. HIRSUTUM
 2   FLOWER DIAMETER 6 TO 10 MM,                                                                  3
  3   STEM WITH SPREADING SIMPLE HAIRS THROUGHOUT, LEAVES SESSILE OR SUBSESSILE, GLANDULAR HAIRS PRESENT ON STEM,
      LEAVES HAIRY ON BOTH SIDES, LEAVES OBLONG-LANCEOLATE,                           E. PARVIFLORUM
  3   STEM SUBGLABROUS OR WITH APPRESSED SIMPLE HAIRS,                                            4
   4   BASE OF LEAVES ROUNDED, LEAVES OVATE TO OVATE LANCEOLATE, STEM MORE OR LESS TERETE, LEAVES ALL OPPOSITE,
                                                                                         E. MONTANUM
   4   BASE OF LEAVES CUNEATE, LEAVES OVATE- TO LANCEOLATE-ELLIPTIC, STEM WITH RAISED LINES, LEAVES
       OPPOSITE AND ALTERNATE,                                                        E. LANCEOLATUM
1   STIGMA ENTIRE,                                                                                5
 5   FLOWERS AXILLARY,                                                                E. NERTEROIDES
 5   FLOWERS TERMINAL,                                                                            6
  6   FLOWER DIAMETER OVER 10 MM,                                                           E. LAMYI
  6   FLOWER DIAMETER 6 TO 10 MM,                                                                 7
   7   FLOWER BUDS DROOPING,                                                         E. ALSINIFOLIUM
   7   FLOWER BUDS ERECT,                                                                         8
    8   GLANDULAR HAIRS ABSENT FROM CALYX TUBE,                                           E. ADNATUM
    8   GLANDULAR HAIRS PRESENT ON CALYX TUBE,                                           E. OBSCURUM
  6   FLOWER DIAMETER UP TO 6 MM,                                                                 9
   9   LEAVES SESSILE OR SUBSESSILE,                                                             10
    10   CAPSULE 5 TO 10 CM,                                                             E. PALUSTRE
    10   CAPSULE UP TO 5 CM,                                                    E. ANAGALLIDIFOLIUM
   9   LEAVES (AT LEAST SOME) DISTINCTLY STALKED,                                                11
    11   FLOWER BUDS DROOPING, FLOWERS WHITE TO PALE PINK, STIGMA ABOUT EQUAL TO STYLE,     E. ROSEUM
    11   FLOWER BUDS ERECT, FLOWERS PINK, STIGMA SHORTER THAN STYLE,                  E. ADENOCAULON
```

Fig. 1. A key for British species of Epilobium.

Since so much effort is expended both on the construction and use of diagnostic keys, it is natural that the possibility of automatic production and editing of keys should begin to interest the taxonomist. A recent paper in *Taxon* (Morse *et al.*, 1968) has given some account of work in progress on editing keys by computer for the Flora North America project and in the present paper we are describing a computer program for automatic key generation.

The advantages of automatic key production can be briefly summarized as follows:

1. *Flexibility*. Several different keys can be generated from the same initial data by introducing known biases (or "weightings") in favour of different diagnostic characters, and the keys can then be tested in the field or in the herbarium against individual specimens.
2. *Ease of editing*. The addition or subtraction of one or more taxa from an existing key, laborious by traditional methods, is very easily accom-

plished by computer. This facility could be especially valuable when converting a large regional Flora into a series of local Floras, where the "best" simplification of the keys could be produced automatically from the original data.

3. *Taxonomic discipline*. Since a computer-generated key is impossible without a minimum diagnostic difference between the descriptions of the individual taxa, the method provides a much-needed check on the "fit" between key and descriptions. It is salutary, if somewhat depressing, to find how badly much published taxonomy shows up when the allegedly diagnostic descriptions are used to generate a key by computer. At the least, the writing of keys by computer means that no character can appear in the key which is not present in the description—and even this elementary requirement is occasionally ignored in published work!

A published Flora or monograph rarely gives alternative keys. There seems to be no reason for this restriction, except the obvious one that the space taken up by keys is precious, especially in a Flora which is designed to be used in the field. Two comments might be relevant here. Firstly we might ask whether the average national or regional Flora, containing keys, descriptions and other varied information, is not too strictly limited by convention—might we not consider producing more often a separate pocket "determinator" containing the field keys, which could be a companion volume to the desk Flora? If so, a variety of dichotomous keys to the same taxa might often be included with advantage. An obvious example would be a key to trees and shrubs based on leaf characters only. Secondly, many keys in published Floras are compromises between the best key for field use, and the best key for herbarium use, when the compromise often means that the key is irritatingly inefficient for either purpose. A good example is the varied use of flower-colour in keys. Perhaps our Floras should where useful include both "field" and "herbaria" keys.

The generation and editing of keys by computer opens up these and many other possibilities. The process liberates the author from the difficult restriction, so often self-imposed, of using as a basis for the new key existing published keys which cover some or all of the same taxa. At the same time the author can, indeed should, express preferences based upon his own taxonomic experience of the groups studied in favour of using certain "leads" early in the key; such preferences can be "written in" to the key by a diffierential weighting. Finally, the products of the computer can be tested and edited by the author just as any other key, and he must exercise the choice, bearing in mind the purpose or purposes for which the key is designed.

THE DESCRIPTION OF TAXA

Each taxon is described in terms of a set of characters and values. The taxon might be a species, a genus, or any category of specimen. In fact, there is nothing in the program which places any restriction on the nature of the specimen, and this could equally well be non-biological, e.g. a rock sample, or non-tangible, e.g. a disease.

The characters, also known as properties, features, or attributes, may take on two or many different values, or states. These are known as binary or multi-valued characters respectively. If the values are arranged in a table with the characters along the top and the taxa down the side, then we have a matrix which is equivalent to a description of the taxa.

Values may be missing, for two principal reasons. One reason can be that the character is only appropriate for some of the taxa, and not for others. Suppose a group of plants has leaves which are either toothed or not toothed. The value of the character "number of teeth on leaf" must then be missing in some cases. The second reason is that the value has not been recorded. This can be due to oversight, or because the taxon is rare, or because the character is so variable that it is not worth using.

THE LOGICAL NATURE OF A KEY

The nature of a key is best described in terms of a tree-like structure, as in Fig. 2, which is equivalent to the key in Fig. 1. Each lozenge is

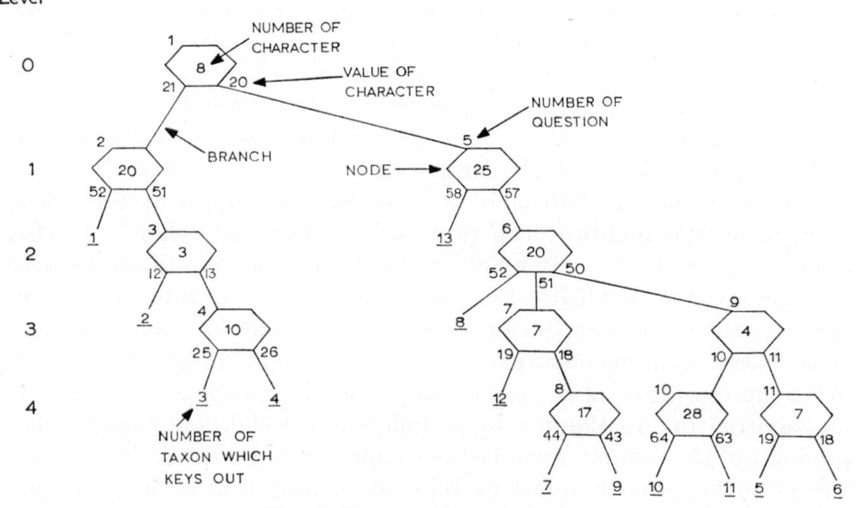

Fig. 2. Tree version of a key.

called a node of the tree and represents a set of leads or questions in the key which have the same lead number. The lines between nodes are the branches of the tree and represent the paths from one lead to another taken in response to the result of the questions asked in each lead. In mathematics, the tree is often called a decision tree, and the convention is to draw the tree upside down. The first lead is the root of the tree, and is said to be at level zero. Subsequent branchings are given increasing level numbers. The taxa key out at the tips of branches, and are shown with the number of the taxon underlined.

WHAT THE PROGRAM CAN DO

(1) The names of taxa, characters and values are represented by strings of letters or digits. The length of a string is arbitrary and its end is marked with a slash. As an example take one of the taxa included in the key of Fig. 1 for which the data are presented in Fig. 3. The first taxon, named *E. hirsutum*, has a character *flower diameter* with the value *over 10 mm*. The user of the program gives numbers to the characters and values. In this case we have character 20 with value 52. The program assigns numbers to the taxa in the order given.

(2) Each different character may have as many values as is convenient—that is, characters may be either binary or many-valued. As an example, character 19 in Fig. 3 represents the teeth on the *leaves*. The possible values are 47, 48 or 49 which are *entire or subentire, denticulate* or *sinuate-toothed*, respectively. If a dichotomous key is desired, this can be ensured by using only binary valued characters. Otherwise, a polychotomous key may result.

(3) Each character that is used can be given a weight. The weight is a positive integer. As many different weights as required may be used. The program compares the weights given and gives priority to the largest. This means that a weight of 10 will be considered before a weight of 1, and not that the first character is ten times more important than the second. The characters listed in Fig. 3 have been weighted: for example, the character number 8 has been given a weight of 4. If the user does not have any special weighting in mind he may give all characters the same weight. This will tend to produce the shortest key. A given character weight may be used more than once. If all characters are assigned different weights then the program will simply produce that key which is determined by this order of weighting.

(4) When due allowance has been made for the distribution of values amongst characters for missing values and for the character weightings, the program will produce an optimized key. The key is optimized in the sense that it should have the shortest identification path for an average

Fig. 3. Input data for the key-generating program.

```
TITLE
                                              KEY TO BRITISH EPILOBIUM SPECIES
CHARACTERS
  30    2
   1    1STOLONS/             PRESENCE
   2    1STEM/                HEIGHT
   3    1STEM/                SIMPLE HAIRS
   4    1LEAVES/              PETIOLE LENGTH
   5    1LEAVES/              AMPLEXICAUL
   6    1LEAVES/              DECURRENT
   7    2FLOWER BUDS/         HELD
   8    4STIGMA/              LOBES
   9    2STIGMA/              LENGTH RELATIVE TO STAMENS
  10    1BASE OF LEAVES/      SHAPE
  11    1GLANDULAR HAIRS/     ON STEM
  12    1LEAVES/              HAIRINESS
  13    2FLOWERS/             COLOUR
  14    2STIGMA/              LENGTH RELATIVE TO STYLE
  15    1LEAVES/              SHAPE
  16    1LEAF/                TEXTURE ABOVE
  17    2GLANDULAR HAIRS/     ON CALYX TUBE
  18    1LOWER LEAVES/        LENGTH
  19    1LEAVES/              TEETH
  20    2FLOWER DIAMETER/
  21    1STEM/                LINES
  22    1STOLON LEAVES/       COLOUR
  23    1STEM/                ROOTING AT NODES
  24    1FRUIT STALK/         LENGTH
  25    3FLOWERS/             POSITION
  26    1PLANT/               HABIT
  27    1LEAVES/              ARRANGEMENT
  28    2CAPSULE/             LENGTH
  29    1STEM/                WIDTH
  30    1LEAVES/              COLOUR
VALUES
  68
   1     PRESENT IN SUMMER OR AUTUMN/
   2     ABSENT/
   3     UP TO 20 CM/
   4     BETWEEN 20 AND 80 CM/
   5     OVER 80 CM/
   6     MORE OR LESS TERETE/
   7     WITH RAISED LINES/
   8     ROOTING AT NODES/
   9     NOT ROOTING AT NODES/
  10     SESSILE OR SUBSESSILE/
  11     (AT LEAST SOME) DISTINCTLY STALKED/
  12     WITH SPREADING SIMPLE HAIRS THROUGHOUT/
  13     SUBGLABROUS OR WITH APPRESSED SIMPLE HAIRS/
  14     SEMIAMPLEXICAUL/
  15     NOT AMPLEXICAUL/
  16     DECURRENT/
  17     NOT DECURRENT/
  18     ERECT/
  19     DROOPING/
  20     ENTIRE/
  21     FOUR-LOBED/

  22     LONGER THAN STAMENS/
  23     ABOUT EQUAL TO STAMENS/
  24     SHORTER THAN STAMENS/
  25     ROUNDED/
  26     CUNEATE/
  27     PRESENT ON STEM/
  28     ABSENT FROM STEM/
  29     HAIRY ON BOTH SIDES/
  30     SUBGLABROUS (EXCEPT PERHAPS FOR MARGINS AND VEINS)/
  31     WHITE TO PALE PINK/
  32     PINK/
  33     ROSE/
```

Fig. 3.—Continued.

```
    34      SHORTER THAN STYLE/
    35      ABOUT EQUAL TO STYLE/
    36      OBLONG-LANCEOLATE/
    37      OVATE TO OVATE LANCEOLATE/
    38      OVATE- TO LANCEOLATE-ELLIPTIC/
    39      LINEAR TO LINEAR-LANCEOLATE/
    40      BROAD OVATE TO SUBORBICULAR/
    41      MORE OR LESS SHINY ABOVE/
    42      DULL ABOVE/
    43      PRESENT ON CALYX TUBE/
    44      ABSENT FROM CALYX TUBE/
    45      UP TO 5 CM/
    46      OVER 5 CM/
    47      ENTIRE OR SUBENTIRE/
    48      DENTICULATE/
    49      SINUATE TOOTHED/
    50      UP TO 6 MM/
    51      6 TO 10 MM/
    52      OVER 10 MM/
    53      GREEN/
    54      YELLOWISH/
    55      0.5 TO 2 CM/
    56      2 TO 5 CM/
    57      TERMINAL/
    58      AXILLIARY/
    59      DECUMBENT TO ASCENDING/
    60      PROSTRATE/
    61      ALL OPPOSITE/
    62      OPPOSITE AND ALTERNATE/
    63      UP TO 5 CM/
    64      5 TO 10 CM/
    65      .1 TO 2 MM WIDE/
    66      2 TO 3 MM WIDE/
    67      YELLOW-GREEN/
    68      BLUE-GREEN/
OBJECTS
    13
E. HIRSUTUM/
     1     5    12    10    14          18    21    22    26    27    29    33          36
          46    48    52     6           9    55    57    18    62    64
E. PARVIFLORUM/
     1     4    12    10    15    17    18    21    23    25    27    29    32          36
                48    51     6           9    55    57    18    62
E. MONTANUM/
     1     4    13    11    15    17          21    24    25    28    30    32          37
                48    51     6           9    55    57    18    61
E. LANCEOLATUM/
     1     4    13    11    15    17    19    21          26    28    30          34    38

          45    48    51     7           9    55    57    18    62    64
E. ROSEUM/
     1     4    13    11    15    17    19    20          26    27    30    31    35    38
                48    50     7           9    55    57    18
E. ADENOCAULON/
     2          13    11    15    17    18     2          25    27    30    32    34    36
                48    50     7           9    55    57    18    62
E. ADNATUM/
     1     4    13    10    15    16    18    20          26    28    30          35          41
    44          48    51     7           9    55    57    18    62    64
E. LAMYI/
     1     4    13    11    15    16    18    20          26    28    30          35          41
    44    45    48    52     7           9    55    57    18    62    64
E. OBSCURUM/
     1     4    13    10    15    16    18    20          25    28    30    33    35    36    42
    43          48    51     7           9    55    57    18    62
E. PALUSTRE/
     1          13    10    15    17          20          26    28    30          34
                47    50     6           9    55    57    18    62    64
E. ANAGALLIDIFOLIUM/
     1     3    13    10    15    17          20          26    28    30    33          38
          45    49    50     7    53     9    56    57    59    62    63    65    67
E. ALSINIFOLIUM/
     1     3    13    11    15    17    19    20          25    28    30                37    41
          45    49    51     7    54     9    56    57    59    62    63    66    68
E. NERTEROIDES/
           3    13    11    15    17          20          25    28    30    32          40
          45    47    50     7           8          58    60    61    63
END
FINISH
```

taxon. This means that wherever possible the key divides the taxa into two groups of as near as possible equal size.

(5) The program can produce either of the two popular forms of key. The term "indented" key is misleading, since either form of key can be indented. Hence the program permits either type of key to be printed with indentation. If no options are specified the program assumes that the "indented" type of key with indentation is required.

(6) The program searches for unusual taxa and if it finds them it keys them out first. The program computes an estimate of the average likeness of every taxon given and detects those taxa which are unusually unlike the others.

(7) The user of the program can specify the maximum number of characters to be used in each question or lead. In practice it is found that questions composed of combinations of more than two characters are very rare because such questions tend to divide the available taxa into a large number of groups, which does not favour an optimised key. In addition, the use of a large maximum number of characters per question tends to take a large amount of computer time without producing a substantially different key.

(8) Missing character values are tolerated at all stages of the program. However, the more missing characters there are the harder it is for the program to produce a properly optimized key. Worse still, if there are too many missing characters, the program may find it impossible to construct any key at all. In such a case the user will have to improve his data before the program can produce a key.

(9) The program checks when the data are read in that none of the characters are redundant and that all the taxa can be distinguished. A redundant character is one whose values are either all identical or all missing or which does not have more than one different value. Indistinguishable taxa are such that for each pair of values, the value is the same, or that one or both of the values are missing. If some of the taxa are indistinguishable it is impossible for the program to construct a complete key, so it lists the indistinguishable taxa in pairs and stops.

(10) As an option to help to check the input data the program will print a conspectus of the data, that is to say for each taxon all the pairs of characters with their values are printed in a readable form. Characters with ill-matching values then show up clearly. The data can then be corrected for a key-producing run.

(11) The key is printed as output in a ready-to-use form. It is possible that a little editing may be needed to improve the style before publication. However, in most cases the key can be taken straight from the line printer to be used in the field.

WHAT THE PROGRAM DOES NOT DO

The following remarks do not imply that the facilities mentioned cannot be implemented, or that they have not been implemented since the submission of this paper.

(1) There is no provision for "weighting" the taxa directly. This might be useful to bring common species to key out at an early point. The effect of taxon weighting can sometimes be achieved by careful choice of a character which has an unusual value for the taxon of interest and giving that character a higher weight.

(2) The program makes no direct allowance for characters which are variable. If the user wishes to include them he may give them a low weight because he feels that they are unreliable, or he may choose to omit them if sufficient other characters are available. A third possibility is that he will repeat the character value data for the taxon using the same name but altering the value of that character which is variable. It is then possible for a taxon with a given name to key out in more than one part of the key.

(3) If for the reason mentioned above, taxa appear in the data which are different but similar but which have the same name, then it is possible that these taxa will key out near to one another, in which case the program does not recognize that they have the same name. Hence, different taxa which happen to have the same name do not get lumped together where they key out. If this happens, however, it is a sign that for that particular manner of constructing the key the repetition of these similar taxa with the same name was unnecessary.

DATA PREPARATION

The data for the key program are generally prepared from a written description of the taxa or species, for example, a Flora or a geological handbook. A more convenient form for the data is a table of taxa against their properties, i.e. a matrix. For convenience in using this program, the taxa are listed at the left of the table while the characters are listed across the top. If the table is not already prepared the user will have to prepare the table by hand on a suitably large sheet of paper. The input data which were used to prepare the key in Fig. 1 are shown in Fig. 3. The data are divided into three principal sections. The first section is for the characters, the second for the values and the third for the taxa as represented by the matrix. The user is expected to assign indices (positive integers) to each character and value. The matrix describing the taxa then takes on numerical values for its elements. Missing values of characters are represented by a value of zero. The weights to be given to the characters are chosen subjectively according to which characters are thought to be easy to observe or to be important in the classification. The matrix must be sufficiently complete; that is to say, at least one character must have all its values present. If this is not the case the program cannot begin. Experience so far suggests that the traditional manner in which taxonomists prepare descriptions of their taxa are often only just sufficiently complete for a small number of alternative keys to be constructed. It should be realized, that if weights are assigned to characters, then these characters have to be well-scored for the weighting to be effective in the construction of a key. These are two reasons for working with a complete a matrix as is possible. There are questions of style to be considered when choosing character-value pairs. For example, suppose we wish to represent "stem glabrous" the character in question here is "stem hairiness" and the value is "glabrous". If these two strings were used in conjunction, the result would be "stem hairiness glabrous". A more stylish phrase would be "stem glabrous". The program allows comments to be inserted in such a case as this, so that the user might be reminded that the character "stem" concerns stem hairiness.

THE PROGRAM METHOD

The first stage is for the data to be read and checked. The strings of letters and digits which represent characters, values and taxon names are saved in a free storage area. The free storage area is a large array, which can be split into numbered blocks of variable size. There is a package of programs to manage this free store; in particular, routines are provided for making available more store, or for returning a block which has already been used.

When the program considers how to make the first lead of the key, all characters and taxa can be considered. Characters which do not have a complete set of values have to be rejected. However, at a later stage when a smaller group of taxa is being considered, such characters may again become useful. Characters and taxa are marked with flags in order to show whether they are "interesting" or not. An interesting taxon is one which belongs to that group of taxa being considered at this node of the key. The interesting characters at a node of the key are those characters which have not been used already and which are completely scored. The next step is to choose the interesting characters which have the highest weight. If there are no characters which are interesting which have been assigned the highest possible weight than a lower weight is considered. A test is made at this early stage in order to detect unusual taxa. (The technique for doing this is explained below.) If any unusual taxa are found the program jumps to the point where a new key branch is stored. Next, the program selects subsets of interesting characters, first one at a time, then two at a time and so on up to the maximum number of characters per line that is allowed. Each of the possible divisions of the key that is considered in this way is evaluated numerically (as explained below), and only the one which has the best score is retained. The new leads for the key which are constructed in this way are stored in a list structure (see below). So far only one stage of dividing the key has been described. The program systematically selects all the groups of taxa as the key divides them into smaller and smaller groups and repeats this until all taxa have keyed out. At each stage the program remembers to exclude all taxa already keyed out and all characters already used. When the indented type of key is wanted, each new set of leads which is added to the key is recorded so that the lead which corresponds to the smallest number of taxa appears first and the lead with the largest number appears last. Once the key has been completely generated in the form of a list structure or tree, this structure is interpreted in order to produce the key in printed form. If the bracketed type of key is required this amounts to following the tree first of all across and then down, and if the indented type of key is wanted then the structure is first followed down, then across.

Unusual taxa are detected by the following technique. (This cannot be used when the taxa concerned have missing values, since the method compares taxa meaningfully only when they all have an effectively equal number of characters.) For all the relevant characters the average similarity is found. The similarity between two taxa is the count of those characters for which they both have the same value. It is assumed, without any theoretical justification, that this average similarity has a uniform distribution. On this assumption the variance is calculated from simple statistical arguments. Any taxon whose similarity is less than the average by more than this artificially calculated variance is said to be unusual.

The function which is computed in order to distinguish between different groupings of taxa is defined as follows. It is a function of the number of different groups K, number of taxa in each group n_i where $i = 1$ up to K, and where the total number of taxa is N. Since the best key branches dichotomously with equal sized groups in each branch (Osborne, 1963), we wish this function to be zero for $K = 2$, and we also want it to be zero for groups of equal size. Since some compromise is necessary in practice, we seek to minimize this function. The function could be defined in many different ways, but the following has been found satisfactory. The function is defined as $F = F_1 + F_2$, where $F_1 = (K - 2)^2$ and

$$F_2 = \sum_{i=1,K} \left| 1 - \frac{n_i K}{N} \right|$$

It is clear that the function F_1 is zero for $K = 2$ and increases rapidly with K. The explanation of the choice of the function F_2 is as follows. Ideally, each of the K groups should be equal in size, in which case the ideal number of taxa in each group is N/K. In fact, the number in each group is n_i. Hence the ratio n_iK/N equals 1 in the ideal case. The absolute difference between this ratio and 1 is summed over all the groups. It is possible to take a short cut and avoid some of these calculations. Since the program begins by considering divisions of the taxa based upon the values of only one character, it tends to find the smallest value of F fairly soon. Once all the possibilities from considering one character at a time have been covered, one has a minimum value of F, from which one may deduce the largest value of K (the number of groups) that can be worth considering. This is found using the function F_1, which increases rapidly with K. Hence many of the possible divisions which come from dividing groups of taxa according to their characters taken severally can be rejected without further consideration.

The list structure that is used by the program is a representation of the key regarded as a decision tree. At each node of this tree are two blocks of storage of arbitrary size. The first block contains pairs of characters and values, and the second contains a list of taxon numbers which have these properties. These blocks are linked together in a chain which branches at each node. Each block has a space reserved in it for the number of the next block on the chain. When numbers are being assigned to the leads at the stage just before the key is printed, extra links are put into the list structure. These extra links in the structure connect nodes which were not adjacent in the original tree.

The program makes use of a push-down stack. The stack is an array containing numbers of blocks belonging to the list structure, together with the index or number of the stack array, which refers to the number of the block currently being processed. The stack is useful in keeping track of the growing key, and is also used in rearranging the list structure when different types of key are to be printed.

PERFORMANCE

The key program has been written in ASA Fortran (also known as USASI Fortran). It was implemented on the University of Cambridge Titan computer. The program occupies 16K, i.e. 16 000 words of storage, and additional 16K of store is sufficient for producing a key for approximately 150 taxa. For an example with 134 taxa with 32 characters the program took 3 minutes to compute a key. This is on the assumption of one character maximum per lead. This particular key was for the European genera of the Umbelliferae, based upon data compiled by Professor T. G. Tutin in Volume 2 of the "Flora Europaea". The computer generated key had approximately 20% fewer leads, used fewer characters, and was line by line more compact.

Another successful application of the program was to 86 cultivated varieties of potato described by Mr. T. Webster of the National Institute of Agricultural Botany. This key was generated from 25 characters and took approximately 2 minutes. No key for these varieties had previously been composed on account of the difficulty of doing this by hand.

Fortran was chosen for writing this program as a matter of convenience for prospective users and not because it was an ideal choice by programming considerations. Because Fortran is used, the program is cumbersome in the following respects. Fixed format data input proved rather inflexible, and the handling of letters and digits in the strings was most inefficient. In addition the coding of the free storage routines required the use of an array referred to by an index, whereas such routines are normally written by using an absolute address as a pointer, hence saving the repeated addition of the first address of the array to the index.

The Fortran program and a writeup are available on request from R. J. Pankhurst at the Cambridge University Mathematical Laboratory.

OTHER WORK

Morse *et al.*, (1968) describe a computer program for editing keys. The leads of the keys are punched on cards and the program simply reorders them. This program cannot be said to generate keys automatically, since the user has to choose the order of the leads himself.

Niemela *et al.* (1968) were interested in generating keys for microbes. They only consider two-valued characters, which have to be coded as zero and one. Their character matrix has to be complete, i.e. no missing characters are allowed. There is no weighting of characters, and only one character per lead is considered. Further, the program does not produce a key in a printed form. They experimented with different methods of dividing the key. One of these is equivalent to a simplified function F, as above, and the other was based on an estimate of the number of possible further keys which could be derived from a division at the current node, and by trying to minimize this quantity.

Morse (1969) has a more advanced program, which actually generates keys automatically. The characters are only permitted binary values and there is no character weighting. Missing values of characters are tolerated and two types of key may be produced. Only one character per lead is permitted, but weighting of taxa is included. An unusual feature of this system is that the character value pairs are not split, but treated as an indivisible couplet.

The above remarks apply to attempts by biologists to write computer programs to generate keys. In addition, within the discipline of computer science, experts on artificial intelligence have been working in parallel on similar problems. For instance, the book written by Hunt *et al.* (1966) has some relevance. The authors here are concerned with what they call "concept learning". In their case, they consider a given group of objects which have been identified into one of two categories. For example, they quote data recorded from hospital patients who either have suffered infection from surgical wounds, or who have not. These are referred to as positive and negative instances of a concept, and the authors are concerned to make use of the characters available to distinguish positive and negative instances (e.g. risk of infection or no) by a decision tree (equivalent to a key). The manner in which computer programs are written to handle this decision tree as it is constructed is similar to the key constructing program. However, they give great importance to positive instances as opposed to negative instances. In biological problems of identification, one is not concerned with deciding for or against a concept in this sense. When a key is used, the purpose is to distinguish between the members of a large group of taxa, while the key user has to assume that the taxon to be identified is one of those covered by the key.

Further references to artificial intelligence work are contained in a companion paper (Pankhurst, 1970) which presents computer generation of keys for computer specialists and non-biologists.

CONCLUSIONS

A computer program for the automatic construction of keys has been implemented, and shows great success with its early applications. During this work, the authors have gained the impression that there may exist in biology and allied sciences many applications of computer methods which are not yet being exploited. On the other hand, there exist techniques in artificial intelligence and computer science generally, which could find immediate application to biological problems, if only the computer experts were aware that such problems existed. It appears therefore that there are two communities of experts who ought to be talking to one another, but who are not.

REFERENCES

Hunt, E. B., Marin, J., Stone, P. J. (1966). "Experiments in Induction". Academic Press, New York and London.

Morse, L. E. (1969). Private communication.

Morse, L. E., Beaman, J. H., Shetler, S. G. (1968). A computer system for editing diagnostic keys for Flora North America. *Taxon* **17**, 479–483.

Niemela, S. I., Hopkins, J. W., Quadling, C. (1968). Selecting an economical binary test battery for a set of microbial cultures. *Can. J. Microbiol.* **14**, 271–279.

Osborne, D. V. (1963). Some aspects of the theory of dichotomous keys. *New Phytol.* **62**, 144–160.

Pankhurst, R. J. (1970). A computer program for generating diagnostic keys. *Comput. J.* **13**, No. 2, 145–151.

Voss, E. G. (1952). The history of keys and phylogenetic trees in systematic biology. *J. scient. Labs Denison Univ.* **43**, 1–25.

15. Automatic Data Handling in Geochemistry and Allied Fields

P. WILKINSON

Department of Geology, University of Sheffield, England

ABSTRACT

The nature of geochemical data is examined together with the problems which it presents. Progress in the automatic collection of data is reviewed. Computer processing of data and the problem of storage and retrieval are discussed at greater length.

I. NATURE OF GEOCHEMICAL DATA

The multivariate and stochastic nature of most geological relationships necessitates a multiplicity of observations. The use of automatic data processing, especially by electronic computers, in interpretation is not only desirable, but almost indispensible in the long term. The fundamental data of geochemistry, as also of geophysics, are almost exclusively quantifiable and hence numerical and ideal for automatic handling.

Geochemical data are derived from two main sources:

(a) "Observational" data, that is from the analysis of natural minerals, rocks, waters, volcanic gases, atmosphere, organic and biological matter, extraterrestrial materials, etc.

(b) Experimental data from laboratory experiments simulating aspects of geological processes and conditions, for example, solid-liquid-vapour relationships in crystallizing silicate melts, diffusion experiments, precipitation from saline solutions, etc.

More fundamentally geochemical data can be characterized as follows.

Systematics Association Special Volume No. 3. "Data Processing in Biology and Geology", edited by J. L. Cutbill, 1970, pp. 205–234.

1. Concentrations: elemental, molecular, isotropic. They can be expressed as absolute or relative abundances, ratios or more complex functions.
2. Positional: (a) atomic: crystallographic data. (b) distribution within phases, that is disequilibrium zoning and diffusion gradients. (c) distribution between phases: partition coefficients. (d) "geographical", that is the three-dimensional distribution of elements within polycrystalline, often polyphase, rock and mineral masses, homogeneous or heterogeneous. This also applies to geochemical exploration surveys.
3. Temporal: the variation of elemental concentrations with time. All evolutionary data fall in this category, including diffusion and geochronological models.
4. Thermodynamic and kinetic data and other constants, for example, atomic weights.

Intrinsically, all such data, except atomic positions have exact values, but sampling error and analytical error limit the precision and accuracy with which these can be known. Consequently, study of the nature and size of these errors with a view to determining the limitations on the data and to the improvement of experimental procedure is of utmost importance.

II. AUTOMATIC DATA COLLECTION

Although it is not the purpose of the present article to present a detailed account of modern geochemical analytical methods, a brief review of the changes of the last twenty years is necessary in order to understand the exact nature of the data-handling problem.

Up until the 1950s, the classic gravimetric techniques were still supreme and, although new methods were being developed, many petrologists and geochemists looked at the accuracy and precision of the latter with grave suspicion. Even an experienced analyst could generally produce at most half a dozen standard rock analyses for a hard week's work. Since the war years, a whole battery of techniques often involving extensive and expensive instrumentation, have so speeded data production that, in some cases, an analysis of up to maybe forty elements can be produced in a matter of minutes. The comparison is not direct since questions of relative precision and accuracy arise. In the following brief survey, which does not pretend to be comprehensive, the types of development are presented.

Geochemical analytical procedures fall into well-defined stages:

1. Sample collection
2. Sample preparation
3. Analysis
4. Data presentation.

1. Sample collection

With the possible exception of soil and water samples, the collection of geological samples is, and probably must remain, the most time-consuming operation in the chain. There is no short-circuit for the geological fieldwork which is necessary to provide meaningful samples, and few labour-saving devices, except the hand-held rock drill for providing fresh, unweathered specimens in awkward situations, have been evolved. In the writer's view, this has meant that many highly instrumented laboratories are either under-utilized or produce an excessive number of analyses of comparatively little geological significance or inadequate documentation. The rationalization of a geochemical laboratory should imply the servicing of sufficient field units to utilize its resources economically.

2. Sample preparation

This generally falls into three phases:

(a) crushing and screening;
(b) (i) separation for single mineral preparations; (ii) homogenization and splitting for whole rock samples;
(c) specimen preparation.

Of all the laboratory procedures, sample preparation is almost invariably the most time-consuming and the weak link in critical path analysis. Two approaches to improvement are possible: acceleration of the actual procedure, and multiplication of the units of preparation, although the latter may be expensive in terms of staff and equipment, especially where platinum-ware is involved.

(a) Crushing and screening. Most of the hard work has been taken out of these procedures by laboratory scale jaw-crushers, hydraulic and fly presses, and high speed vibratory and disc mills. Crushing can usually be kept to the order of ten minutes per sample, although cleaning of equipment between samples may considerably increase the time necessary.

(b) (i) Mineral separation has always been a slow and tedious procedure and although improvements in equipment have been made, these have resulted usually in greater efficiency in separation rather than in reduction of time. Any time-saving has resulted from a decrease in the amount of final hand sorting necessary.

(ii) Homogenization is another procedure which cannot be hurried, and although improved equipment has probably somewhat shortened the time necessary, multiple sample homogenizers seem to be the best answer for improving throughput.

Splitting the total sample to provide the analytical aliquot is one of the most neglected aspects of many laboratory procedures, and probably contributes considerably to total analytical variability. The writer is not aware of any radical innovations in equipment or procedures over the last twenty years except in the field of on-line flow sampling.

(c) Specimen preparation. A great variety of processes are employed depending on the analytical technique to be used. In "wet" chemical techniques this is still the most time-consuming unit operation. Since it normally involves dissolution in hydrofluoric and strong acids or alkali fusions (except in the case of carbonates, water-soluble minerals, such as borates, and hydrocarbons) little can be done to speed this process although operation under pressure in polytetrafluorethylene bombs has been used successfully. The usual approach to speeding overall operation is to perform large numbers of simultaneous dissolutions, although this leads to heavy investment in platinum and other expensive metal-ware.

Other analytical techniques, such as infrared and X-ray fluorescence spectrometry employ pelletised powders or fused beads. These procedures are usually very much quicker than dissolution and much has been done to accelerate the procedures, even to the extent of automatic glass bead-producing equipment. Nevertheless, specimen preparation still occupies a substantial proportion of total analytical time and there is room for improvement.

3. Analysis

It is in this field that the most revolutionary development has taken place, and it is probably a fair assessment that little can now be, or needs to be done to improve analytical time for most methods.

Classical "wet" methods were considerably speeded-up, first by the introduction of so-called "rapid" methods using a variety of titrimetric, spectrophotometric and flame photometric methods. These have since been further accelerated by techniques of automatic pipetting, automatic transmission and mixing of reagents by use of variable-bore capillary tubes, the automatic recording of absorption spectrophotometers and flame photometers, and the introduction of new techniques such as atomic absorption spectrophotometry.

The major innovations in geochemical technique include mass spectrometry, isotope and radioactivation techniques, resonance spectroscopic techniques (including infrared and Mössbauer spectrometry), X-ray fluorescence spectrometry, electron microprobe and direct reading emission spectrometric techniques. For descriptions of these techniques, the reader is directed to such works as Smales and Wager (1960), Zussman (1967). Mainly they employ solid specimens and some are non-destructive. Most, although not all, are extremely rapid compared to classical methods, sometimes requiring only seconds to determine an elemental concentration.

From the point of view of speed, a distinction of major importance is between multi-channel and sequential analysers. This is best illustrated in the field of X-ray fluorescence spectrometry. In the sequential type of instrument, a single detector unit (usually a gas flow proportional counter and a scintillation counter mounted in line) moves from angle to angle, determining one element at a time. In the multi-channel type of instrument a detector unit is present at the Bragg angle of each element to be determined, sometimes up to forty positions, so that all elements are determined simultaneously, thus drastically reducing the analytical time per element. The additional cost of detectors and counting equipment is, however, very high and the saving of time is rarely utilizable or the cost justifiable, except in industrial production situations.

Multi-channel techniques are more or less essential in emission spectrography in order to utilize the relatively short duration of arcing. In the classic technique photography provides simultaneity, whereas in the "direct-reading" spectroscopes photomultiplier cells are used as detectors, their output being accumulated on capacitors and measured by a digital voltmeter.

A further time and labour saving refinement is utilized, especially in X-ray fluorescence spectrometry. A small on-line digital computer or built-in logic circuits permit the fully automatic execution of a complete programme of element analyses. To make full use of this facility automatic specimen changing devices are necessary, thus allowing untended operation for long periods of time. Such automated specimen changing is also employed in other methods, such as β- and γ-ray spectrometry and fully automated wet chemical techniques.

4. Data presentation

Most of the instruments mentioned have direct displays of instrumental readings. However, in order to utilize fully the rapidity of analysis, especially in multi-channel instruments, some form of a chart-recorder for scanning instruments and an electric typewriter, or similar data printer, for set-position instruments is necessary. Such a device is essential for fully automatic instruments of course.

There is a further problem in that many of the techniques depend on interelement interference and other complex correction procedures, normalization to reference standards and the use of calibration curves, often non-linear. These are usually best carried out by digital computer and are, indeed, sometimes so lengthy as to be only feasible, as a routine procedure, if so handled. Some manufacturers have incorporated small computers in their instruments or provide facilities for on-line interfacing with locally available computers. More normally such a relatively costly device is not justified and off-line computation is performed on the local computer. In this case digital tape output, punched paper or magnetic, is normally provided in addition to, or in place of, print-out.

Relatively few geochemical analytical techniques employ continuous functions (for example, diffraction line-profiles) but in such cases, if analogue techniques of interpretation are not used, analogue to digital converters may be used to provide an output suitable for digital computer processing.

Summary

Although classical analytical techniques are still extensively used, especially for reference analysis and mineral analysis, the whole structure of geochemical analysis has been revolutionized since the war years and many phases have been fully automated.

Speed: The time needed for analytical determinations has been decreased, depending on circumstances, by one to three orders of magnitude.

The speed of collection of samples has, however, hardly changed and that of sample preparation has in some cases only been increased by a factor of 2 or 3. These factors are for practical purposes somewhat increased by the greater number of persons employed in earth sciences today.

Precision: Instrumental precision is readily determined because of the ease of making replicate analyses. It is also a function of time spent in determination and is controllable within wide limits. Unquestionably analytical precision is generally much higher than, say, twenty years ago.

Accuracy: This is a hotly debated question. Undoubtedly accuracy has been raised generally over the years, but the difficulty of producing absolute standards for naturally occurring substances makes quantification problematic.

Detectability: The introduction of new methods has vastly improved the limits of detectability of many elements and for some this is now in the region of 10^{-9}.

Rate of production of data: The natural growth of the earth sciences has produced a marked increase, since the Second World War of personnel engaged in all aspects of geochemistry, and in the number of laboratories devoted to geochemical determinations. Together with the great increase in instrumentation, some fully automated, and the increased rapidity of analysis, this has increased the output of geochemical determinations by two to three orders of magnitude.

III. GEOCHEMICAL DATA PROCESSING

Geochemical data processing may be subdivided into five categories for ease of treatment:

1. Analytical data reduction
2. Function evaluation
3. Data evaluation procedures
4. Simulation
5. Data storage and retrieval

Simulation is included here for convenience although it could equally well be considered as a form of data collection or generation.

Only the first four will be treated in this section since storage and retrieval are dealt with in the following section on Data Banking.

1. Analytical data reduction

In this category are included all computational procedures which are used to reduce raw instrumental readings to final, corrected analytical results. They include such various procedures as dead-time corrected rates from total counts and times, correction for zero errors, calculation of calibration curves, correction for interelement effects, prediction of concentrations from regression equations, normalization to reference specimens, evaluation of experimental equations and the calculation of instrumental errors. In general the individual algorithms are fairly simple, but a single result may have to undergo many stages, some iterative, and the whole process may be very time-consuming when undertaken manually. A large number of laboratories reduce their analytical data by computer, but since the programming is dependent on the particular instrument and analytical procedure adopted there is little easy interchangeability of programs between laboratories, except where a complete technique is imported, and moreover, computers are compatible.

Since such procedures have been in use for many years now, and most laboratories have their own program packages for experimental data reduction, there is little point in describing any individual package or trying to achieve uniformity or compatibility. In any case, it is probably fair to say that competence in programming at this level should be part of the stock-in-trade of any modern geochemical laboratory.

Crystallography was one of the first disciplines in which the great potentialities of electronic digital computation was realized and many years of use have greatly extended the accuracy, power and range of crystallographic methods, as well as removing the major part of the sheer drudgery of manual methods of calculation. Computer methods have for many years been standard in X-ray, electron, and neutron-diffraction laboratories. For description of program packages the interested reader is referred to such works as Rollet (1965) and Shoemaker (1966).

2. Function evaluation

Procedures in this category generally fall into two types:

(i) Geochemical interpretation commonly requires the evaluation, tabulation and plotting of standard thermodynamic, kinetic and other functions (such as radioactive decay functions), over wide ranges of values in the variables. Electronic computation is certainly advantageous since it not only greatly reduces labour and time, but may also extend the range of what is possible. Automatic graph-plotting of functions is not so commonly used, but has considerable potential for time-saving.

(ii) All geochemists, mineralogists and petrologists use standard methods of recalculating analyses. These include:

- (a) Conversion of oxide percentages to atomic proportions, combining of elements, formation of ratios, calculation of coordinates, etc., for plotting as binary, ternary and quaternary diagrams.
- (b) Projection of quaternary diagrams onto ternary sections.
- (c) Recalculation of mineral analyses as norms, Niggli or Zaravitsky values, Barth's standard cell, and so on.

Most of these procedures are arithmetically simple, but often logically complex and leave plenty of room for calculation errors. Moreover, by their nature they are usually required to be calculated for large numbers of analyses at a time. The electronic computer thus saves a very great deal of time, and eliminates operator error, except at the input stage. For example, a norm may require 20 minutes to calculate manually, with the use of tables, for one analysis, but depending on the individual unit, only a few seconds at most by computer. To be realistic a minute per analysis must be added to computer time for punching and verifying input data. However, a well-designed program package will utilize the same input record for a number of the procedures outlined above so that a large proportion of data preparation time can be discounted.

Normative calculations seem to have provided a useful toothcutting exercise for novice programmers and a high proportion of geochemists already have their own programs and packages for processing analytical data. A number of these covering all the usual program languages have been mentioned in the mineralogical literature and informal contact is sufficiently good that there seems to be little need for further measures aimed to achieve greater availability of programs and packages. However, experience with programming standard procedures shows that errors, inconsistencies and ambiguities exist in many published accounts and the interests of accuracy and comparability would be served by revision of many of these, together with some general agreement on conventions.

As an illustration one may take the case of standard C.I.P.W. norm calculations. The original description provides for the calculation of normative minerals from usually minor constituents such as ZrO_2, BaO, SrO, SO_3, Cl, F and some more recent versions are even more comprehensive. However, for many practical petrochemical problems it is necessary to deal with bodies of data in which a component has been determined for some rocks, but not for others. This may lead to significant differences in the major normative constituents in nearly identical rocks. For many comparative purposes, particularly evolutionary trends, it is preferable to calculate norms only for the dozen or so major constituents that are almost invariably determined. A similar problem is posed by the increasing number of analyses, for which X-ray fluorescence analysis is responsible, in which only total iron is known.

3. Data evaluation procedures

No strict division can be drawn between these procedures and those of the preceding category. It is here used to define the methods, mainly statistical, which enable an evaluation of the nature and significance of variation in geochemical populations to be made. Most available mathematical methods of data evaluation seem to have been used at some time in geochemistry, but listed below are only the more important or those that are used with any degree of regularity.

(*a*) *Analysis of error:* All geochemical data are subject to error—sampling error, specimen error, instrumental error, operator error and so on. Any valid evaluation of the significance of the data must therefore entail, as a prerequisite, an analysis of the nature and magnitude of such errors. Analysis of variance of experimental results and calculation of the propagation of error in derived functions is too often not performed or at least not stated in publications. Handling of errors is often implicit. For example, the errors in major element determination in rock analysis are now fairly well known, and the petrologist can for many purposes content himself with screening sets of data to remove analyses which are sub-standard by generally accepted criteria, for example Manson (1967). For a general discussion of the theory of geochemical error see Miesch (1967).

(*b*) *Analysis of interdependence of variables:* This section must be prefaced by the remark that many geochemical data suffer from the restriction that chemical analyses are normally expressed as weight percentages. Since the components of such a system have a constant sum there is a loss of one degree of freedom. For two component systems this means that only one variable is independent. Even for multi-component systems there will be a degree of interdependence of all component percentages and simple bivariate analysis is liable to generate spurious correlations. There is no simple universal solution to this problem, which has been widely discussed especially by Chayes (1960) and Vistelius (1967).

Some part of geochemical research involves the search for and description of *deterministic associations.* This commonly consists of the solution of sets of simultaneous equations or various types of curve-fitting—linear, polynomial approximation, hyperbolic, logarithmic, exponential and Fourier analysis.

More often the search is for *stochastic associations* and the following types of evaluation are employed.

Single variate analysis:

- One-dimensional, by linear regression and forms of curvilinear fitting as for deterministic association.
- Two-dimensional, trend-surface analysis.
- Three-dimensional, trend analysis.

Bivariate analysis:

- Correlation coefficients.
- Regression lines. It should be noted that in geochemistry both variables are commonly subject to error and considerable care should be taken in the choice of type of fit.

Multivariate analysis:

- Variance-covariance matrix.
- Correlation matrix.
- Principal component analysis, see for example, Le Maitre (1968).
- Factor analysis, see for example Manson (1967) and Prinz (1967).
- Multiple regression.
- Multiple and partial correlation coefficients.
- Discriminant analysis, see for example Chayes and Velde (1965) and Le Maitre (1968).
- Cluster analysis, see for example Chase and Mackenzie (1961).

For more general information on statistical applications in earth sciences reference should be made to standard texts, such as Miller and Kahn (1962), Krumbein and Graybill (1965), Smith (1966), Vistelius (1967), Griffiths (1967) and Howard (1968).

Hypothesis Testing: Most stages of statistical analysis can be subjected to significance testing and this is common, although not invariable practice in geochemical publication.

Miscellaneous techniques: Time series analysis by moving averages, or harmonic analysis, see for example Vistelius (1961), Schwarzacher (1964) and Anderson and Koopmans (1963); theory of games, for economic decision making, see Vistelius (1967); cartographic techniques, including automatic contouring, using formatted printout and graph-plotters (techniques which are widely used, especially in economic exploration and regional geochemistry, to provide easily assimilated visual displays of processed data, including trend surfaces).

The application of statistical methodology is very uneven as judged by published papers and personal communication. Some techniques are over-used, many others under-used. There is little general consensus of opinion at present as to the relative merits and demerits of many techniques, for example, principal components, factor and cluster analysis. Much could be done to raise the minimum standard of statistical usage.

Program packages: Most of the data evaluation procedures named are general in nature and need no special adaptation for geochemical purposes. Consequently computer manufacturers' software packages which are designed for specific systems, usually suffice for all the simpler curve- and surface-fitting and statistical needs.

Programs for the more sophisticated procedures, still not specifically geochemical have been designed and written, and are available in a number of libraries. Normally geochemical laboratories acquire those that fill their own specific needs and incorporate them in their own libraries. No comprehensive, generally available, specifically geochemical package is known to the writer. However, there exist large program packages for general geological purposes of which STATPAC and GEOPAC of the U.S.G.S. and ROKDOC of the Sedimentological Research Laboratories, University of Reading (see Loudon, 1967) may be cited. Special mention must also be made of the excellent series of publications of the University of Kansas, Kansas State Geological Survey, which contain many sophisticated programs with specifically geological orientation.

Evaluation of crystallographic data is a special field and many programs and packages in many high-level languages exist and reference is made to the I.U.C. publication (Shoemaker, 1966).

4. Simulation

Because of the multivariate and complex nature of most geological processes and the general difficulty, on account of time and other scale factors, of producing realistic laboratory models, the use of computer simulation models is growing in the geological sciences. In particular much work has been done on sedimentation and fossil population dynamics models, and use of the technique is growing in geophysics. In geochemical problems, however, many of which one might have been thought particularly susceptible to this approach, practically no significant work has been done.

Electronic computer simulation is especially suitable for:

(i) deterministic models where there are a large number of significant variables; and
(ii) stochastic models where all the dependent relationships are not known or thoroughly understood, or where randomizing processes are at work.

The evaluation of such models involve so much arithmetical work in order to investigate usefully any degree of variability in the system that manual methods are quite impracticable.

A prerequisite for simulation is that deterministic or stochastic mathematical models are available for all the major processes affecting the system to be investigated. This is now generally true, or at least practicable, for a great deal of geochemistry. The rapid growth of experimental techniques in the last half-century has provided a vast body of geochemical data so that where exact relations between variants are imperfectly understood and deterministic models are unavailable, polynomial or other fits can be made to empirical lines and surfaces, as in phase equilibrium diagrams.

A second requirement is that the interdependence of physical factors, such as pressure and temperature, in the geological environment are sufficiently known so that realistic use of the computer model can be made. In particular, the dependence of physical and chemical variables on time in what may be called *evolutionary* models is of fundamental importance. Thus rate of cooling in rate-controlled processes will determine whether equilibrium is achieved and hence the subsequent course of evolution. In such cases the kinetic factors must be quantitatively known where disequilibrium conditions are likely to be approached under geological conditions. Where the kinetic constants are not known, but observational data is adequate, the model could alternatively be used to explore the limiting values of these kinetic factors. For example, in a given magma, zoning of plagioclase felspar is dependent essentially on the relationship between the rate of cooling, hence of crystal growth, and rate of lattice diffusion of sodium and calcium atoms in the plagioclase. Since the diffusion constants are not sufficiently known, the evolution of the residual melt composition under such disequilibrium conditions cannot be directly simulated. If, however, the dimensions and compositions of these zones and sufficient other chemical data are known for an igneous body for which a realistic cooling history can be calculated, the model could alternatively be used to explore the values for the diffusion constants of sodium and calcium.

A number of problems that could probably be simulated readily suggest themselves.

(i) *Magmatic crystallization.* Sufficient is probably now known about phase relations, thermodynamics and partition coefficients in basaltic systems to simulate quantitatively the course of crystallization and liquid residuum composition under varying pressure regimes, rates of cooling, with and without crystal–liquid segregation. Such a computer model would be useful in following the detailed progress of crystal fractionation and in testing current theories of the evolution of magma series, which are mostly based on semi-quantitative estimates or on the gross chemistry of widely separated stages. It would be particularly valuable in studying quantitatively the behaviour of minor and trace elements, still imperfectly understood. Shimazu (1959a,b; 1961) has provided many of the mathematical models required, and work is in progress on a simulation program. The main weakness at present lies in our fragmentary knowledge of rate constants, especially of diffusion processes, thus making it possible only to treat equilibrium conditions of cooling with any degree of rigour.

(ii) A logical extension of such work would be the simulation of fractional melting of the mantle although the data on high-pressure phase relations is much less abundant at present and would limit the flexibility of the model.

(iii) The thermal history of many metamorphic terrains, especially in contact aureoles, can be realistically simulated. Knowledge of the P-T conditions of specific metamorphic reactions and the position (e.g. distance from igneous contact) of the relevant zonal boundary in particular cases would permit an evaluation of the minimum value of the rate constant.

(iv) Simulation of the development of oceanic and atmospheric compositions over long periods of geological time. It would probably be logical to commence with the present composition and use the model to extrapolate backwards in time.

(v) Simulation of precipitation in evaporite facies.

(vi) Radioactive decay models can be readily calculated at present, but with the addition of relatively little fundamental data, it should be possible to simulate the effects of various types of thermal event, thereby improving the interpretation of age data.

These are only some of the possible geochemical applications of computer simulation and are suggested in the hope of stimulating more work with this powerful, but underused tool.

Analog computers: Most of this survey concerns the use of the digital computer, but it must be remarked that in some fields, especially that of simulation, the analog computer has considerable potential and its use has been little explored. This is probably due to a much more restricted availability of analog computers and hence lack of familiarity in geological circles. This is not the place for an extended discussion of the nature and potential use of the analog computer, and only the main characteristics will be summarized.

Advantages:

(i) the analog can handle continuously variable functions;
(ii) effectively instantaneous evaluation of the most complex functions or solution of equations;
(iii) can operate in real time.

Disadvantages:

(i) limited precision, often only 1% and 0·1% can only be improved on at excessive cost;
(ii) very limited logical ability;
(iii) no real memory.

It seems clear that in the future the application of the hybrid computer—varying configurations of analog and digital computers combined—will considerably increase.

One potential use of the analog computer is in the complex corrections for interelement effects in various types of instrumental analysis.

A special type of analog is the curve synthesizer, which has considerable value, for example, in analysing line-profiles in diffractometer traces.

IV. GEOCHEMICAL DATA BANKING

(1) The size of the data problem

Increasing sophistication and automation in analytical methods, linked with a fairly high level of capital investment in equipment over the last decade or so, has resulted in a geochemical data explosion, which has by no means reached a maximum yet.

In the area of X-ray fluorescence analysis alone, there are of the order of 30–40 instruments, mostly automatic or semi-automatic, representing more than £½ million of investment in university and similar geological departments in Great Britain. Each instrument has a high potential annual output of determinations and in order to assess the magnitude of the problem the Electronic Data Processing Sub-Committee of the Geochemistry Committee of the Mineralogical Society asked Dr. J. G. Holland in 1967 to conduct a survey with a view to assessing the probable rate of data-production in geochemistry in the United Kingdom for the following decade.

Response to the survey was good (20 replies out of 24 institutions circularized) and the analysis of results was detailed and sophisticated. However, since many of the departments were only just in the process of setting up new techniques of analysis, the estimates were speculative and the conclusions necessarily tentative. Retrospectively it may be guessed that many departments were over-optimistic about the speed of establishing new instrumental methods and the efficiency of utilization. It is the writer's view, however, that the conclusions were broadly valid but that one or two years should be added to the time estimates.

The main conclusion was that by 1970 we may expect an annual output of the order of 10 000 analyses, each consisting of 15–20 elemental determinations on average, from geological departments in Britain. This prediction refers to rock and mineral samples only and does not include water analyses, isotope determinations, geochemical exploration surveys and so on.

These figures can be compared with estimates from other countries. For example, in the U.S.G.S., geochemical prospecting accounts for some 10^5 determinations per year, and at the beginning of the fiscal year 1967, the file of chemical and physical properties contained records of 805 000 determinations representing 36 000 samples, the result of about 5 years' accumulation. In May 1966, D. R. Davidson reported that the Geochemical Census Branch of the U.S.G.S. has for several years analysed 15 000–20 000 specimens a year and counts on analysing 50 000–60 000 specimens in 1967. The number of elements determined in each specimen is in general from 15–20.

In the Canadian Geological Survey, at the end of about 4 years of operation, the Geodat system contains some 70 000 records filed on magnetic tape (K. R. Dawson, personal communication). In 1966 S. C. Robinson reported that the Geological survey completed some 10 000–15 000 analyses per year. At the same conference M. K. Roy-Chowdhury stated that the Geological Survey of India analysed 15 000–20 000 specimens per year.

It can be seen that the output of rock and mineral analyses from each of these Geological Surveys is of a similar order to each other and must, to varying degrees, represent an underestimate of the total national outputs. This fact tends to support the reasonableness of the estimate for Britain and, at the same time, underlines the magnitude of the data problem. It is perhaps also reasonable to estimate that the world "store" of geochemical data is being increased annually by more than 10^5 analyses representing considerably more than 10^6 elemental determinations. If the data from geochemical prospecting surveys by mining companies were included this number would be greatly increased. However, since most of these latter data tend to remain confidential it ceases to be a general problem.

A factor not so far raised is the decay rate of geochemical data and this is not easy to assess. The fact that many pre-1950 analyses would be rejected by many petrologists as inferior or not comparable with modern analyses suggests that at present the average life of geochemical data is of the order of 20 years. However, innovations in analytical techniques mentioned earlier have greatly improved precision and sensitivity and are gradually improving accuracy, and it may well prove that the decay rate may be considerably slower in the future. This implies that a British National Data Bank would have to contain analytical data on more than 20×10^4 rock and mineral samples alone, and might easily exceed this by an order of magnitude.

(2) The need for data banking

Since there exists scepticism in some quarters about the practicability or desirability of data banking it is advisable to review the arguments that may be put forward to justify large-scale, and maybe expensive, data-banking projects.

(a) Owing to the stochastic nature of many geological processes, alluded to earlier, many major advances of recent years in our understanding of geological phenomena have come from the statistical study of large masses of data. One could instance many examples in the fields of sedimentology and palaeontology and, in the field of geochemistry, specifically in our understanding of the variability and origin of basalts and their differentiates and of granitic rocks. It is certain that the study of regional variability in geochemistry will be a source of major advances in knowledge for a long time to come. Since the processing and evaluation of large amounts of data is most rapidly and economically handled by computer the data has to be rendered into machine readable form, which is half of the labour of data banking. In addition most data require selecting and screening before processing, again more readily accomplished by computer.

(b) With the vast increase in geochemical information the time is rapidly approaching when data obtained for a specific project will become buried in the mass of published data, or increasingly in unpublished work such as M.Sc. and Ph.D. theses or local records. This immediately raises economic and socio-ethical questions. The central problem is whether data, once used for the project for which they were obtained, should be preserved. One view is that the problems and cost of data storage and retrieval are such that it is frequently quicker and cheaper to repeat work when occasion demands. The opposing view is that retrieval problems can be overcome satisfactorily and at an economic price and that it is unethical to discard data used once only, and sometimes not at all, especially when expensively purchased by public funds. It is certainly true that nowadays the major consumer of time and money is often the collection of material rather than the analysis of it. Unfortunately it is difficult to obtain valid data to adjudicate between the economics of the two standpoints.

However, other questions apart, it is true that some specimens and data are difficult or impossible to repeat and there is room for the compromise view that some data must certainly be retained and that storage and retrieval studies must be seriously prosecuted until valid feasibility and economic conclusions can be drawn.

(c) It has been the practice in geochemistry in the past that all new data supporting the thesis of a paper should be published with it. With increasing speed and ease of analysis projects are being executed involving several hundred analyses in some cases and it is already clear that it is becoming impractical for all new analyses to be published in the conventional sense and some journals are refusing to do so. This situation raises two problems in its turn:

(i) Geochemical information will have to be published in summary form and it is desirable that some conventions and standard statistical procedures should be generally adopted.

(ii) It will still be vital that the raw data should be readily accessible so that other geochemists may satisfy themselves about the validity of the methods of data reduction and of the conclusions drawn. One way of solving this is for the data to be deposited with an institution—geological survey, museum, university, library—or the publisher, but the work and cost involved in making the data available is likely to make this procedure unpopular and, in the end, unworkable. The ideal would be inclusion of the data in a National Data Bank and publishers could be encouraged to make such deposition a condition of publication.

(d) In the sphere of economic geology the rapid availability of organized data, often of a nature not easy to anticipate, is essential. Computer search is the only efficient way of dealing with this requirement.

(e) Sterling and Pollack (1968) have recently suggested that "the scientist has a moral obligation to produce for the same amount of investment an optimal amount of information—even if he himself is not the person requiring it. A serious moral question can be raised about the scientist who does not avail himself of techniques that exist so that the results of his work are relatively meagre when compared to what they could and should be." The writers were advocating the use of the computer for data processing and evaluation, but the argument could equally well be applied to the ready availability of data in national banks or by machine communication.

3. Types of Data Bank

Data banks may conveniently be classified as:

(a) Domestic banks
(b) National banks
(c) Supranational banks, or an equivalent.

By domestic it is intended to characterize a bank which is internal to a single institution and which caters more or less exclusively for, and hence is designed for, internal needs. The institution may range in size from a single university chemical laboratory to the largest geological survey.

The difference between a domestic bank and a national or supranational bank is not merely one of size. Design of a data bank is very sensitive to the exact requirements from it, that is to say the type of enquiries it will be expected to answer. Thus, an institution may have limited spheres of enquiry so that its files may contain specialized or limited records, whereas a national bank will be expected to fill a much more general role. On the other hand an institution may demand detailed description and complex coding, if not as a condition of employment, at least as a pre-condition for analytical services, whereas a national or supranational system must depend almost entirely on goodwill. The latter will depend heavily on the simplicity of its record structure and hence determine the amount of work which the individual will be expected to contribute, which will inevitably, in its turn, be set against what he may expect to benefit from it.

A national bank, which aims to file all geochemical data produced within the country, as well as all data relating to the country (the two sets are not necessarily completely coincident) may be an unattainable ideal. An alternative is the concept of a *National Index* which registers the contents of domestic banks, thus making the task of locating data relevant to a problem greatly easier (see Brisbin and Ediger, 1967, p. 23).

The reality of a supranational bank in which global data is accumulated is probably, for a long time to come, an impossible ideal, and the economics of searching such a system non-viable. However, the alternative concept of ready international exchange of data, so far as politics permits, is a vastly important one and raises difficult problems to which a solution must be sought.

None of the different systems defined need be mutually exclusive and, indeed, it is the writer's contention that the design of even a domestic bank should take into account the problems of the other types. To be more specific, it is highly desirable, and certainly practicable, to design the record structure of a domestic bank so that records compatible with the country's national bank or of an international communication format, may readily be derived from it, preferably by the computer itself.

4. The design of data banks

(A) GENERAL REQUIREMENTS

Lovering and Davidson (1961) have stated some of the requirements that a geochemical storage/retrieval system must meet. These may be amplified as follows.

(i) Documentation must be sufficiently comprehensive to provide satisfactory answers to a wide range of geochemical enquiries.
(ii) On the other hand, the record structure should be sufficiently simple to permit data contributors to furnish descriptive information quickly and easily.
(iii) Each sample should be uniquely identified in such a way that it is always possible to determine where more detailed information is to be obtained.
(iv) The storage system must permit easy reorganisation and updating of records.
(v) The retrieval system must be sufficiently flexible as to allow searching over a wide range and combination of pertinent parameters. It must also be capable of delivering data in a wide range of useful output formats.
(vi) It is desirable that it should be possible to modify file and record structure in the light of experience. This implies that file structure should be internally described, or self-defined and that records should preferably have a free format, or at least, be of variable length.

(vii) Record length and storage medium should be such that searches through files of the order of 10^6 records are economically viable. This may include the possibility of multiple simultaneous enquiries.

It should be clear from the foregoing that file and record format design cannot be divorced from the supporting storage and retrieval software.

(B) TYPES OF GEOCHEMICAL ENQUIRY

The usefulness of a geochemical data bank will be measured almost entirely by the efficiency with which data can be recalled in a useful form. The successful design of the retrieval system must, in its turn, depend on the ability of the designers to anticipate the types of enquiry that may be presented to the system. Some of the general categories of geochemical enquiry are relatively easily predicted.

(i) *Provincial:* that is, listings of all analysed rocks and minerals from a defined geographical or geological area—map-sheet county, tectonic province, sedimentary basin, etc.
(ii) *Statistical:* for example, "how many analysed rocks designated as 'granite' or 'granodionite' have potash contents falling within concentration limits x_0–x_1, x_1–x_2, . . ., x_{n-1}, x_n?"
(iii) *Geochemical screening:* for example, "list separately all analyses of oceanic and continental 'basalts' with Si, Al, Na and Na/K values within specified limits and that also have analytical values for V, Cr, Co, Ni or Cu."
(iv) *Matching:* that is, listings of rocks whose major element concentrations lie within certain defined limits of those of a given target analysis.

All are essentially forms of data-screening, normally prior to more sophisticated processing. A good example is provided by Manson's (1967) work on the major element variation of basalts. The main difficulty is selecting descriptive parameters such that future geochemical enquiries, of a type perhaps not yet anticipated, may be answered.

Notwithstanding the last comment, records may not be of such inordinate length as to preclude efficient and economic searching. This leads to the concept of a record containing the minimum information required for efficient enquiry, which could also be identical with, or very close to the content of a supranational or communication format. Depending on requirements a domestic bank could also have a record infrastructure of secondary statements amplifying the primary minimum statements on which searches would normally or exclusively be prosecuted.

Existing data-banks appear to agree that the record structure should consist of four elements:

(i) *A label:* This uniquely identifies the sample which the record represents and should imply where further information is to be found. It should consist of three parts, the first identifying the country of origin of the data, the second identifying the institution from which it originates, and the third, allotted by the institution, uniquely identifying the sample or record within its collection.

(ii) *Type of material:* Igneous, sedimentary or metamorphic rock, mineral (possibly divided into silicate, oxide, carbonate, etc. or into rock-forming, economic), water analysis, geochemical exploration survey, extraterrestrial material, soil, organic fuel, etc. This essentially defines the file or sub-file into which the record is placed.

(iii) *Descriptive parameters:* Geographical, geological, petrological, which are required to give meaning to the analytical data.

(iv) *Analytical data.*

(C) DESCRIPTIVE PARAMETERS

Examination of descriptions of existing or proposed geochemical data-banking schemes reveals a general agreement on the categories of information necessary:

(i) Location.
(ii) Geological environment, regional and local.
(iii) Age, stratigraphic and, where available, radiometric.
(iv) Petrography/lithology/associated minerals.
(v) Sampling specification.
(vi) Literature reference, optional.

It is obvious that such a format cannot entirely satisfy all files, e.g. water analyses, but it is nevertheless, fairly basic. There is less agreement as to what should be included under each heading and how. The writer considers that the idea of two levels of statement, suggested above, would solve many problems, but would require suitable software.

(i) *Location:* There are many conflicting interests in selecting suitable parameters (see, for example, Brisbin and Ediger, 1967, 38–42). It seems clear that the only method which would uniformly serve for oceanic samples and areas such as Antarctica, as well as the better surveyed land-masses is latitude and longitude. It would also help many searches to include the country of origin of the sample. Secondary statements may include "province" (e.g. Oceania, Antarctica), state, county, town, height/depth, complex or formation name (e.g. Skaergaard, Torridonian) or local coordinates. The latter would be available for information or output, but not necessarily for searching. The problem here is that for efficient search names must be spelt uniformly or coded and this would lead to an undesirable increase in work for the submitter.

(ii) *Geological Environment:* This is conceived as consisting of two parts:

Regional: a simple statement of the regional setting of the sample, e.g. Oceanic, continental, regional metamorphic, lacustrine, cratonic, island arc, orogenic (pre-, syn-, post-), etc.

Local: a concise indication of the local geological provenance of the sample, e.g. "margin of dyke in matamorphosed limestone", "sandstone at base of cyclic unit". This is perhaps the most difficult of the parameters to specify since, to a large extent, it will depend on the type of material involved (that is, from file to file) and the purpose for which the file has been created. Even if much of the detailed description is relegated to a secondary statement, much work will be needed to obtain general agreement about the minimum acceptable description for a particular class of material, desirable for the primary or search parameter. There is no doubt that in the long term a system of semantic coding, similar to that proposed by Dixon elsewhere in this symposium, may provide the ultimate answer.

(iii) *Age:* It is probable that for most geochemical purposes stratigraphical age can be restricted for enquiry purposes, to about a dozen terms, such as Tertiary or Upper Palaeozoic. More detailed stratigraphic age determinations can be relegated to secondary statements. Where radiometric ages are available provision should be made to include this as well, together with an estimate of precision.

(iv) *Petrography/lithology/associated mineralogy:* The content of this entry will obviously vary from file to file. Once again it seems desirable and entirely practical to restrict the primary statement to a very few statements. For example, in the case of the igneous rocks, where agreement about nomenclature is still a long way off, the use of some twenty to thirty terms should be more than adequate for search purposes. The classes should be recognizable in the field and the definitions have no genetic or chemical connotations. Where detailed petrographic or mineralogical description exists it is desirable that it be incorporated in the record and this can be done as a secondary statement or in a comment field. Textures and structures should also be described. The best way of handling this seems to be the use of a limited glossary of standard adjectives from which a choice of two or three can be made.

(v) *Sampling specification:* This should define the sampling technique, for example for rocks and economic minerals, spot sample, grab sample, channel sample, composite sample, borehole chippings, etc. There should also be an estimate of the representativity of the sample.

(vi) *Literature reference:* Where there is a published description of the analysis, sample, or sampling locality it is helpful, although not essential, to include a reference to it in the record.

The selection of suitable terms has been much discussed and the interested reader is directed to the systems described by the U.S.G.S., the Canadian Geological Survey, and the Committee on Storage, Retrieval and Automatic Processing of Geochemical Data of the I.U.G.S. (U.S.G.S., 1969; Brisbin and Ediger, 1967; I.U.G.S., 1967 and 1968—see note at foot of references).

(D) ANALYTICAL DATA

Although the essential data of the record is numerical a number of problems exist:

(i) Provision must be made to record all naturally occurring elements. This may be done in a fixed format, which is wasteful of memory space, since some records may have only one or two elements determined and provision must be made for all, or by the use of identifying symbols in a variable length or free format.

(ii) In addition the format must permit the identification of isotopes and other forms such as different valency states (ferrous and ferric, as well as total iron), H_2O^+, H_2O^-, loss on ignition, SO_3 (as well as elemental or sulphide S), free silica, organic as opposed to carbonate carbon, and so on.

(iii) The form in which concentration is stated (%, p.p.m., p.p.b., gm/gm, etc.) must be decided or provision made to indicate this in the data.

(iv) A measure of precision, such as the overall coefficient of variance, should ideally be associated with each concentration.

(v) It is desirable that some measure of accuracy should also be incorporated with each concentration. How best to state this is open to debate.

In addition to the above problems, there should be attached to the record an indication of the technique of analysis, the laboratory or analyst, and the date of analysis. It is considered by some geochemists that the last three items provide the only satisfactory indication of accuracy at present available.

(E) CODING

Brevity of record and ease of search of a fixed format has been achieved in many data banks by means of coding information. The choice has often been largely dictated by the use of punched cards as the input medium. The main disadvantage is that coding of a record of a useful length is a relatively lengthy business requiring the use of glossaries or coding charts, which can rapidly become bulky and a disincentive to use. Coding, even where mnemonic codes are used, also increases the possibility of error and virtually excludes the filling-in of data forms in the field.

With advances in programming techniques the retrieval from variable format records has become far more efficient. Although still less efficient than fixed format searching the gap is sufficiently small now to consider whether a loss in computer time is not outweighed by the gain in ease and accuracy at the input end. Limited vocabularies and hence glossaries, with accurate spelling are still necessary for efficient retrieval, but they are easier and more natural to use and the psychological factor is considerable.

(F) CONVENTIONS

Standardization on modes of expressing concentrations, measurements (metric), gazeteers for the spelling of place and stratigraphic names, decimal or sexagesimal expression of latitude and longitude, and so on, are necessary and conformity with international practice, where this exists should be practised.

(G) COMPARABILITY OF DATA

One of the criticisms of the concept of geochemical data banking is that analyses relevant to a particular problem gathered from many sources, one of the main purposes of banking, usually differ too much in precision and accuracy to be treated as a single population from which valid conclusions may be drawn. Unfortunately, this is probably true for many purposes. However, to use this as an argument against data banking is a counsel of despair. The more positive approach is the incorporation in all records of more rigorous estimates of precision and accuracy and the wider dissemination of inter-laboratory reference standards on a national and, preferably, international scale.

(H) SOFTWARE PACKAGES

The success of computer-based data banking must ultimately depend on the quality of the storage and retrieval systems available. Some packages, such as the U.S.G.S. RASS (Rock Analysis Storage System), are already in operation. However, more versatile systems, permitting much more flexible file and record formats, and a greater variety of output formats, need to be developed. Such an operation is a major one, involving a mass of intricate and often tedious programming in machine language for efficiency of operation.

Such a comprehensive package for the life sciences is under development by Dr. John Cutbill of Cambridge University, sponsored by the Office of Scientific and Technical Information, as described elsewhere in this symposium. A pilot study of storage and retrieval of geochemical data is to be made jointly by Dr. Cutbill and the Electronic Data Processing Sub-Committee of the Geochemistry Group of the Mineralogical Society. A limitation of such systems is that the software has to be developed for a range of computers which share a fundamental programming compatibility. They will not be modified for use on all computers.

(I) MATCHING ANALYSES

An important aspect of data retrieval is the recovery from a file of samples which match a target analysis within defined limits. A similar facility is often required in data evaluation and simulation when it is required to know how closely a deduced or simulated composition approaches a target analysis. This is often done by inspection, but it is desirable that a criterion should exist so that closeness of match of two multicomponent compositions can be quantified automatically so that a logical decision, such as whether or not to terminate a routine, can be made. As far as the writer is aware no criterion has been proposed. It is suggested here that, following statistical methodology, the logical procedure is to use the sum of squares of the deviations of the individual concentrations of components of the contending analysis from those of the target analysis. Deviations may be weighted according to the enquirer's estimate of the relative importance of each of the components in achieving a match. The closest match is achieved when the S.S.D. is minimized.

(J) COMMUNICATION FORMATS

If geochemical data are to be exchanged internationally on any large scale, as is highly desirable in many types of enquiry, the problems of communicating data in machine readable form must be solved. This means that a communication format must be agreed internationally. The requirements are:

(i) *Simplicity:* otherwise it will not be used. This means the use of the absolute minimum of descriptive data consistent with usefulness. The record label will ensure that further information can always be obtained from source if required.

(ii) *Should overcome language problem:* This could be achieved in three ways—(A) common language; (B) multilingual glossary; and (C) coding.

(A) seems improbable for political reasons, and (B) is too cumbersome. Coding seems the only practical solution at present. This does not necessarily mean that domestic or national banks need adopt coding. A computer based translator could quite readily be developed to convert the relatively limited vocabulary of primary record statements into an international code.

(iii) *Standard conventions:* similar to those discussed above in paragraph (f) would need to be agreed. It is suggested that domestic and national banks should adopt any internationally agreed conventions (for example, those of the I.S.O.)

(iv) *Agreed medium and specifications:* magnetic tape would appear to be the most practical medium, although punched paper tape might be an acceptable alternative. It would also be highly desirable to reach an international compromise agreement on such specifications as standard widths, numbers of channels, blocking systems, and a limited number of character codes.

International discussion of this problem is considered as one of the most urgent in the field of geological communication.

5. Existing systems

A number of large-scale geological storage and retrieval systems are in operation, including those of the U.S.G.S., the Canadian Geological Survey, the Australian Bureau of Mines and Resources, the Bureau des Recherches Géologique et Minières and others in a pilot stage such as those of the Institute of Geological Sciences and the École des Mines in Paris under P. Laffitte.

Since geochemical data presents intrinsically one of the easier problems in geological data banking it is surprising that few institutions have purely geochemical files. The Rock Analysis Storage System of the U.S.G.S. and that of the Canadian Geological Survey are the best examples, although development work is in progress in India, Italy, France, U.S.S.R. and U.K. A number of universities have simple computer-based storage systems, but retrieval is mostly simple listing. The most advanced one described in the literature appears to be that of Manson (1967).

6. International action

Following the setting-up of a Committee of the I.U.G.S. concerned with the codification, storage and retrieval of geologic data at the International Geological Congress in New Delhi, 1964, a "Committee on Storage, Retrieval and Automatic Processing of Geochemical Data" was formed at Nancy in 1966 by officers of the I.U.G.S. The Committee had global representation, but asked M. Roubault to form a Working Group at the Centre de Recherches Pétrographiques et Géochimiques at Nancy. The first Committee produced a "list of specifications to be added to geochemical data" and the first project of the Working Group was to define and codify these specifications. The first group of specifications was produced and circulated in April 1967, just prior to a meeting held in May during the Paris meeting of the International Association of Geochemistry and Cosmochemistry when views on the project were exchanged between thirty delegates from twelve countries. The project was completed in November 1967, and in January 1968 a meeting to discuss the completed work was held in Nancy and attended by sixty participants from eight European countries. Subsequently, an operational booklet for reporting "observations and data concerning geochemical samples" was distributed in July 1968. The various, unpublished reports of this Working Group are listed in the bibliography under I.U.G.S.

The "Specifications to be added to Geochemical Data" and the "Operational Booklet" are a very useful source of ideas for institutions designing a data banking system. They do not seem to form an immediate basis for any international agreement, being too comprehensive and oriented rather towards documentation than communication. The writer considers that the most urgent objective should be the discussion of the specification for a communications format along the lines described above, with a view to international agreement.

V. CONCLUSIONS

It is suggested that certain general conclusions can be drawn from the foregoing survey.

(1) Our ability to produce good quality geochemical data has outstripped our facilities for processing and making optimum use of the data.

(2) The large increase in geochemical data production is leading to a situation where:
    (a) only a small percentage of analytical data will be published.
    (b) a high percentage of data may be used only once.
(3) This raises important scientific, economic and social-ethical problems. It is suggested that:
    (a) large-scale data banking with efficient and flexible retrieval systems is the only available solution to the data explosion problem.
    (b) it is highly desirable that there should be generally accepted conventions and minimum statistical standards for summarizing geochemical data or publication.
(4) Free exchange of data, especially internationally, is highly desirable and will lead to much more rapid advance in many geochemical problems. It is urgently necessary to discuss and agree on acceptable standard conventions and formats for the communication of geochemical data. The International Union of Geological Sciences has a Committee on the Storage, Retrieval and Automatic Processing of Geochemical Data which is the appropriate body to take action.

REFERENCES

Anderson, R. Y. and Koopmans, L. H. (1963). Harmonic analysis of varve time series. *J. Geophys. Res.* **68**, 877–893.

Brisbin, W. C. and Ediger, N. M. (Eds) (1967). A National system for storage and retrieval of geologic data in Canada. Geological Survey of Canada.

Chase, K. E. and Mackenzie, F. T. (1961). A statistical technique applied to the geochemistry of pelagic muds. *J. Geol.* **69**, 572–582.

Chayes, F. (1960). On correlation between variables of constant sum. *J. Geophys. Res.* **65**, 5185–5193.

Chayes, F., and Velde. (1965). On distinguishing basaltic lavas of circum-oceanic and oceanic—island types by means of discriminant functions. *Am. J. Sci.* **263**, 206–222.

Griffiths, J. C. (1967). "Scientific Method Analysis of Sediments". McGraw-Hill, New York.

Howard, J. C. (1968). "Bibliography of Statistical Applications in Geology". C.E.G.S. Programs Publication Number 2, American Geological Institute.

Krumbein, W. C. and Graybill, F. A. (1965). "An Introduction to Statistical Models in Geology". McGraw-Hill, New York.

Le Maitre, R. W. (1968). Chemical variation within and between volcanic rock series—a statistical approach. *J. Petrology* **9**, 220–252.

Loudon, T. V. (1967). "The ROKDOC Package". Sedimentology Research Laboratory, University of Reading.

Lovering, T. G. and Davidson, D. F. (1961). Storage and retrieval of analytical data on geologic materials. *Colo. Sch. Mines Q.* 247–257.

Manson, V. (1967). Geochemistry of basaltic rocks: major elements. In "Basalts—The Poldevaart Treatise on Rocks of Basaltic Composition" (eds. Hess and Poldevaart). Vol. I, pp. 215–269. Interscience, New York.

Miesch, A. T. (1967). Theory of error in geochemical data. U.S.G.S. Professional Paper 574–A, U.S. Government Printing Office.

Miller, R. L. and Kahn, J. S. (1962). "Statistical Analysis in the Geological Sciences". John Wiley, New York.

Prinz, M. (1967). Geochemistry of basaltic rocks: trace elements. In "Basalts—The Poldervaart Treatise on Rocks of Basaltic Composition" (eds. Hess and Poldevarrt). Vol. I, pp. 271–323. Interscience, New York.

Rollet, J. S. (1965). "Computing Methods in Crystallography". Pergamon Press, Oxford.

Schwarzacher, W. (1964). An application of statistical time-series analysis of a limestone-shale sequence. *J. Geol.* **72**, 195–213.

Shimazu, Y. (1959a). A thermodynamical aspect of the earth's interior—physical interpretation of magmatic differentiation process. *J. Earth Sci.* **7**, 1–34.

Shimazu, Y. (1959b). A physical interpretation of crystallisation differentiation of the Skaergaard intrusion. *J. Earth Sci.* **7**, 35–48.

Shimazu, Y. (1961). Physical theory of generation, upward transfer, differentiation, solidification, and explosion of magma. *J. Earth Sci.* **9**, 185–223.

Shoemaker, D. P. (1966). "World List of Crystallographic Computer Programs" (2nd Edn.) I.U. Cryst.

Smales, A. A., and Wager, L. R. (1960). "Methods in Geochemistry". Interscience, New York.

Smith, F. G. (1966). "Geological Data Processing". Harper and Row, New York.

Sterling, T. D., and Pollack, S. V. (1968). "Introduction to Statistical Data Processing". Prentice-Hall, Englewood Cliffs, New Jersey.

U.S.G.A. (1969). "Sample submittal manual". (3rd Edn.). U.S. Dept. Interior.

Vistelius, A. B. (1961). Sedimentation time trend functions and their application for correlation of sedimentary deposits. *J. Geol.* **69**, 703–728.

Vistelius, A. B. (1967). "Studies in Mathematical Geology". Consultants Bureau.

Zussman, J. (1967). "Physical Methods in Determinative Mineralogy". Academic Press, New York and London.

In addition to the above references there are a number of unpublished reports of the Committee on Storage, Retrieval and Automatic Processing of Geochemical Data of the International Union of Geological Sciences (I.U.G.S.) available from the Secretariat of the Committee at CNRS–CRPG, BP 682, Nancy, France. The more important of these are:

"Specifications to be added to Geochemical Data" (1967) French and English;

"Observations and Data Concerning Geochemical Samples—Operational Booklet" (1968) French and English;
"Structure du fichier géochimique" (1968). French only;
"Un système descripteur des échantillons géologique" (1968) by H. de la Roche et Ph. Grandclaude, French only;
"L'Evolution des fichiers géochemiques et leur adaptation à la communication" by H. de la Roche (1968) French only;
"Report of the Committee: Proposals Concerning its Future Activities" (1968) by M. Roubault and H. de la Roche;
"Quelques points importants concernant le stockage des données géochimiques" (1969) by H. de la Roche, French only;
"Recommendations for the Recording of Geochemical Data on Exchangeable Computer Files" (1969) by H. de la Roche.

16. Implications of Data Processing for Museums

DONALD F. SQUIRES

Marine Sciences Research Center, State University of New York, Stony Brook, New York, U.S.A.

This symposium is addressing the subject of data processing, museums and museum research, an area having relevance far beyond the community of repositories, and with special implications for library sciences. Objects, or specimens, have an increasing importance in a world which is experiencing dramatic and rapid changes. It is by the collecting of objects that mankind can most rapidly and effectively store data about his natural world, data which cannot otherwise be recorded completely for posterity. The problems to which we are applying the technology of computers are: (1) the management of the vast supply of these discrete data bases called specimens, a phase of application we term Collection Management; and, (2) the need for making research on these specimens or objects more meaningful and expeditious, the phase called Museum Research.

Through collection management, curators must make the resources of the collections available to the scientific community, not only by means of the limited cross-indexing permitted by collection arrangements and card files, but also by the various other desirable means of entry created in answer to the demands of scientific advance. Further, curators must bear the responsibility of making available both the actual specimens in their care and all of the existing information associated with them. These data are traditionally factual items relating to the occurrence in time and space and facts related to the acquisition of the specimens. Increasingly imperative is the capture of information enrichment, the knowledge which successively accrues to individual specimens, over the course of re-study and re-examination by various investigators.

This relationship between objects and information is analogous to that of museum collections and libraries. Libraries may be thought of as

Systematics Association Special Volume No. 3. "Data Processing in Biology and Geology," edited by J. L. Cutbill, 1970, pp. 235–253.

collections of objects (books). If indexing and cataloguing available to the user consisted only of the descriptive characteristics of the books (height, width, color, number of pages, data and mode of acquisition, etc), the library would be a useless place indeed. But indexing does reach deeper, with title, author, date, subject, etc. given, with cross-indexing as an additional aid. These clues to the informational content of each book are the stuff of which the value of the library as a resource is made. When fuller treatment of the indexing is made possible by means of data processing and indexing such as that provided by the KWIC indices, or by more sophisticated abstracting and indexing techniques such as Medlars, the real content of the library becomes immediately available to the user. Libraries have engaged in research on improvement of information and indexing systems for decades. Museums are just now embarking upon the same activity.

The role of the museum as an information bank is not totally new. In the course of their existance natural history museums have served this function in a limited manner, but it should be enlarged in a significant proportion. In fact, the increase of the scope of this role is vitally important, both at present and in the future, if museums are to remain as useful components in the structure of science.

The delicate subject of the place of the Curator in the museum has been questioned (e.g. Washburn, 1967; Squires, 1969), and even the role of the museum in contemporary science (e.g. Crompton, 1968). Is a natural history museum a vital part of the natural science community, or is it a vestigial heritage? Are museums in "left field" because of their emphasis upon studies of material culture and objects? Can their present despair be reversed, and can they be revivified as the important and significant resources they should be? These are survival level questions for the traditional museum.

As Project Director for the development of "An Information Storage and Retrieval System for Biological and Geological Data" now being implemented at the Smithsonian's Museum of Natural History, I have seen the consequences of the application of computer techniques become more than speculation. The realities with which we have been dealing are my credentials for speaking of the impact of this new tool, computer technology, upon museum administration, upon the scientific resident of the museum, and upon The Museum as an institution. I wish to speak to the questions of What? Why? and How? as they apply to data processing and the museum utilization thereof.

WHAT?

The concept of museums as informational resources or data banks addresses itself to one aspect of the application of computer technology to research, collections management. It demands the enhancement of informational resources within the museum community. This aspect has been emphasized in our project as it is one in which I have a deep commitment. In the early stages of developing a program of data processing for the Smithsonian Institution, we made a clear distinction between data-bank like activities and research utilization of computers. This distinction is dependent upon recognition of the time framework for the two activities. Research is a real-time situation requiring immediate access to data and ability to develop extended amounts of data into complex treatment such as statistical analyses, etc. While the volume of such data may be large, it does not reach the proportions of the total resource of a large museum. Establishment of a data bank, on the other hand, is an enormous undertaking requiring a longer period of time for development, and probably while in a semi-completed form, not susceptible, except at great expense, to real-time manipulation of data. Research data files become a part of the larger data base. Data retrieved from a larger data base are a start for a research data file. One complements the other and thus each enhances the other.

We are discussing larger data files (in the instance of the Smithsonian, millions of entries) which commence with the ordinarily available data recorded with specimens, and which have the capacity to incorporate new data as it is recorded in the course of study of the specimens. We are considering data files which not only give access to the collection through a diversity of entries, but which also have the capacity to retrieve data about and associated with specimens and to print or manipulate selected items in specified arrays, or to otherwise make such data available to investigators. We are interested in information storage and retrieval systems which have the characteristics of complementing the data bank inherent in the specimen, and through its existence to make possible increased and more effective utilization of collections, and further, to make possible more advanced research upon specimens and collections.

WHY?

Collections of specimens or objects are rich and important resources for study and research. But increasing costs of upkeep have forced many educational institutions to abandon the study collections amassed during years of collecting by faculty and students. As a result, students in many universities do not have the first-hand access to museum information as had been the case when the university museum was not only a "place to visit" but was also a resource for study. Faculty researchers must now travel to distant museums to select, and/or study materials, or to simply record the data associated with the specimens.

Museums are faced with the necessity for more detailed data to be recorded if their collections are to continue to serve a useful scientific function. Numbers of specimens being housed by museums are sharply increased in response to the growth of science.

Already badly behind in documenting their collections, museums are facing larger backlogs of cataloguing at the same time they are being called upon to supply more information to the academic community than ever before.

Students who find their way to collections are often discouraged by the problems of manually extracting information from the collections and are dissuaded from following a line of research or a potentially rewarding career in systematic studies or in museology.

In science there is a growing awareness of the importance of museum collections and the information associated with specimens in the documentation of both direction and magnitude of environmental changes. Growing use of collections for use in analytical or faunistic studies creates new problems for the museum curator. New combinations of data are desired at a time when the museum profession, because of reduced work forces, has generally retreated into maintaining only narrowly structured reference files principally for the use of systematic biologists. Renewed awareness of the importance of collections as a part of the information resource of the biological sciences in particular, has come late to a museum profession still proudly conscious of its pre-eminent role of 60 years ago.

In over a decade of responsibilities as a museum curator in two American museums and as a visiting researcher in countless other museums about the world, I have been increasingly impressed with the amount of data not being recorded because of the inability of existing retrieval systems (the curators) to capture and retrieve data in its potential volume. Data acquisition and registration will improve when the working scientist can utilize an adequate information storage and retrieval system, because he will know that these data can be recorded in a fashion which will be useful to him and his colleagues. Museum curators will better manage the data acquisition procedures in their own units so that more information is obtained. Finally, and perhaps most significantly, greater and better use will be made of museum collections in education and research endeavors through the availability of the information contained in museum collections.

HOW?

As there are a multitude of means, there are multitudes of answers to the question posed, how? For some museums established and working paper files of one sort or another will suffice for some time into the future. For most museums, present indices to the collections are not sufficient and even the most unreliable of all information storage and retrieval systems, the Curator, is worked beyond capacity. Edge punched cards, or more effectively, the multiple drilled cards of the Termatrex system provide an effective, efficient, inexpensive, and instantaneous means of accessing collections through a variety of parameters (Van Gelder and Anderson, 1967). But to meet the qualifications previously set forth, to be able to retrieve not only data about the specimens, but also the data associated with them and subsequently accumulated about them, in a form which allows for manipulation in either printed form or within the computer, only one technique will suffice—computerized data processing. Its disadvantages are discouraging: it is expensive, it is slow to build, it will probably be costly to operate. But, we are convinced that it will provide a whole new dimension to the world of natural science.

The system described in here is not the only solution, nor is it necessarily the best. It is the approach to the development of an informational storage and retrieval system developed for the Smithsonian Institution in a pilot project now in its terminal phase. Thus far it has met our specification but we have much more to learn about it. An interim report describing the development of the system over the first half of the project has been prepared (Squires, in press) and final reports will be issued following completion of the project. Currently the system has limited retrieval capability with a data base consisting of 41 000 specimens records of birds, crustacea and rocks. The rationale for the project and some of its implications are described in a number of publications among which are: Creighton and King (1970); Galler *et al.*, 1968; Manning, 1969; Piacesi and Creighton (1970); Squires, 1966; Squires, 1968 and Squires, in press.

Data preparation and input has been discussed extensively (Squires, 1966) and need not be detailed. We have followed the ensuing general principles in establishing data input:

(1) Input should be coupled with the generation of hard copy documentation similar to that traditionally utilized in the originating unit so as to maintain complete, in so far as possible, existing files and records, until such time as a complete, function of, computerized data base exists.

(2) For purposes of implementation of the pilot project, existing data catagories would be utilized for input records, with the potential for enlargement of the systems design to incorporate new data catagories as required.

(3) Input (and output) should require as little codification as possible in order to reduce the time required in data preparation, and to eliminate in so far as possible, potential for operator error.

Punched paper tape is utilized as the input medium. Two Control Data Corporation, SCM typetronic systems are used (Birds and Crustacea) to generate labels and file cards (Crustacea) labels and catalogue pages (birds) as well as input tape; a Frieden Flexowriter is used for the rock collection to generate labels and file cards as well as input tape.

Our experience with source data automation has been mixed. While great efficiencies had been anticipated, experience has demonstrated

TABLE I Specimen data recorded for pilot project collections

Crustacea	Birds	Petrology
Catalog Number	—	—
Generic Name	—	Petrographic name
Sub-generic Name	—	
Species Name	—	
Subspecies Name	—	
Author Name	—	
Locality (in four levels)	—	—
Latitude and Longitude	—	—
Collector	—	—
Collector's numbers	—	—
Date collected	—	—
Depth	Altitude	Depth/Altitude
Number of specimens	Collection code	Collection code
Sex	Description code	Number of specimens
Preservative	Donor's name	Donor's name
Collecting gear	Preparator's name	Chemical analysis (by percentage, element)
Identifier	Age of specimen	
Nomenclatorial type	Sex	Trace elements
Publication information	Fat	Radiogenic isotopes
	Skull ossification	Non-radiogenic isotopes
	Reproduction anatomy	Petrographic description
	Soft part color	Modal analysis
	Molt condition	Thin section(s)
	External measurements	Density
	Stomach contents	Associated rocks
	Ecological notes	Geological age
	Parasites	Mineral composition
	Related specimens	Exhibition information
	Nomenclature type	Texture
	Disposition data	Analytical methods
	Publication information	References

that the greater rigor required by computer technology requires more cataloguer time than the sloppiness accommodated by previously existing file systems. Conversion of the latter to the former has been painfully expensive because of the frequency of errors incorporated into previously existing files. Presently we utilize cataloguers to prepare data for a machine operator or typist, but such a system is abnormal and designed to meet our requirements of establishing a data base rapidly. Under normal operating conditions, the cataloguer and the operator are one person.

Table I lists the data catagories presently being utilized for the various collections in our experiment. As mentioned above, this list includes existing data catagories and can be expanded beyond the present form as new data are recorded.

Two other steps are required beyond the input of data in the form of punched paper tapes: (1) preparation of the numericlature data, entailing the preparation of the hierarchy (phylum, class, order, family, genus, species, author and date, and such synonyms as are to be recorded) for incorporation into the Numericlature thesauri (after the first entry of a name combination, or higher taxon, such a compilation is not necessary, as machine lookup is a part of the input program); (2) assignment of latitude and longitude coordinates to locality information. These need be recorded only the initial time for an insertion of locality information and are thereafter recorded automatically through a table lookup system of geopolitical terminology.

During the first 18 months of the project, a total of 137·5 man-months were expended in establishing a data base for specimen records on over 23 000 specimens. The distribution of this effort is indicated on Table II.

Systems design and programming

A number of constraints were placed on the original design of the system. (1) the system must be expandable to large capacity, for the present holdings of the Smithsonian's Museum of Natural History alone are in excess of 50 million specimens. It is estimated that there are over three million known species of plants, animals and rocks and minerals, each having an average of ten synonymous names, for a total of 30 million names. Although at the outset we could reckon on recording an average of 14 data catagories for each specimen, this number did, and is still rapidly expanding, with the result that the record length associated with each specimen has increased dramatically to over 30 data catagories. (2) The system should not require large numbers of new personnel, but once developed should be adaptable within the museum structure and staffing. (3) The system, must be compatible with existing card files and other visual records, none of which will be

TABLE 2. Manpower requirements during first 18 months of pilot project

	Specimen record preparation	Specimen record entry	Preparation of numericlature files	Preparation of global reference Codes	Source data automation programming	Supervision and other miscellaneous tasks
Grant supported man-months	—	58·0	22·0	4·5	7·0	16·5
Smithsonian contributed man-months	21·0	7·5	—	—	2·0	—
Production achieved	— —	11 867 processed 8 000 backlogged 19 867 lots	40 389 names	1965 localities	—	—
Productivity	—	303·3 lots per man-month	1836 names coded per man-month	436 codes assigned per man-month	3·0 man-months per division	—

replaced within the present (and perhaps next) generation of museum curators. (4) Because of the immense literature and the classification schemes deeply imbedded in both literature and the philosophy of practitioners of systematic biology, it was important that the system use for input, names, not codes, retaining the flexibility of structural classifications of those names, and recognizing the necessity for dealing with geo-political terminology and other geographic references rather than an arbitrarily selected grid reference scheme.

Although important strides are being made in the technological problem of machine interfacing, we have been from the beginning concerned with facilitating communication between a non-computer oriented scientist and the data processing system and between various types of computers located at different centers. For these reasons, COBOL (Common Business Oriented Language), a subset of the English language was selected as the query language for the system.

Data input in the form of punched paper tapes are batched and run through a reader and accepted by the computer. A number of checks are made against the input data as a part of the entry program including verification of fields, validation of names, conversion of units of measurement to standard expressions and calculation of mean values when ranges are expressed (the original form of the data are preserved), and so forth. The data are reformatted and placed in the Work in Process File. A copy of the input data are returned to the originating unit for verification before further operations are performed. Because the Work in Process File may contain up to 10 000 specimen records, a query capability has been developed permitting questions such as the following:

(1) If Country equals "United States" and state equals "Alaska" and locality equals "Bristol Bay" and breeding status equals "breeding" perform search through retrieval. Yield: a listing of these bird specimen records (printed in full in the Work in Process format) satisfying these requirements.

(2) If genus equals "*Gonodactylus*" perform selection through retrieval. Yield: a printing of the specimen records of all species of Crustacean.

(3) If collector-name equals "Ecklund, C. R." perform selection through retrieval. Yield: a listing of specimen records of specimens collected by C. R. Ecklund.

(4) If genus equals "Dacite–0148"* and island-group equals "Mariana Islands" perform selection through retrieval. Yield: specimen records of Dacite–0148 from Mariana Islands.

* Rock terminology has been developed parallel to biological nomenclature. "Generic" equivalents are rock names adapted from Troger's Classification.

The function of this initial state then is to accept input data, reformat from the various forms in which originating units submit the data, to edit *and* to add new information. Through the Numericlature Input Thesaurus, a numerical code for handling names internally within the system, is assigned. A Numericlature Output Thesaurus relates author and date citations of the name and links synonymous names with the valid name. Input data are also examined and an Abstract Search Index prepared.

Upon validation of the material held in Work in Process, the data are released to the main storage for the system, the Data Bank, in which specimen records are ordered by Numericlature (or, put another way, taxonomically) and within each species, sequentially by catalogue number. Specimen records are stored in their complete text.

Because taxonomic sequence is utilized for the main file structure, it is important to clarify the methods used to construct this sequence. No fixed system of classification exists. We have adopted the principle that, just as curators will order specimens on shelves by some arbitrarily selected scheme, so should they utilize the same arbitrariness in selecting a classification for the data bank. We have followed the general principle of utilizing a published, comprehensive classification wherever possible so that other users of the system can be aware of the groupings involved. A scientist may change the higher taxa assignments of any and all species as he so desires, constructing a classification to his own needs and changing concepts. This flexibility is vital to the system in order that it does not prematurely harden the arteries of systematic biology.

Retrieval from the data bank is accomplished as follows:

1. A request. The requester writes simple conditional statements identifying the taxonomic unit of interest and other restricting parameters. For example: if information on the American Pelican is desired, the request would be written as follows.

If species equals *Pelicanus americanus*, then perform record selection. The computer will search the Nomenclature Input Thesaurus and associate both the numericlature code for the name *Pelicanus americanus* and the address for the specimen abstract record in the Abstract Search Index. A negative response from the Nomenclature Input Thesaurus means that the name has not yet been placed in the system or that the input name was misspelled. A negative response from the Abstract Search Index would indicate that although the name had been included, no specimen records have been incorporated into the data bank. If records are present, they are then selected from the Data Bank.

2. A request containing a geographic term. The requester would write the additional qualifying term containing the geographic terminology. For example: if information was desired on the American pelican from Key West, the request would be written as follows.

> If species equals *Pelicanus americanus*, and locality equals Key West, then perform record selection. The query cards containing the geographic term will initiate selection of the appropriate Global Reference Code polygon functions from the Location Query Thesaurus. The abstracts of the taxonomic group selected from the Abstract Search Index by the numericlature search will then be matched in Global Reference Code to nominate those records which will be retrieved.

3. A request containing a collection date. This particular parameter was selected to provide the third dimension to the collection, a characteristic of specimens which is of increasing importance. Should the requester be interested in *Pelicanus americanus* from any locality, but only those which had been collected in 1855, then the request would be written as follows.

> If species equals *Pelicanus americanus*, and if date collected equals 1855, then perform record retrieval. Expression of date collected, entered in time-span query cards would be utilized to select those record abstracts which had survived the first screening.

At this step in the query process, the essential elements of what, where and when have been met. More specific requirements relating to special data fields will be met by first searching the Abstract Search Index which will indicate those records containing data in specific fields, and then through a match of the data involved in the specimen records. The Abstract Search Index is expressed in the form of dichotomous yes-no choices recorded against the various data catagories, and serves principally to eliminate those records in which information is lacking in the requisite fields.

Specimen records meeting the requirements of the search and which have been located in the three major files, for which further restricting statements have been issued, met and retrieved, will then be copied onto an output magnetic tape in preparation for printing as output. Before the final printing is accomplished, the requester will have the opportunity to specify the printing format (for example, see Manning, 1969) and to associate additional taxonomic information including synonyms, authors, dates of publication of the species, and such other taxonomic remarks which are incorporated from the Nomenclature Output Thesaurus.

Examples of interrogations presumed to constitute the genre of questions of the future are more complex and include the following.

Print the geographical distribution of the genus *Gonodactylus* collected at depths greater than 20 fathoms, listing the species alphabetically.

What species of the family *Squillidae* collected between 50 and 100 fathoms occur together in the Gulf of Mexico? Arrange list as cross-index listed alphabetically by genus and by species.

List all records of the *Dendroica caerulescens* collected above 3000 feet, in May, June or July from the Appalachian Mountains. Arrange by months and alphabetically.

List all records of the orders Procellariiformes and Pelecaniformes which have recorded soft part colors or external measurements. Arrange list by catalogue numbers and cross-reference to an alphabetical listing.

WHEN?

The system described above has been undertaken as a pilot project to test the feasibility of a data retrieval system within a museum of natural history and particularly to determine the costs involved (more about this aspect of the project follows). Has the system, in its preliminary demonstrations affected the Museum? It has indeed. The immediate problem facing the museum administrators is that of implementation, a factor tied closely to the question of cost, but also involving questions of where such a program should be imbedded within the framework of the administrative establishment, how it should be operated, and with whom it should be staffed. These questions will be dealt with later, and for now I will consider only the temporal question. Obviously, we who are engaged in the project feel that the time is now, and that delay only worsens an already bad situation.

Few of the indices to the collections of most museums are being maintained and the level of accessions continues to be at or higher than the post-war rate. With each year the museum becomes further behind in its own chores and the acquisition of information associated with specimens suffers. Now, with many studies of environments being geared to the ecological crisis, there is a pressing need for full documentation to be available with specimens. Further, the system once implemented is not static and susceptible to being put on the shelf until the museum problem has worsened to a level which makes some form of information treatment mandatory; it is dynamic and must respond to the needs of the museum and the scientific community at the immediate time of utilization. As museum curators change their views in response to contemporary science, so must the system.

Traditionally oriented curators will argue, persuasively, that other museum needs are more pressing: bottles, trays, cabinets, and above all, more curators and technical staff who will process the massive amounts of information now accumulating and who will examine the backlogs. There is no doubt that these realities exist, but my colleagues and I believe that unless an adequate means of really indexing and making available this flow of information is developed and implemented, all of the curatorial and technical effort put forth into the treatment of the collections will be of little avail. The danger lies in the fact that collections are becoming progressively the domain of the systematic biologist and the remainder of the scientific community is being forced to utilize other sources of information in the default of the museum.

If this argument is valid, museums should begin now to process data in a machine readable form; a step which carries a greater commitment than is immediately apparent. Most source data automation techniques, i.e. the methodologies which produce the paper files which must be maintained until a total replacement is available, and simultaneously makes these documents utilizable as input to a computerized system, are not infallible. They must be checked by actually reading the material into a computer, filing the information on magnetic tape and reprinting it in some form which serves as assurance that the data are actually present in the correct form. Thus, with implementation of the initial stage, the museum must be prepared to go all the way in a series of planned phased steps.

Two alternative procedures are available to undertake the initial steps: (1) implementation of SDA at all specimen processing points so that all new specimens are appropriately indexed; or, (2) processing of certain key collections in depth so that the entire resource of that segment is quickly accessible to the curatorial and other research community. For the small museum, the first alternative is probably workable for the flow is not great and centralization of specimen processing procedures is possible. The great advantage is that the enormous intellectual effort involved in bringing information associated with specimens up to some standard is obviated. I am sure that all curators must question the validity of including all of the collections in their care in an information storage and retrieval system. Older materials, invaluable for a systematic viewpoint, often have limited associated data making their utility to the general community questionable. For larger museums, the second alternative is probably the most workable, and has the great advantage of placing its inhouse research staff in the position of reflecting priorities. By emphasizing those collection areas which are of immediate research importance, the benefits to selected research programs are immediately recognized. Once the data base is established, relatively little effort is involved in keeping it to date. We have suggested that the National Museum of Natural History undertake to establish a cataloguing team which moves into a unit of the museum and assists the existing staff in getting its backlog in hand, and then which moves to another segment of the collection upon completion of its task, leaving existing cataloguing personnel to cope with the incoming flow of materials.

In short, we feel that the question of when has only one dimension—that being as soon as possible within the context of the resources of the museum.

WITH WHAT?

Perhaps no question has been placed with greater fervor by visiting museum curators and administrators than "how much will it cost?" As yet we are not prepared to answer for our own pilot project. The answer of the cost is entirely related to all manner of other costs within a museum, none of which are really well understood. As a part of our research in our pilot project we have undertaken a cost study which is broadly based and examines acquisition of specimens and specimen related data, the processing of both specimens and specimen data, the maintenance of both specimens and data in the collection and its associated files, and finally the costs of making this information available to curators and the scientific community through information requests, visits, use of collections and associated files, loans and exchanges. This study is now underway and I cannot as yet report any significant results.

Through interviews with visitors and staff of the museum, we are attempting to determine which museum personnel are involved in the steps enumerated above. As with any modern scientific effort, the most significant costs are those which are associated with the people required to do a task, and if we examine our own experiences in museums, we must be painfully aware of how consuming of the time of the staff the present human information retrieval system is—and hence how costly.

As mentioned above, our experience has been that insertion of data within the computerized file through automated cataloguing techniques has not resulted in the dramatic savings originally postulated. In large part this reflects upon the quality of data formerly input to retrieval systems and the great tolerance for ambiguity which we as museum curators have learned to accommodate. Thus one of the benefits to be obtained through data processing is a higher standard of exactitude, but only at a price. But one can argue significantly that if the cataloguing is to be done at all, be it by a curator or by a cataloguer, that the additional effort required for the difference between insertion of minimal data and the full data available is not significant in the general enhancement of the value of the collection to the scientific community. The availability of the information associated with specimens to the user, both inhouse and visiting on a demand basis, is another positive aspect.

But, the point may not be avoided that a program of mechanization of museum data is essentially a new program and will require new funding and personnel increasing the strain on already meager museum budgets.

Thus far I have mentioned only development costs, but what of the operational costs? These are matters deserving immediate attention by museum administrators. Who will pay for the services provided by the computer? Will individual departments budget for their computer time so that the curatorial staff will be able to utilize the system, and if so, will these costs be sufficiently high to drive the museum staff back to traditional catalogues? Should such costs be assessed by some annual proration of the total operational costs of the computer facility or should there be hourly charges assessed? Who will determine the priority of requests and assess the costs, and who will determine which inquiries should be processed with those funds remaining in the budget? Will the visiting scientist pay, or will the service be rendered gratis? If the latter, to what extent should visitors be given free rein?

WHO?

Manpower to accomplish automation of museum reports is a point of some fundamental dispute between various groups engaged in research on this activity. Some argue that unless museum personnel, probably at the curatorial level, are deeply involved in the establishment of the system and the programming required to accomplish the final goals, that the system will not be appropriately responsive. We, on the other hand have taken the position that the extensive systems design and programming required is beyond the probable limits of time available from the curatorial resource and the professional computer personnel are required. If these persons work in close relationship with curators, they can be taught the required principles and requirements perhaps more quickly than can be museum curators the disciplines of computer techniques. We feel that the indications of our own system and its requirement for eight man-years of programming and systems design represents an incredibly large investment few museums could make from their own inhouse resources.

In our own program, the Smithsonian's Information Systems Division has been the responsible agency for the design and implementation of the computerized program. Participating curators have worked closely with systems analysts and programmers either directly or indirectly through a liaison man. Thus we have been able to accomplish the design and implementation of a system in a relatively short time, but still feel that it is responsive to the needs of the curator. Future requirements will keep at least one programmer occupied in the tasks of introducing the new units into the system (our requirement that each unit have the ability to format its own data in its own way requires customized input routines for each unit) and to assist curators in the interrogation of the file.

Perhaps the key to the issue is the position of liaison man who, in our scheme of things, has had the responsibility of getting new units started with source data automation, instructing curators and cataloguers in the techniques and protocols required and interpreting their concepts to the programming staff. Such a person must have both a scientific background and an extensive knowledge of computer techniques—but he need not be a professional in either field. Among the responsibilities of such a person are the implementation of systems, validation of numericlature and global reference codes validation of specimen records is a responsibility of the originating unit and its cataloguing staff), general monitoring of the system, and screening of the query statements.

The last statement bears amplification. We are concerned with the cost of operating the system, a value as yet unknown, and are presently operating under the principle that requests will be batch-processed. The system itself is amenable to remote keyboard query, but realities of the computer available and demands upon its time dictate that we go the other route for the present. We are developing query instructional leaflets for distribution to curatorial staff and visiting scientists which outline the informational content of the files and the procedures by which their questions are transformed into COBOL query statements. While in flow charts and descriptions such a system would appear to work, the realities are that few curators have ever been able to ask significant questions of their informational resources and get responses and they have lost the knack of asking questions in the most searching way. We are prepared to assist both visitors and scientists for some interim period by having expertise available to them and to scan all query statements to assure that file dumps are not included among them.

Many problems associated with the operational aspects of a computerized museum catalogue are not yet thought through. Should visiting scientists have the privilege of printing significantly large segments of the file to carry off as a reference to the collection, and if so, what of the unpublished data incorporated into such a file through the process of research but prior to publication?

WHERE?

Finally, I should like to ask where the facility should be located within the framework of the museum. Not many museums will have the good fortune to have a computing center as a part of their own organization. However, a survey made some years ago demonstrated that of the 100 largest natural history museums in the United States, all were within easy distance of a major computer, either located on a university campus or in a local industry. Since then the advent of time-shared computers has made possible the linking of museums in the United States through commercial telephone lines, independent of the computer capability of any of the museums.

In the absence of a computer, the museum will have to acquire rights to the use of one through some financial arrangemement. Service bureaus are now to be found in many major cities who will offer computer time on a rental basis. More probably, consortia of museums will acquire a computer and share the time jointly, and remotely. The consideration of the physical location of the computer thus becomes a second order importance. The placement of the intellectual resource required for the development of the program is more difficult.

We have argued that professional computer competence is a requirement for museums. Such personnel should be centrally located, probably within the Office of the Director, for the activity will soon become a major budgetary item which will occupy his careful attention. For most museums, centralization of source data automation equipment will be feasible, and mandatory for efficiency of operation. We also urge the development of a team of cataloguers who roam the museum, supplementing the efforts of the regular staff as needed and on a priority basis established by the regular staff of the museum and an advisory committee to oversee the operations, is probably desirable.

SUMMARY

The list of implications of the introduction of data processing to the museum can be protracted indefinitely as it ramifies through the structure and philosophy of the museum. It is apparent that many museums are not administratively structured to cope with the problems induced by such an activity and that management practices must be carefully considered and restructured as the role of the museums enlarges through the advent of the new tool. No pat answers are available to the problems presented, for each solution will depend upon the local environment and the resources both human and financial available to the museum. Actions must be undertaken with full awareness that the larger consideration of the role of the museum in contemporary society and science must be considered simultaneously. There has been little discussion of the utility of data processing to museums, or of the multitude of purposes to which this tool will be put. Such a catalogue will be as outdated as are most published museum catalogues, for given the ability to interrogate the informational resource of the collection, the scientific mind will develop whole new activities beyond my ability to project. Museums are potentially entering into a new era, emerging from a passive resting stage as a repository for the relics of scientific inquiry to a new dynamic existence as an informational resource as utilitarian as the library, and as vital to scientific research as is the written record of scientific achievement.

REFERENCES

Creighton, R. A. and King, R. (1970). The Smithsonian Institution's Information Retrieval System (SIIRS). Proc. 6th Annual Colloquium on Information Retrieval, pp. 31–50.

Crompton, A. W. (1968). The Present and Future Course of our Museum. *Museum News*, **46** (5), 35–37.

Galler, S. R., Oliver, J. A., Roberts, H. R., Friedmann, H., and Squires, D. F. (1968). Museums Today. *Science, N.Y.* **161**, 548–551.

Hey, M. H. (1966). Catalogue of Meteorites. British Museum (Natural History), London.

Manning, R. B. (1969a). A Computer Generated Catalogue of Types: A By-product of Data Processing in Museums. *Curator*, **12**, 134–138.

Manning, R. B. (1969b). Automation in Museum Collections. *Proc. biol. Soc. Wash.* **82**, 871–878.

Piacesi, D., and Creighton, R. A. (1970). An Approach to the Geography Problem in Museums. Proc. 6th Annual Colloquium on Information Retrieval, pp. 441-456.

Squires, D. F. (1966). Data Processing and Museum Collections: A Problem for the Present. *Curator*, **9** (3), 216–227.

Squires, D. F. (1968). Collections and the Computer. *Bioscience* **18** (10), 973–974.

Squires, D. F. (1969). Schizophrenia: The Plight of the Natural History Curator. *Museum News*, **48** (7), 18–21.

Squires, D. F. (in press). An Information Storage and Retrieval System for Biological and Geological Data: An Interim Report.

Van Gelder, R., and Anderson, S. (1967). An Information Retrieval System for Collections of Mammals. *Curator*, **10**, 32–42.

Washburn, W. E. (1967). Grandmotherology and Museology. *Curator*, **10**, 43–48.

17. A Format for the Machine Exchange of Museum Data

J. L. Cutbill, A. J. Hallan and G. D. Lewis

Department of Geology, Sedgwick Museum, University of Cambridge; Imperial War Museum, London; Sheffield City Museums, Sheffield, England

With our evident enthusiasm for data processing we shall soon be producing hundreds of files of machine readable data—data laboriously assembled to serve some specific research project and then set aside with the thought that it might prove useful again, or to someone else. The advantages of exchanging tapes are too apparent to need elaboration but the obstacles to this are numerous. This paper attempts to solve one of these obstacles: compatibility between record formats. To achieve this, a Museum Communication Format is proposed which for convenience here is called MCF. Such compatibility should neither restrict the range or complexity of data which can be held nor produce losses in transmission from originator to user. However, before describing the communication format, a brief history of the project will not be out of place together with a general discussion of the problem of compatibility and our approach to this.

HISTORY OF PROJECT

The Information Retrieval Group of the Museums Association (IRGMA) is responsible for this proposal. Just over four years ago one of us drew attention to the failure of museum cataloging practice to meet the demands of research and museum administration (Lewis, 1965). In addition, the need for indexing to satisfy interdisciplinary search requirements was stressed. It was suggested that there was an urgent need to provide a nationally recommended standard for the cataloguing of museum collections which would be a first step towards a national index of museum holdings: an aim, if then a somewhat pious one that the Museums Association had had at its inaugural meeting in 1888.

Systematics Association Special Volume No. 3. "Data Processing in Biology and Geology," edited by J. L. Cutbill, 1970, pp. 255–274.

In May 1966 a number of biologists and geologists met at Leicester to plan a common approach to the cataloguing of biological specimens and to allow the exchange of specimen data between institutions through computer operated files. This was the beginning of the Leicester Group which produced recommendations in March 1968.

Another of us (J. L. C.) carried out a feasibility study for the computer handling of geological specimen data three years ago which was financed by the Office of Scientific and Technical Information (OSTI). This work has, of course, now developed into the Cambridge Geological Data System and is described elsewhere (Cutbill and Williams, this volume, p. 105).

The third author of the paper (A. J. H.) has been concerned with setting up a deep indexing system for the film archive at the Imperial War Museum, London; this will be operational in 1970.

In the spring of 1967, a colloquium was held at the Sheffield City Museum under the auspices of the Museums Association to which all known to be concerned with the problems of information retrieval in museums were invited; from it IRGMA was formed (Lewis, 1967) IRGMA now comprises both technical and subject committees as well as an *ad hoc* body of interested people who receive an occasional *Newsletter* and attend lectures on related problems. It is perhaps significant to the potential advantages of the inter-disciplinary approach that the Leicester Group amalgamated with IRGMA a year later and for that matter that the authors of this paper have diverse specialist interests being from museums with geology, military history and general collections.

PROBLEM OF COMPATIBILITY

It was not until January 1969, however, that we found the problems leading to the idea of a communication format. A rough draft of this has been published by the Museums Association (1969) but the present paper elaborates and improves on these draft proposals. The overriding problem that we had to face, of course, was compatibility. This may be grouped under three headings: machinery, systems and data.

Machine compatibility includes such things as the reading by one machine of a magnetic or paper tape written or punched by another, and the running of programs originally tested on one machine, on another. The problems here are numerous, annoying, unnecessary and almost all may be overcome, though sometimes at great expense. At the outset it was realized that machine compatibility for all museums and other workers in this field could not be dictated by a professional body. There are over 200 computers in use in British local government—

which administers many of the Country's museums—to say nothing of those used by national government and in the universities.

System compatibility amounts to the ability of one museum using a particular machine and suite of programs to read and process files produced by another museum using a different machine and programs. This is seldom possible, but if both systems are well designed special programs can be written to translate a file from one storage format to another. This may be expensive and usually some direct editing is necessary which adds to the cost. If several systems wish to exchange data then special programs may be needed for every pair of systems and the cost becomes prohibitive.

Data compatibility necessitates the use of words with the same meanings and using the same conventions in recording information. It is necessary to distinguish clearly between compatibility of formats, which is often trivial, and real compatibility of data, which is not. Data compatibility includes the problems of vocabulary control, standardized indexing languages and consequently all the differences of terminology and classification both within and between disciplines which can be overcome only by strong incentives.

Of course this is not a new problem but people are much better than machines at recognising and allowing for such incompatibilities. Attempting to process data by machine brings out these difficulties rather clearly. In the long run true compatibility of data will only be achieved if the benefits to the producer of museum data give sufficient incentives to encourage the adoption of general standards. In the meantime the effects of data incompatibility are minimized if the conventions used in cataloguing are recorded along with the data.

The IRGMA proposal is designed to achieve system compatibility which will provide, we feel, a valuable incentive to data compatibility. At first sight the easiest way to achieve compatibility of systems might be through agreement on the storage format to be used for museum data, but there are good reasons why this is not possible. For example, the standards of representation of data vary from machine to machine. Also, for efficiency, the storage format must be chosen in each case to match the characteristics of the actual computer and the data to be processed. A format which can handle all museum data on any machine would be inefficient. In addition, development of new computing techniques will soon make obsolescent any particular storage format and computer system.

COMMUNICATION FORMATS

The solution consists in providing a satisfactory method of communicating between systems by machine; that is a method for translating one storage format into another. Since this operation will seldom be performed in comparison with operations on data held in a storage format, it can afford to be relatively inefficient.

As it would be wasteful to have separate translation programs between every pair of storage formats, a single communication format has been devised which can take all museum data without exception. Each system wishing to send and receive data must possess programs capable of translating those data in which it is interested from the communication format into its own storage format and *vice versa*. Compatibility is potentially achieved once a Communication Format is agreed and the test of the compatibility of any particular system is the existence of the necessary programs for translation between its storage format and this communication format.

A communication format is not dependent upon a particular machine or machine code. It is, in a sense, a layout for an abstract document. Since data in such a format can be viewed as a document, we must consider using ordinary language for communication and merely encoding letter by letter existing catalogues, labels and indexes. This has attractions but there are serious obstacles. Language conveys information through the combination of syntactical structures and the knowledge and experience of the reader. Much is conveyed implicitly rather than explicitly and the process of understanding is difficult to reproduce on the machine. Since the major advantage of machine processing is that the date can be manipulated, it is essential that the machine can recognise and isolate from the text the data elements it has to process.

Much can be achieved at the cost of severe restrictions on vocabulary, grammar and syntax. How great a price must be paid will be known to anyone with even a slight knowledge of computer programming languages. Techniques for the analysis of language by machine are improving and the severity of the restrictions will decrease in the future. However at present museum data must be analysed so that implicit information can be expressed explicitly in the communication format and so ensure that the machine can recognise the elements in the data.

It should be clear from our emphasis on the analysis and manipulation of data that we are concerned with data systems rather than indexing systems. That is, all information should be in a fully analysed form allowing independent access to each data element. A typical indexing system, for example the computer operated MEDLARS medical indexing service, abstracts pertinent data from the body of the source documents as a first stage of retrieval. The source document however, is unaltered, is not held in machine readable form, and is reached only at the second stage of a search. Indexing systems are particularly appropriate to a library situation where source documents are long and originate independently from the library systems which store and retrieve them. In Museums and in other institutions which originate files of specimen-based data, there is potential control over the recording of data and the volume of data per item is small in comparison with library materials. The interdisciplinary nature of museum material and the fact that much of it is primary evidence in the various disciplines increased our belief that a data system was required.

You will note that the design of the MCF was required to be interdisciplinary. In other words it had to accommodate museum data whether from the biological sciences, geology, fine arts, archaeology, military history and so on. An additional requirement was that it could be used to allow curators and other research workers to record different selections of data about a similar subject. For example, a study of the decorative motifs on a class of artifacts may take little account of material composition, geographical occurrence, etc.; another's work may provide quite different information about the item. In other words the MCF had to be structured to take all types of information and impose no restrictions on researcher or recorder. To compel the recording of specific categories of information would limit its use and indeed limit its usefulness: the compulsion would cause either a rejection of the system or else demand information which the compiler may not be competent to provide.

METHODS OF STRUCTURING DATA

It was necessary, therefore, to find some principle by which we could divide and sub-divide the data, which was compatible with any subject in the sciences and the humanities, which placed no restrictions on the information to be recorded, and which yielded an intelligible, manageable format.

An obvious principle is subject classification. This, however, means dividing the format by disciplines with a general section for specimen number, location in museum, etc. For each subject area it would be necessary to repeat a large number of specialist headings (e.g. find-spot material composition, date) and so on. This would result in a cumbrous, inelegant format with all the data elements required in each subject area included.

In library and similar cataloguing, where the item has a relatively homogeneous physical form, data about the item (as opposed to its contents) can be treated in this way. However, as we know, subject classifications require bulky cataloguing rules to deal with the exceptions amongst their material. Any attempt at subject classification for museum material, therefore, would be likely to result in volumes of, cataloguing rules.

Also a subject classification, in the context of museum material, fails because the principle conflicts with the idea of a museum data system in which data elements are individually accessible. Such a classification also hinders access because it contains implicit as well as explicit information both in the data elements and the cataloguing rules. It would also fossilize categories of information, preventing easy amendment of the structure as knowledge changes. Instead we need an approach which makes all the information explicit and, therefore, manipulable: a flexible, open-ended data system which permits economical interdisciplinary processing.

Earlier in the paper the use of natural language was rejected because not enough is known about its structure to manipulate it effectively. However, as we are exercising a degree of control over the origin of the data, could we not employ a non-natural language with an artificial structure which permits information so written to be manipulated successfully? The vocabulary could be in English or any other language but the grammar of the language would be something quite different. The Museum Communication Format incorporates such an artificial grammar.

DESCRIPTION OF THE MUSEUM COMMUNICATION FORMAT

Before describing the structure of MCF, it is necessary to introduce some terminology. A collection of information for communication forms a *file* which may be thought of as a *box* into which the information is placed. The file is divided into a number of smaller boxes, each of which contains one *record*; a record might consist of information from one entry in a catalogue or refer to a single specimen.

A box containing a record, however, is an over simplification for the record may be made up of a series of inter-related but quite distinct pieces of information. In order to gain access to these individually, the *record box* itself is divided into a further series of boxes one for each kind of information or *data element*. Thus a record box could contain boxes for the date elements specimen number, identification, date of manufacture, date of collection and so on. However, if the box for date of collection is taken as an example, it will be noted that it can be further sub-divided into day, month and year. In this way the structure is built up, each sub-division forming a *level* within the record. Boxes such as date of manufacture and date of collection need the same structure to record similar information although in a different context; therefore, each kind of information recurring within the structure, e.g. year, is known as a *common element.*

The conventions for separating and labelling the individual boxes in a record will be common to many communication formats and will not be chosen by the user. The technique will probably involve the use of special characters or tags to break the document up into records and subdivisions of each record. At the same time the data in each record will be preceded by a directory listing those boxes actually present and giving the position in the record at which they may be found. This device is used to speed machine processing when the data is complex.

One advantage of this system is that it allows great flexibility for variable data Thus, in a specimen record which includes two separate identifications, the identification box will occur twice. If an identification was made jointly by two people, then the identifier box will occur twice within the identification box This idea is referred to as a *multiple entry.*

The design of the MCF, therefore, consists of an arrangement of boxes adequate to receive any data which museums might wish to communicate. It consists of:

(a) the main data structure;
(b) a number of boxes to hold common elements; and
(c) rules to assemble together the main structure and common element boxes.

The main structure divides initially into: a key by which the item record is identified; statements by which the item is described, and administrative data which is concerned with the management of the item and record. For the purpose of this paper, however, we shall be concerned only with the statements (Fig. 1).

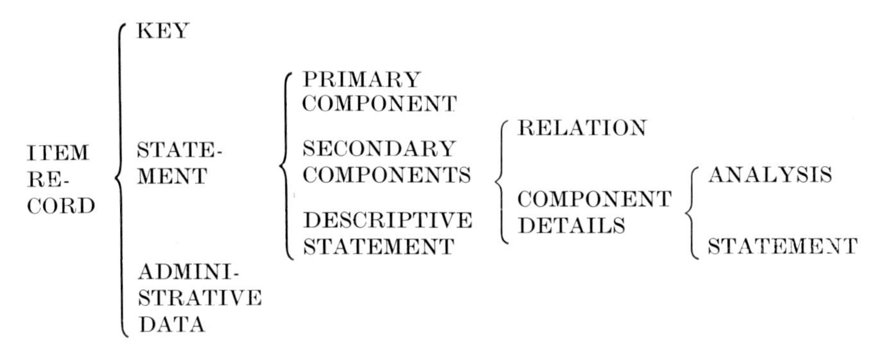

Fig. 1

The entry in the statement box may be in the form of free text as a descriptive statement or further broken down into one primary component with any number of secondary components. The primary component box contains an identification of the item which the statement is about. Thus a statute made of fossiliferous limestone could appear as a work of art in one statement, a type of rock in another and a fossil in a third. The secondary components break down as shown in Fig. 1. There is here, however, a vital rule to be followed:

> In any one entry under Component Details, the Component Analysis box must be converted to take any one of the following common elements: class identification, part, physical form, time/date, place, person, reference, record key, object or dimension, part.

COMMON ELEMENTS

It is these common elements that form the bulk of the data. Each of the boxes holding them has a structure several levels deep and although the use of any particular box is optional, the structure itself is fixed. For example, the structure for Class Identification (Fig. 2) divides first to define the classication being used and then provides the identification in terms of the classification used: this may be simply the class name or number, but can also be supported by a fuller description of the class. In using the term classification and class, this should not be confused with analytical classifications with a formal notation such as Dewey or UDC. The Special Systems category is to accommodate such things as taxonomic names the hierarchical structure of which is incorporated directly into the data structure.

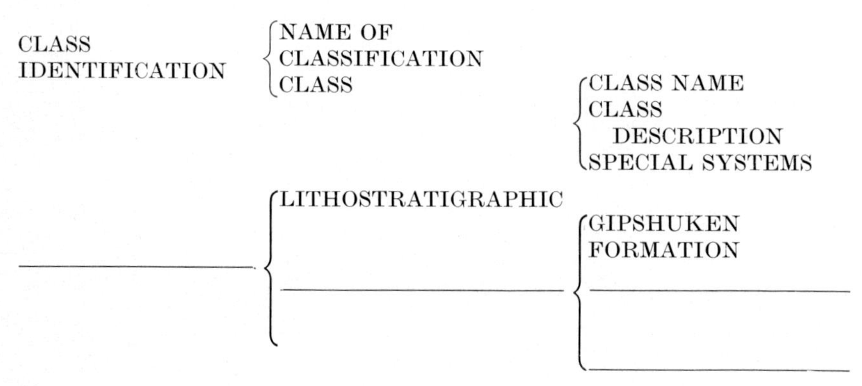

Fig. 2

The example (Fig. 2) shows as the Name of Classification the term lithostratigraphic, this referring to a body of reasonably well organized and widely recognized terminology. Where an accepted classification is not available, it would be necessary to refer to common usage and use uncontrolled keywords from ordinary English; this, however, lowers retrieval efficiency.

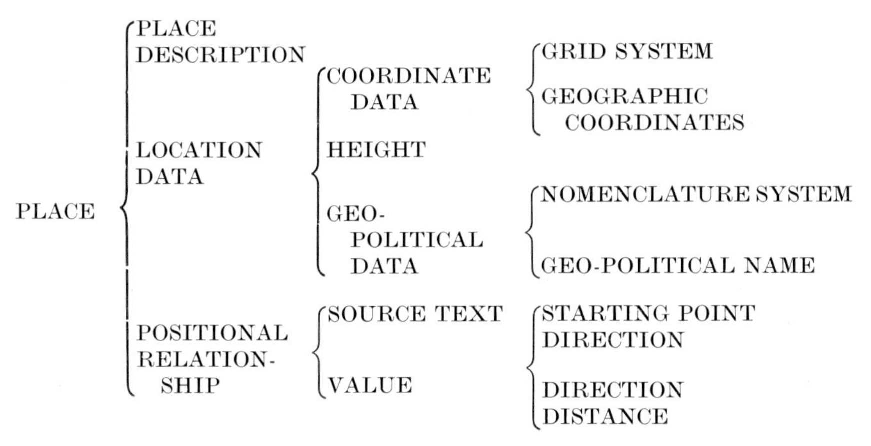

Fig. 3

Two other examples, one showing the analysis of Place (Fig. 3) and the other of Dimension (Fig. 4) show further the basic approach. In this way the boxes can hold a great range of highly structured data.

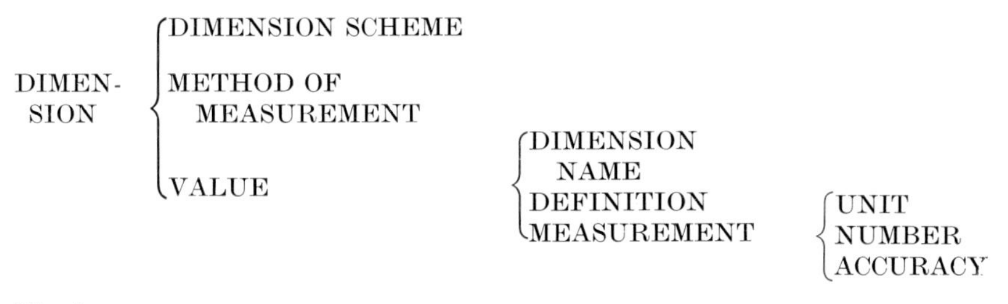

Fig. 4

COMPOUND STATEMENTS

So far simple statements have been described with the data structure. To give another one, this time from an archaeological context, we might say: here is a helmet (primary component); Thomas Bateman found it at

Benty Grange, near Monyash, Derbyshire (secondary components). However, museum data is notoriously far from simple; one statement will give rise to a chain of further statements. In order to analyse a complex statement, recursion has been introduced into the structure in such a way that a subordinate statement can be associated with any one of the secondary components of a higher statement.

STATEMENT NUMBER	PRIMARY COMPONENT	SECONDARY	COMPONENTS	DEGREE OF RECURSION
I	PICTURE A	PATRON P	COMMISSION ARTIST X	Zero
II	ARTIST X	DRAWING	PICTURE B	1st
III	PATRON P	COMMISSION	ARTIST Y	1st
IV	ARTIST Y	PLACE K	PERIOD L WORKING	2nd

RELATIONSHIPS OF STATEMENTS AS EXPRESSED IN STRUCTURE:

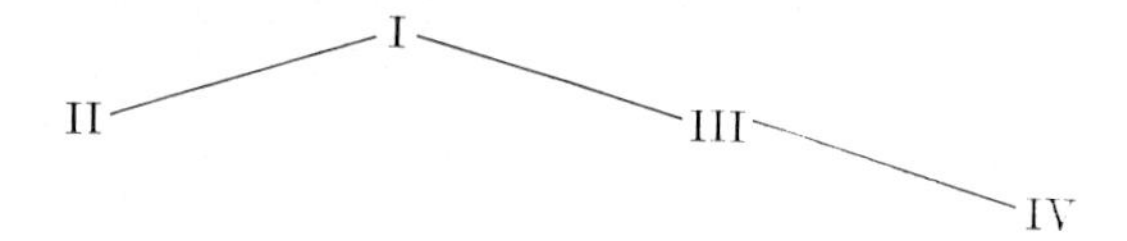

Fig. 5

Figure 5 is an analysis of the statement that picture A was commissioned by patron P from artist X who also drew picture B; another commission was given by patron P to artist Y who during period L worked at place K. Here there are a primary statement, two independent secondary statements and a tertiary statement related to one of these. The structure ensures that the data is unambiguous by placing picture B in the same statement as artist X who executed it and in a different statement from artist Y who had nothing to do with it. Thus, when information is retrieved from the structure, combinations of vocabulary are valid only when found within a statement and its, associated component. By preventing false combinations of vocabulary, the structure performs the same task as links in coordinate indexing systems. These are annotations added to index terms to show valid combinations (Fig. 6).

SMALL_1 / $\text{MUSEUM}_{1,2}$ / DISPLAY_2 / TECHNIQUE_2

TO BE READ AS: "DISPLAY TECHNIQUES FOR SMALL MUSEUMS"

NOT AS "TECHNIQUES FOR SMALL DISPLAYS IN MUSEUMS"

Fig. 6

However, links do not overcome all ambiguities in co-ordination indexing. Consider the statement: James is the father of Bill and Harry. This could be written:

JAMES/FATHER/BILL/HARRY

To find the father it is necessary to assume that it is written first; this, of course, is the first step towards voluminous cataloguing rules. To overcome this type of ambiguity, the idea of relation has been introduced into the structure (Fig. 1). It is possible to associate with any secondary component its relationship to the primary component or any other secondary component in a statement. Since the relationship takes the form of an identification any system, including ordinary language, may be used. However, a formal system of logical operations is likely to be more effective. For instance Derek Austin (1969), working in the field of bibliography, suggests the use of number prefixes with logical meanings to show inter-relationships. Thus we could relate the secondary component FATHER to the primary component JAMES by relation (5) and the secondary components BILL and HARRY to the secondary component FATHER by relation (4). Here (5) means "effect produced on related component" and (4) "that which produces the effect". Or in less formal terms—"James is a father and the existence of Bill and Harry is why he is a father!"

USE OF PART BOX

A statement may apply to only part of the primary component by specifying the part as a secondary component. A part may itself be subdivided and the subparts described by using subordinate statements attached to the appropriate part box. This provides a powerful technique for organising descriptions of things.

SET LISTS

The format described above is adequate in theory but one additional device may be used to make the communication records easier to use. All the common elements may be collected together in lists, one for each kind, at one place in the record and a pointer to the position in these lists substituted at the original place. Such Set Lists economise on space where there is a lot of duplication in a complex record, but more importantly make it easier to find particular kinds of data in the record without having to follow through the structure used by the original cataloguer.

USE OF THE COMMUNICATION FORMAT

Now we must turn to the use of the structure itself which, as we have seen, is intended to promote communication. The MCF can be used for holding data to be communicated between two otherwise incompatible systems. It can also be used as a tool for designing local input and processing formats compatible with it. As a design tool it ensures that those designing local formats analyse their data to the same specification, although it is necessary to use only that part of the structure applicable to the local requirements.

The crucial test of the compatibility of such a design with the MCF is to devise a translation procedure by which data in the local format is expanded and re-structured into the MCF. Fig. 7 shows a small part of the local format used by an art museum:

LOCAL BOX NAME	EXAMPLE OF DATA
SPECIMEN NUMBER	P.12–1966
ARTIST'S SURNAME	ROBERTSON
ARTIST'S FORENAMES	HENRY ROBERT
ARTIST'S DATES	1839–1921
TITLE	AMBER BEADS
SUBJECT	YOUNG GIRL

Fig. 7

The translation procedure would first insert data implicit in the local box names and in their supporting cataloguing rules. The result is shown in Fig. 8. You will notice the amount of implicit data that has been inserted.

MCF BOX NAME	IMPLICIT DATA	EXPLICIT DATA
ORGANISATION	VICTORIA AND ALBERT MUSEUM	
ITEM IDENTIFIER		P.12–1966
CLASS NAME (PRIMARY COMP.)	PAINTING (NOUN)	
CLASS NAME (SECONDARY COMP.)	MANUFACTURE	
TYPE OF PRIMARY NAME (PERSON, SECONDARY COMP.)	SURNAME	
PRIMARY NAME (PERSON, SECONDARY COMP.)		ROBERTSON
TYPE OF ELEMENT (PERSON, SECONDARY COMP.)	FORENAME	
SECONDARY ELEMENT (PERSON, SECONDARY COMP.)		HENRY ROBERT

Fig. 8

Both the implicit and explicit data is now structured into the MCF (Fig. 9).

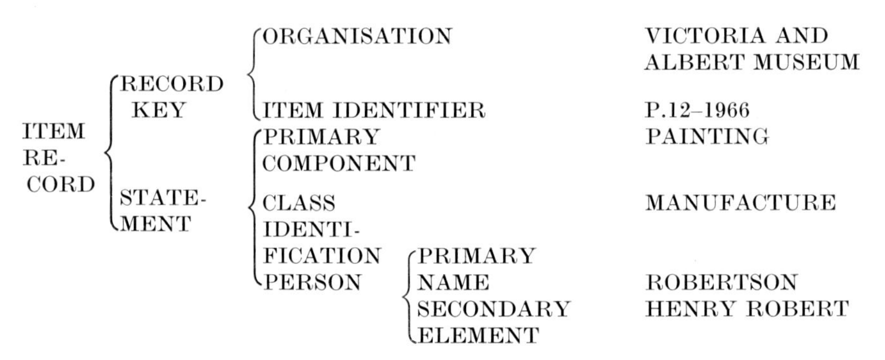

Fig. 9

When a museum receives a request for information, the first step is to translate the required data into the communication format. It would be uneconomic to use MCF to store the data; for example, it would be obvious from this particular entry that Robertson was an artist, but a request to use the file in a different context requires that all information must be equally accessible. Further, as the request may be made of a number of different data sources it is far more reliable and economic to the client for the implicit data to be supplied at source. In response to his request, then, the client receives all files in a single, standard format and consequently needs only one set of programs to process the data.

Successful manipulation of data in this form is not guaranteed solely because it is in the MCF. These records have originated from different sources with differing local formats and cataloguing rules. Two characteristics distinguish the different sources:

(a) their "minimum contents list", i.e. the data elements supplied for all records;
(b) the classification used to control the vocabulary of their data elements. (Vocabulary control typically consists of eliminating synonyms, distinguishing homonyms, defining terms and establishing generic relationships between terms).

Joint filing and retrieval on records from different sources will be hindered seriously if data elements are missing from such records or if the vocabulary varies considerably.

FUTURE WORK ON THE FORMAT

The MCF is not yet in a final working form and has been presented here in in the hope of stimulating discussion. Parts of the format cannot be altered. The ideas of using a structure of statements and subordinate statements and of using concepts such as people, place and identification, which are not subject based, as components in these statements are fundamental to the format. On the other hand the list of these common elements, and particularly the detailed structures necessary to specify them satisfactorily, could certainly be improved.

One final point, although devised for specimen information, the structure of the communication format can be used for any indexable item. In addition it is possible to interrogate the record, ignoring the item indexed, should the answer lie in a statement of the second or higher degrees of recursion: the item indexed simply provides a starting point for a series of related statements. The aim of MCF is both simple and ambitious: it is intended to manipulate data by algorithm, in contexts other than that in which the data originated and this has applications far beyond museums.

DETAILED STRUCTURE AND RULES OF THE MUSEUM COMMUNICATION FORMAT

Notes:

1. All boxes are given an identifying label consisting of a unique number preceded with a letter prefix: boxes in the main structure are indicated by the letter S (e.g. S101) and those containing common elements with the letter C (e.g. C200).
   NB. The identifying label has no classificatory or structural significance.
2. Some of the boxes have an identical structure to receive similar information, e.g. TIME/DATE—C50. Where this occurs the label of the other identical boxes is placed in brackets after its identifying label, e.g. TRANSFORMATION DATE—C324 (C50).
3. A basic box, that is one which cannot be further sub-divided, is marked with a letter b, e.g. NAME OF CLASSIFICATION—C2b.
4. The structural level of each box is indicated on the left-hand side of the list by a numeral for the main structure and by letters for those containing common elements. *NB.* Structural levels should not be confused with levels of recursion.
5. Set lists are a means of cross-referencing to economise on data storage.

MAIN STRUCTURE

Level	*Box name*	*Box number*
0	ITEM RECORD	S0
1	RECORD KEY	S1 (C53)
1	STATEMENT	S101 (C200)
1	ADMINISTRATIVE DATA (under discussion)	S2
1	SET LISTS	S3
2	PART LIST	S15
3	PART (SET LIST)	S180 (C7)
3	REFERENCE TO SOURCE (PART)	S111 (C11)

(and likewise for Person, Event, Reference, Place and Time lists)

S3—SET LISTS

Level	*Box name*	*Box number*
1	SET LISTS	S3
2	PART LIST	S15
2	PERSON LIST	S620
2	EVENT LIST	S36
2	REFERENCE LIST	S37
2	PLACE LIST	S621
2	TIME LIST	S622

The substructure of each list is the same, for example:

Level	*Box name*	*Box number*
2	PERSON LIST	S620
3	PERSON (SET LIST)	S183 (C52)
3	REFERENCE TO SOURCE (PERSON)	S184 (C11)

C200—STATEMENT

Level	*Box name*	*Box number*
A	STATEMENT	C200
B	PRIMARY COMPONENT	C201 (C1)
B	SECONDARY COMPONENT	C203
C	RELATION	C224
D	RELATED COMPONENT	C225
D	RELATIONSHIP	C226 (C1)
C	COMPONENT DETAILS	C227
D	COMPONENT ANALYSIS	C205 (see Rule 3)
D	STATEMENT	C200
B	DESCRIPTIVE STATEMENT	C204b

Rules

1. In any one entry under STATEMENT—C200 there must be either:
   (a) At least one entry under DESCRIPTIVE STATEMENT—C204b, or
   (b) one entry under PRIMARY COMPONENT—C201 (C1) and at least one entry under SECONDARY COMPONENT—C203.
2. In any one entry under SECONDARY COMPONENT—C203, there must be at least one entry under COMPONENT DETAILS—C227 in which there must be an entry under COMPONENT ANALYSIS—C205.
3. In any *one* entry under COMPONENT DETAILS—C227, COMPONENT ANALYSIS—C205 must be replaced by one of the following:

C1 —CLASS IDENTIFICATION
C300—PHYSICAL FORM
C51 —PLACE
C57 —REFERENCES
C97 —DIMENSION
C7 —PART
C50 —TIME/DATE
C52 —PERSON
C53 —RECORD KEY
C12 —OBJECT

C1—CLASS IDENTIFICATION

Level	*Box name*	*Box number*
A	CLASS IDENTIFICATION	C1
B	NAME OF CLASSIFICATION	C2b
B	CLASS	C3
C	CLASS NAME	C4b
C	CLASS DESCRIPTION	C5b
C	SPECIAL SYSTEMS	C6 (C145, etc.)

C145—TAXONOMIC NAME (special extension of C1—Class Identification)

Level	*Box name*	*Box number*
C	TAXONOMIC NAME	C145
D	NOMENCLATURE RULES	C146b
D	QUALIFIER	C147b
D	SUBNAME	C148
E	TAXONOMIC LEVEL	C149b
E	LATIN NAME	C150b
E	ORIGINAL AUTHORITY	C151
F	ORIGINAL AUTHOR	C154b
F	YEAR OF AUTHORITY	C155b
E	SENSE OTHER THAN ORIGINAL	C152
F	AMENDING AUTHOR	C156b
F	YEAR OF AMENDMENT	C157b
E	SUBNAME QUALIFIER	C153b

C7—PART

Level	*Box name*	*Box number*
A	PART	C7
B	PART IDENTIFICATION	C104 (C1)
B	PART POSITION	C105
C	POSITION IDENTIFICATION	C260 (C1)
C	RELATIVE PART	C7
B	PART COUNT	C263b
B	PART KEY	C262b
B	PART STATUS	C108b

C300—PHYSICAL FORM

Level	*Box name*	*Box number*
A	PHYSICAL FORM	C300
B	PHYSICAL DESCRIPTION	C301b
B	PHYSICAL DETAILS	C302
C	COLOUR	C303 (C1)
C	MATERIAL	C304
D	MATERIAL TYPE	C305 (C1)
D	CHEMICAL ANALYSIS	C306 (C97)
C	SHAPE	C337 (C1)
C	PHYSICAL DIMENSIONS	C307 (C97)
C	SURFACE TEXTURE	C308 (C1)
C	OPERATING SYSTEM	C309 (C1)
C	SOUND	C310
D	SOUND CATEGORY	C311 (C1)
D	VOLUME	C312
E	SUBJECTIVE VOLUME	C313 (C1)
E	NUMERIC VALUE (VOLUME)	C314 (C97)
D	TONE	C315 (C1)
C	SMELL	C316 (C1)
B	INSCRIPTION	C317
C	INSCRIPTION PRESENCE	C318b
C	INSCRIPTION TEXT AND ORIGIN	C503
D	INSCRIPTION TEXT	C190b
D	TEXT STATUS	C319b
D	SOURCE MATERIAL	C320 (C57)
D	TRANSFORMATION PROCESS	C321 (C1)
D	TRANSFORMER	C323 (C52)
D	TRANSFORMATION DATE	C324 (C50)
C	LANGUAGE	C326b
C	ALPHABET	C328b
C	LETTERFACE	C329b
C	LAYOUT	C330b
B	MARK	C331
C	MARK PRESENCE	C332b
C	MARK DESCRIPTION	C334b
C	MARK MEANING	C335b
C	MARK IDENTIFICATION	C336 (C1)

C50—TIME/DATE

Level	*Box name*	*Box number*
A	TIME/DATE	C50
B	QUALIFIER (TIME/DATE)	C121b
B	SOURCE TEXT (TIME/DATE)	C93b
B	TIME/DATE VALUE	C120
C	STANDARD DATE	C66
D	CALENDAR	C122b
D	ERA	C123b
D	DAY OF MONTH	C124b
D	MONTH	C125b
D	YEAR	C126b
D	DAY OF WEEK	C128b
D	ACCURACY (DATE)	C127 (C49)
C	STANDARD TIME	C130
D	TIME TYPE	C131b
D	TIME VALUE	C132
E	HOUR	C133b
E	MINUTES	C134b
E	SECONDS	C135b
E	ACCURACY (TIME)	C136 (C49)
E	PART OF DAY	C129b
C	CLASSIFIED TIME	C266 (C1)

C51—PLACE

Level	*Box name*	*Box number*
A	PLACE	C51
B	PLACE DESCRIPTION	C106b
B	LOCATION DATA	C95
C	CO-ORDINATE DATA	C77
D	CO-ORDINATE SYSTEM	C81b
D	GRID VALUES	C82
E	REGION	C84b
E	NUMERIC VALUE	C85 (C97)
D	GEOGRAPHICAL CO-ORDINATES	C83
E	ORIGIN FOR LONGITUDE	C72 (C97)
E	VALUE OF CO-ORDINATES	C73 (C97)
C	GEO-POLITICAL DATA	C78
D	NOMENCLATURE SYSTEM	C91b
D	GEO-POLITICAL NAME	C92b
C	POSTAL ADDRESS	C79b
C	HEIGHT ABOVE SEA LEVEL	C80 (C97)
B	POSITIONAL RELATIONSHIP	C74
C	SOURCE TEXT (POSITION)	C96b
C	VALUE OF RELATION	C94
D	STARTING POINT	C98b
D	DIRECTION	C100 (C97)
D	DISTANCE	C101 (C97)

C52—PERSON

Level	Box name	Box number
A	PERSON	C52
B	TYPE OF PRIMARY NAME	C118b
B	PRIMARY NAME	C119b
B	SECONDARY ELEMENTS	C140
C	TYPE OF ELEMENT	C141b
C	SECONDARY ELEMENT	C143b

C57—REFERENCES

Level	Box name	Box number
A	REFERENCES	C57
B	ITEM REFERENCE	C59 (C53)
B	BIBLIOGRAPHIC REFERENCE (sub-structure under discussion)	C60
B	PRIVATE REFERENCE	C71
C	TYPE OF SOURCE	C99 (C1)
C	PERSON (REFERENCE)	C8 (C52)
C	PLACE (REFERENCE)	C9 (C51)
C	TIME/DATE (REFERENCE)	C10 (C50)
C	ADDITIONAL INFORMATION	C67b
B	REFERENCE TYPES	C48b

C53—RECORD KEY

Level	Box name	Box number
A	RECORD KEY	C53
B	ORGANISATION	C87b
B	COLLECTION NAME	C88b
B	ITEM IDENTIFIER	C89b
B	ADMINISTRATIVE NAME	C90b

C12—OBJECT

Level	Box name	Box number
A	OBJECT	C12
B	OBJECT IDENTIFICATION	C13 (C1)
B	OBJECT REFERENCE	C14 (C57)

C97—DIMENSION

Level	Box name	Box number
A	DIMENSION	C97
B	DIMENSION SCHEME	C58b
B	METHOD OF MEASUREMENT	C68b
B	DIMENSION VALUE	C103
C	DIMENSION NAME	C109b
C	DIMENSION DEFINITION	C110b
C	MEASUREMENT	C111
D	MEASUREMENT UNIT	C112b
D	MEASUREMENT VALUE	C113b
D	MEASUREMENT ACCURACY	C114 (C49)

REFERENCES

Austin, D. (1969). Prospects for a new general classification, *J. Lib.* **1** (3), 149–169.

Lewis, G. D. (1965). Obtaining information from museum collections and thoughts on a national index, *Museums J.* **65** (1), 12–22.

Lewis, G. D. (1967). Information retrieval for museums, *Museums J.* **67** (2), 88–120.

Museums Association (1969). Draft proposals for an interdisciplinary cataloguing system. Museums Association, London.

18. Pilot Data Processing Systems for Floristic Information*

S. G. Shetler,[1] J. H. Beaman,[2] M. E. Hale,[1] L. E. Morse,[2]
J. J. Crockett[3] and R. A. Creighton[3]

Department of Botany, Smithsonian Institution, Washington, D.C.[1]*;*
Department of Botany and Plant Pathology, Michigan State University, East Lansing, Michigan[2]*;*
Information Systems Division, Smithsonian Institution, Washington, D.C., U.S.A.[3]

INTRODUCTION

As a tool for statistical and classificatory studies, the computer is by no means new to the discipline of plant taxonomy. In fact, since 1963 when the provocative book of Sokal and Sneath appeared and challenged openly the time-honored subjectivity of orthodox taxonomy by introducing computer methods for large-scale data analysis, a whole new sub-discipline, "numerical taxonomy", has developed, and already a vast literature on computer taxonomy in plants and animals has accumulated. Numerical taxonomy, which was a logical, inevitable consequence of the arrival of modern computers, has come of age and is here to stay; slowly but surely its methods are being incorporated into a new orthodoxy and are making contributions to standard taxonomic investigations (e.g. Johnson and Holm, 1968; Stearn, 1968).

Receiving scant attention from the computer taxonomists until quite recently, however, has been the staggering problem of information handling that the taxonomist, whatever his brand, faces daily in conducting his research and communicating his results. This oversight is

* The paper was presented at the Symposium by S. G. Shetler, Secretary, Flora North America (FNA) Program. The program, explained in the paper, is being administered by the American Institute of Biological Sciences (AIBS). M. E. Hale is a member of the AIBS–FNA Advisory Council. J. H. Beaman is a member of the AIBS–FNA Editorial Committee.

Systematics Association Special Volume No. 3. "Data Processing in Biology and Geology", edited by J. G. Hawkes, 1970, pp. 275–310.

the more surprising when one considers the stereotyped, descriptive nature of much of the information dealt with by taxonomists. Such information should be very amenable to machine handling.

The advent of third-generation computers and interactive time-sharing systems that permit remote access has brought us to the threshold of an exciting new era in information science and technology, and scientific communication surely will soon undergo a dramatic revolution. Fortunately, the number of taxonomists who foresee this revolution and some of its implications is growing (e.g. Sokal and Sneath, 1966; Crovello, 1967; Soper and Perring, 1967; Squires, 1966 and in this volume, p. 235). Some taxonomists clearly understand not only that the way in which we order our data affects our very perception of reality in nature but also that the computer will be a powerful new tool for ordering our data which could in fact lead to fundamental re-evaluations in science (Heywood, 1968).

The first information retrieval system to be developed specifically for plant taxonomists is the TAXIR (Taxonomic Information Retrieval) System of David J. Rogers and associates, University of Colorado, Boulder, U.S.A., which represents a significant achievement (descriptive materials distributed in 1968). A more comprehensive system is the Smithsonian Institution Information Retrieval (SIIR) System for Biological and Petrological Data (Creighton and King, 1969; Manning, 1969; Squires, 1970). These are systems with great potential for systematic biologists.

It is not possible to review all the pilot systems that have been developed thus far for the computer which potentially have application to floristic information, and we are confining ourselves to a report of systems developed until now for the Flora North America Program or in coordination with it.

FLORA NORTH AMERICA PROGRAM

In 1965, the American Society of Plant Taxonomists took steps that were to culminate in the organization a year later of a cooperative program of broad scope intended to result eventually in a Flora of North America comparable to the Flora of the U.S.S.R., completed in 1964, and the *Flora Europaea*, now about half completed (Shetler, 1966; Taylor, 1969). This effort has the official designation "Flora North America (FNA) Program". Initially, FNA was conceived in largely conventional terms, as a concise, diagnostic manual of the vascular plants of North America north of Mexico, to be completed through the cooperative labors of many taxonomists over a period of about 15 years. It soon became apparent, however, that to write a Flora in the 1970s and 1980s without using methods of electronic data processing (EDP) surely would prove at least in retrospect to be indefensible.

Presently, Flora North America is visualized as a comprehensive floristic information system. The rationale for this approach and the philosophical and methodological implications are discussed at length in the papers of Taylor (in press) and Shetler (1970) presented at the recent, 11th International Botanical Congress in Seattle, Washington, U.S.A., in the symposium, "North Temperate Floristics and the Flora North America Project". These papers constitute a necessary preface to the present paper for those who wish to know the theoretical basis for the development of FNA as an information system. The discussion that follows deals first with the preliminary general systems concept and then with the three pilot systems developed thus far: Type Specimen (TYPES) System, FNA Automated Bibliography, and Taxonomic Data Matrix System.

As already indicated, the FNA Program is by no means the first effort to develop computer systems for storing and retrieving floristic-taxonomic information; it does represent, however, the first attempt to undertake large-scale, cooperative planning, development, and implementation of information systems specifically designed to handle the full range of floristic and related botanical data. The information process is by nature a group process, governed by group conventions and dependent upon a common data base built up by group effort, and the introduction of automation does not change this fact. The widest possible cooperation is being sought, starting with North American plant taxonomists but extending ultimately to the world-wide taxonomic fraternity and botanical, even biological, community in general. At the limit, science is always international and interdisciplinary, and these truths must be kept in focus from the outset by the FNA Program in developing data-banking and retrieval methods. Because cooperation is so essential, we are quite conscious of the fact that the usefulness and success of a taxonomic information system may depend less, in the final analysis, on the sophistication of the computer systems designing and programming than on the scope and depth of the data base or the scale of user participation. Simple systems can provide large payoffs if implemented on a comprehensive scale. Accordingly, we make no special claims of sophistication or novelty in our pilot systems, but openly state that we are concentrating on adapting existing computer science and technology to the peculiar needs of taxonomic botanists and on gaining the acceptance of the botanists in the process.

Concerning the FNA goal of bringing floristics, if not plant taxonomy in general, into the electronic age, Shetler (1968, p. 177) has written as follows:

"Floras and systematic monographs are time-honored, if non-automated, storage and retrieval systems, which through the years the taxonomist has produced as both the means and the ends of his research. In the FNA Project we propose to computerize such systems and thereby to update and extend the concept of the flora or monograph to cover not only the traditional printed book but also today's dynamic electronic data bank. Output from a data bank can take many forms and present many different kinds of data. FNA output, in addition to the basic identification manual, may include distribution maps, illustrations, glossaries, indices, gazetteers, descriptions, ecological summaries, statistical calculations, phenetic diagrams, and research reports. Thus, FNA will not be a single publication, embalming the state of knowledge. Instant update, revision, and cumulation will be cardinal attributes of the automated FNA information system."

The development of an information system is at best a calculated risk. It is costly and there is little basis for predicting in advance who the real users will be or what their needs and demands will be because automation makes possible a whole new user pattern. For these reasons, FNA development is proceeding very cautiously. A two-year planning study, supported by the National Science Foundation of the United States, was initiated in July 1969. In the course of this study, the whole realm of floristic/taxonomic information handling and communication will be investigated thoroughly in terms of modern information systems concepts, and appropriate modules of the conventional information process will be selected for automation where potential user demand and support seem to warrant this. The planning and architectural designing will be done by a competent systems development manager working in close liaison with the AIBS–FNA Editorial Committee and other botanical advisers. Meanwhile, the Editorial Committee has been attempting to gain some "hands on" experience and to test user reaction with pilot systems. During the past three years the Smithsonian Institution has supported virtually the entire activities of the Editorial Committee, including the development of the type specimen and bibliographic systems. The data matrix work has been financed largely by Michigan State University during the same period.

CONCEPT OF FNA INFORMATION SYSTEM

Like every large, multi-purpose information system, the FNA system surely will comprise many smaller systems or subsystems, which together will involve many computer programs and data files. This point must be emphasized because it is important not to think of a "computerized" Flora North America in oversimplified terms, for example, as a single data system or data file or as merely a magnetic tape image of a published book.

The pilot systems developed thus far are mere suggestions of how the computer can be put to work in the realm of floristics under the organizing theme of FNA; there is no guarantee that any of them will survive the more intensive planning study now underway, to become fully operational elements of the permanent, total system. By the same token, we are confident that at least in basic concept and purpose these pilot systems point the way to permanent systems for data storage and retrieval.

In Fig. 1, the proposed Flora North America Information System is schematically depicted. Let us examine this diagram according to the following points:

1. Data banking v. data processing

An information system is both a data bank and a data process. The bank is the composite of meaningfully structured files of reliable, useful data required to support the information function(s) of the system. The process is the composite of manual and electronic procedures required to "massage" the data and to deliver to and from the bank on command so as to perform the information function(s) of the system. The data may be stored as isolated data, but they must be retrievable as organized information, the organization depending on the demands of the user and not being predetermined entirely by the notions of the original author.

In Fig. 1, the data bank forms the core of the FNA Information System, and the data process is represented by the several boxes that form a shell around the bank and collectively identify major aspects of input, output, and processing ("thruput"). The important point to make here is that the FNA Program is concerned not only with the content and form of the final *product*, namely, the concise, diagnostic Flora. It is equally concerned with the *process* of preparing and communicating floristic information, and every possible step will be taken to improve and modernize the process as well as the product.

2. Sources of data

In the broadest terms, one can say that the data for a Flora can be derived from only three sources, as indicated in the figure: (a) herbarium specimens (or, in more general terms, botanical voucher specimens); (b) literature; (c) field and laboratory (unpublished) experiments and observations.

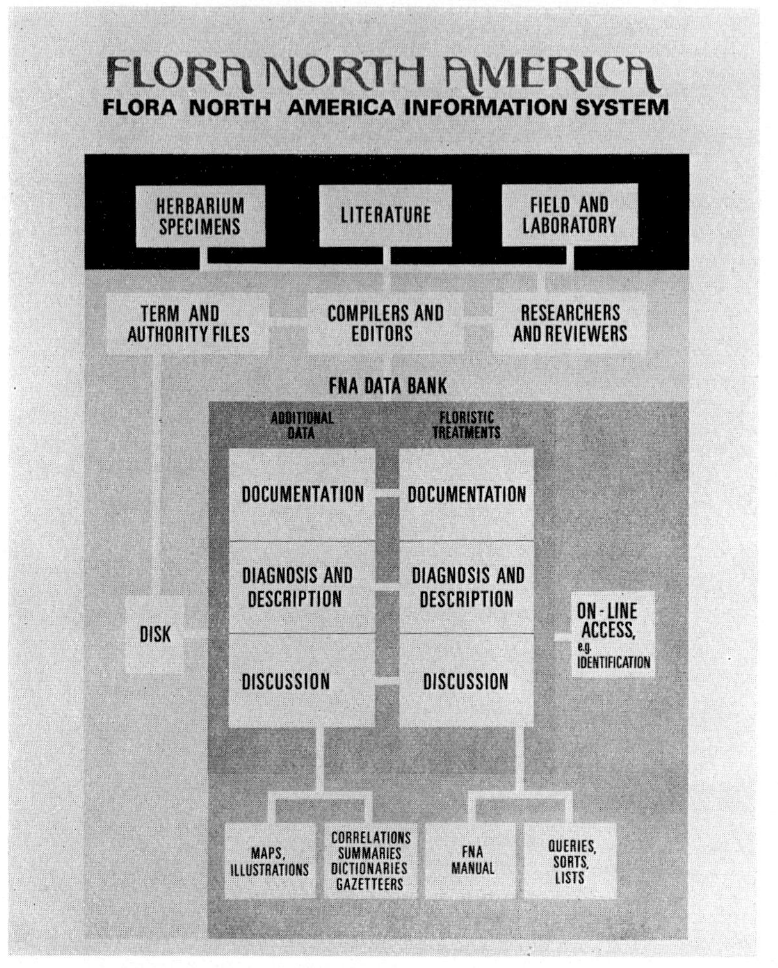

Fig. 1. Schematic representation of proposed Flora North America Information System (from Shetler, 1970).

Floristics, like any science, begins with raw data and through a process of distillation, organization, and corroboration results in authoritative documents, which provide information that is necessary or useful to the scientist or layman. The document may be a published Flora, a voucher specimen, or a field or laboratory note or notebook. Some documents rest only on the authority of their author or collector, while others are cross-referenced to other documents and thereby gain additional authority, although not necessarily greater reliability, a fact that is especially bothersome in descriptive science where authority and reliability tend to be equated by the user unconsciously. Likewise, the concept of "raw data" is as relative as the concept of "documentation". Raw data are unorganized data, but data organized for one purpose may at the same time be totally unorganized for another purpose. Thus one man's finished science becomes another man's raw data.

We emphasize the relativity of the concepts of raw data and documentation because, on the face of it, the three sources of data depicted at the top of Fig. 1 do not seem to be comparable. Ordinarily publication is regarded as the end point of scientific work. The literature, therefore, becomes the ultimate source of organized knowledge, not of raw data. To be sure, the printed page usually represents several to many levels of abstraction, organization, and corroboration beyond the basic level of voucher specimens and field and laboratory notes. The real universe of raw data, however, is nature itself, in all of its temporal and spatial manifestations. Measured against this vast data universe or against the whole range of questions that confront the botanist, a collection of voucher specimens or a set of field notes may constitute just as valid and useful a subset of organized raw data as a published paper or monograph, and in some contexts the specimen or the notebook entry may be an even more authoritative document than the printed page.

In developing the FNA Information System, we are trying to avoid rigid, artificial boundaries between published and unpublished data. Insofar as possible, we hope to build data systems that will enable the user of the future to tap easily the full range of available data, regardless of source. We are trying to take cognizance of the iterative or cyclic nature of the scientific process, wherein there is no such thing as an ultimate, truly perfect, master data file. The file is always a working file, which continuously is open for new cycles of revision and update, no matter how "final" it may seem to be at any given moment. Feedback between the organized and unorganized data levels is or should be continuous, and presumably each new generation of the data bank represents a somewhat more refined and reliable body of scientific knowledge than the last. In one sense, of course, all data, whatever the level of organization or form of storage, reside in a gigantic data bank. The point of distinguishing between the sources of the data and the data bank itself in Fig. 1, therefore, is to emphasize the different stages of data organization and refinement, not necessarily to imply some kind of physical difference or separation in data storage. Thus the diagram is intended to convey the functional rather than physical attributes of the FNA Information System.

It is true that initially the concepts of "sources of data" and "data bank" are being used to contrast the present, non-automated forms of data storage with the proposed, automated forms of data storage. These concepts should still be viable, however, when the day arrives that most scientific information is stored and communicated electronically. Even now, therefore, it is useful to take the completely general and purely functional point of view that the data source and the data bank are one and the same thing, because source data are merely data that have been withdrawn from the bank temporarily for another cycle of organization and refinement and will be returned in some form to the same bank, which after all is nothing more than the cumulative wisdom that science has succeeded thus far in history to capture and put to use.

3. Manpower

Data evaluation and organization is in large part an intellectual process that requires the efforts of scientists and technical specialists no less in the machine environment than in the conventional environment. The use of machines probably will not diminish the labors of the generation of taxonomists who will prepare the first "edition" of the FNA Data Bank. In fact, the development of the FNA Information System, including all necessary standardization and documentation, doubtless will prove to be more tedious and demanding than any floristic work ever undertaken before. Eventually, however, computer methods should reduce some of the routine tasks in taxonomic research to a technology that can be handled by technical specialists. We stress, therefore, that the FNA system is being developed with the needs of the future in mind, and we harbor no illusions about the computer easing the burden of the initial effort, although certainly we expect that this will happen in many individual situations.

The introduction of information retrieval systems inevitably raises questions about the sacred rights of authorship and publication, and clearly certain problems lie ahead for FNA in this area. There seems little doubt that the forms of publication will change fundamentally during the course of the FNA Program because of rapid progress in information science and technology generally. If the FNA Program succeeds in developing automatic or semi-automatic systems for compiling and integrating certain kinds of floristic information, then some of the onerous compiling tasks that ordinarily fall to the specialist can be turned over to a technical editorial staff. This is especially likely in the realm of bibliographic documentation. Thus we anticipate that in the FNA Program the editors and associated staff (i.e. the "editorial mechanism") will play a quite active part in the compilation, screening, and organization of data for the FNA data bank. Accordingly, Fig. 1 shows that data for the bank may enter the research-edit cycle either by way of the researchers (taxonomic specialists) or by way of the editors. In either case, however, the data will complete at least one full cycle before going into the bank; the question becomes only a matter of where data for a taxonomic treatment first break into the research-edit cycle.

4. Standardization

The concept of a *standard* data bank of basic facts about North American plants, from which all future "cutting and pasting" could be done, implies term or vocabulary standardization. This is essential for efficient storage and effective searching for retrieval by machine; it also is essential from the standpoint of the authors and editors who, if they are to cooperate in building a common data bank, must have strict guidelines regarding the use of names and terms. An early phase of the FNA Program, therefore, will be devoted to the preparation of standardized term and authority files. These vocabulary thesauri must be open-ended to allow for growth and revision. To the fullest extent possible, such thesauri will be incorporated into the electronic processing system so that the computer can do some of the routine editing and auditing of data input and output. In Fig. 1, this feature is illustrated by the box marked DISK, which represents a random access memory device onto which the term and authority files can be loaded.

5. Nature of data bank

The generalized representation of the FNA DATA BANK in Fig. 1 symbolizes the totality of files and matrices that ultimately will constitute the computer-stored record. The diagram makes two main points: (a) all data that might be entered into the bank can be classified into one of three categories—(i) documentation, (ii) diagnosis and description (including identification keys), and (iii) discussion (i.e. interpretation); (b) the strictly floristic data, from which the concise manual of North American plants will be generated, will constitute only part of the bank. Eventually, the bank should include a wide range of other data, e.g. ecological, biochemical, and physiological. In theory, the bank should be structured to accommodate n characters and n character states for every taxon or operational taxonomic unit (OTU) included.

What a taxonomist publishes often represents a mere abstract of the full amount of data he has collected and organized during the course of his research. The publication may in fact present a very artificial, biassed record of his real scholarship. This is especially true in floristics because the full supporting documentation, often worked out with great care, seldom can be included in the final Flora. Instead, a Flora tends to present only the end-points of the scholarly research process, i.e. the taxonomic decisions, and the data included are calculated more to define than to support the decisions. Data abstraction in a typical synoptical Flora reaches the level of common-knowledge universals (e.g. "leaves opposite"), which can hardly be claimed under anyone's private authorship. The researcher-author's "copyright", as it were, really covers only the taxonomic decisions, if in fact new ones have been made. The data themselves often have been in the public domain for decades if not centuries. The substance of the scholarship behind a Flora usually falls by the way in some notebook or "shoebox" card file, and the truth is that many potential users of the Flora may have far greater need for this lost scholarship than for what has been preserved in the manual in time-honored form.

Future data banking should permit the capture of the full record of scholarship whenever desirable. At the same time, data banking will tend to discourage certain traditional types of publication. The local floristic list, for example, tends to contain an artificial subset of records that could serve taxonomic needs far better if it were added directly to the gathering universe of records in the data bank without prior publication. Then the user of the bank could request a sub-universe of records according to his own parameters, and he would not have to extract his information tediously from numerous separate documents published according to the parameters of others.

A primary concern in building the FNA Data Bank will be data quality and authority. The editorial mechanism will screen input critically so as to prevent the accumulation of worthless data. No attempt will be made to mount a crash program for "pouring" all herbarium and literature data into the data bank wholesale. The goal is to achieve a new level of data validity and documentation.

Whether the data bank is centralized or decentralized, as a series of linked data banks, remains to be determined, but this question becomes more and more academic as teleprocessing systems, allowing access from remote terminals, are improved.

6. Forms of output

Figure 1 shows some of the main forms of information output anticipated. The aim of the retrieval strategy is maximum user accessibility to the stored data: the user should be able to recall any facts in whatever combination and format and at whatever time and place he needs them. Potentially, time-sharing computer systems offer the ultimate flexibility, and such systems will be exploited as fully as economically possible. The computer can greatly assist some routine taxonomic procedures, e.g. identifying plants, but only if it can be used in an interactive, conversational mode, as is possible with on-line, time-sharing systems.

The output phase is the communication phase. Already it appears that the proposed FNA Information System can transform the whole communication process in plant taxonomy. In the future, much taxonomic information probably will be transferred from one user to another without ever being published in the formal, conventional sense.

TYPE SPECIMEN (TYPES) SYSTEM

1. Purpose

Today there are at least a thousand public herbaria in the world, which altogether house upwards of 200 million specimens (Shetler, 1969). A score of these contain over 2 million specimens each, and several have 4–5 million specimens or more. Unfortunately, data accessibility tends to vary inversely with collection size. The taxonomist embarking on a new study cannot begin to make full and effective use of this vast storehouse of specimens that theoretically is at his disposal.

As Shetler (1969, p. 732) has written, "The constant growth of collections impels us to find a more efficient means of storing the specimens and accessing the data. A computer system provides the ideal answer to the problem of data access . . . Data can be retrieved without necessarily taking recourse to the specimens . . .". Surely after several hundred years of building the herbarium as a physical, specimen data bank, the time has come to spend a few years creating a magnetic image of this data bank so that tomorrow's taxonomist will have an effective electronic key to the massive accumulations of specimen data in our herbaria.

Where do we begin herbarium computerization? With a 200-million-specimen backlog, the first task, obviously, is to define within the total specimen pool a meaningful subset of specimens that presents a feasible automation goal and promises to yield an exceptionally useful data bank. *Type specimens* (types) represent one such subset. These are the original specimens that serve as the vouchers for our nomenclature—to them are attached the names of species and infraspecific taxa. No taxonomist can work without at least occasional recourse to type specimens.

Most of the world's herbaria contain at least a few type specimens. The U.S. National Herbarium (operated by Department of Botany, Smithsonian Institution), for example, has about 65 000 known type specimens, constituting about 2% of its general collection. (About 55 000 are types of phanerogamic species, subspecies, varieties, and forms; the other approximately 10 000 are types of cryptogamic taxa—ferns, mosses, and lichens.) If this percentage were to hold over herbaria in general, then the worldwide total could be upwards of 4 million types.

Some herbaria, including the U.S. National Herbarium, maintain special type files for their own collections, but no one has ever attempted to compile a union listing of type holdings in the world. Such a listing would serve as an invaluable finder's guide to known types. It could provide direct answers to many questions that otherwise can be answered only by seeing the specimens or through time-consuming correspondence. With the ultimate goal of such a guide, the Smithsonian's Department of Botany, under the leadership of Hale, initiated a pilot project to determine the feasibility of producing a computerized Type Specimen Register. Creighton has designed the processing system and supervised the computer programming. The Flora North America Program, while not the instigator, has encouraged and facilitated this effort from the outset, because the Type Specimen Register could form a valuable module of the FNA Information System, which in the final analysis will surely be an international system covering plants as a whole.

2. Procedures

(A) TYPE FILE OF U.S. NATIONAL HERBARIUM

Since the early part of the present century, type specimens have been placed in a special collection at the U.S. National Herbarium as they have been discovered in the general collection. The curators have always checked the original literature before placing a type in the special collection, and the resulting documentation has been recorded through the years in an associated card file (5×8-in cards), containing one card per taxon. As a rule, this card includes: name of new taxon (species, subspecies, variety, or form), author of name, place and date of publication, geographic locality of collection, collector(s) name(s) and number, date of collection, kind of type (holotype, isotype, syntype, lectotype, neotype), Dalla Torre and Harms genus number (by which specimens are arranged), and U.S. National Herbarium sheet number. In both the type collection and the card file only basionyms are used.

Although the type specimens are available for borrowing, the card file can be used only on the premises. Automation will permit dissemination in a variety of ways, so that the thousands of hours of bibliographic research that have gone into the creation of the card file can benefit taxonomists at large.

(B) DATA VERIFICATION AND AUTOMATION

Before data automation, each card is subjected to a quick verification routine, including a check of the specimen it is supposed to document, for obvious discrepancies. If necessary, the geographic data are supplemented by consulting an atlas. As a rule the bibliographic data are accepted at face value, and the literature is not searched again unless a clear mistake or problem requires it. The specimen is given the general designation "Type Material" if the kind of type (holotype, isotype, etc.) cannot be determined from the card or specimen in hand.

The verified cards are passed directly to the operator of a programmatic tape-punching typewriter (in this case an SCM 2816 Typetronic), and she encodes the data on paper tape. After appropriate proof-reading, the paper tape is corrected, if necessary, and sent to the computing center for conversion to magnetic tape, and the data are processed. A preliminary computer workprint is proof-read, and essential corrections are made on the magnetic tape via paper tape before the computer record is declared "clean" and ready for dissemination.

(c) COMPUTER OUTPUT*

The computer file is printed first on two-part cards (Fig. 2). The two parts can be separated easily along a perforated line, and each is by itself a 5 × 8-in filing card. The top part is a catalog card, and the bottom part is a data collection card. The sample in Fig. 2 is largely self-explanatory.

The catalog card is simply a standardized, formatted, and printed copy of the original card in the type file of the U.S. National Herbarium. The sheet number of the Herbarium, 2004769 in the sample, is prefixed by the Herbarium's official acronym, US, as designated in *Index Herbariorum* (Lanjouw and Stafleu, 1964). Space is allowed here for adding the acronyms and identification numbers of other herbaria, as they provide data on duplicate types for the same taxon. Ultimately, an updated version of the card could include a complete listing of the institutions of the world claiming to have type material of the given taxon.

The data collection card is for the convenience of cooperating institutions that wish to add a list of their own type holdings to the master file. Data cards can be returned to the Smithsonian as they are completed; it is not necessary to return them all at once or in any particular order. The two data options explained at the top of this card permit the receiving institution to report not only other types of the same taxon but also types of other taxa not in the Smithsonian collection. The long number in the upper righthand corner of the data

* At present, the data are being processed with a Honeywell 1250, which prints only in upper case. Both upper and lower case letters, as well as diacritical marks, are captured on the paper tape, however, which could be re-converted for a machine with an upper/lower-case printer whenever desired.

```
TYPE REGISTER - CATALOG CARD

SCROPHULARIACEAE

   MIMULUS MULTIFLORUS  PENNELL

      PROC. ACAD. NAT. SCI. PHILADELPHIA 99:161. 1947.

      COLLECTORS: PENNELL, F. W.                 COLL. NO.: 26451
                                                 DATE COLL.: 09 AUG 1940

      US; CALIFORNIA; FRESNO CO.; DUNLAP

      US...2004769   ISOTYPE


                                   THIS CARD MAY BE RETAINED FOR YOUR FILES

                                   RECORD ID. 335


     TYPE REGISTER - DATA COLLECTION CARD.               62649478517531242982

     IF YOU HAVE IN YOUR HERBARIUM A DUPLICATE OF THE ABOVE SPECIMEN,
COMPLETE LINE 12 BELOW ON THIS CARD.
     IF YOU HAVE OTHER VERIFIED TYPES, ESPECIALLY OF TAXA WHICH YOU
HAVE RECENTLY DESCRIBED, PLEASE ENTER DATA AS COMPLETELY AS POSSIBLE ON
THIS OR OTHER CARDS AS NEEDED. RETURN TO DEPARTMENT OF BOTANY, SMITHSONIAN
INSTITUTION, WASHINGTON, D.C. 20560.

01. FAMILY:                                              RECORD ID. 000335
02. GENUS:                             SPECIES:

03. VAR. OR SUBSP.:                    FORMA:

04.   AUTHOR:
05. CITATION:

06. COLLFCTORS:                                        COLL. NO.:
                                                       COLL. DATE:

07.  LOCALITY: COUNTRY:
08.  STATE OR PROVINCE:
09.  COUNTY OR DEPARTMENT:
10.  TOWN OR SUBREFERENCE:

11. REMARKS:

12. INSTITUTION ACRONYM:        CAT. NO:             KIND OF TYPE:
```

Fig. 2. Sample of two-part catalog and data collection card of Type Specimen Register.

collection card serves as a surrogate of the family and name of the taxon listed on the catalog card and thereby identifies this taxon to the computer. This surrogate number is a computer-generated number that does not represent a systematic numericlature. Its purpose is to prevent, as different institutions are reporting their holdings, the introduction into the master file of more than one original record for the same taxon.

Apart from the catalog cards, which can be retained and filed by recipient institutions if desired, many other forms of output are possible as the need arises, including various cumulative listings that can be published in book form, e.g. alphabetical and systematic lists of taxa, giving references to the institutions holding the types; institutional lists of holdings; author and collector lists, giving names of taxa described or collected by each; geographical indices to collecting localities; historical activity profiles of authors and collectors; bibliographic analyses, etc. When circumstances warrant, specific questions can be processed for the benefit of individual users whose needs are not met by published cards or catalogs. Several kinds of catalog listings are now being prepared on a sample basis for user evaluation.

The processing system can handle large volumes of data rapidly, facilitate data validation and correction, and print accurate, timely management reports on the collective holdings of the world's herbaria as reflected thus far in the cumulative master file. The system requires strict bookkeeping, especially because it is designed to accept data continuously from many different sources on a random basis. This bookkeeping is handled more or less automatically by the computer. The system adjusts its transactions according to the status of the records in the data bank; for example, the system automatically prevents redundant mailings and maintains current status reports for each cooperating institution.

3. Evaluation

In March 1969, a trial set of 52 cards, representing the complete type holdings of the U.S. National Herbarium for the genus *Mimulus* (family Scrophulariaceae), was distributed to the 46 largest herbaria in the United States. (Since then a few sets have also been sent abroad.) After three months about 65% of the herbaria had responded, and about half of these had supplied data to be added to the file. The other half had no additions to make, but even this negative information was of positive value. Only three of the 30 responding institutions failed to use the data collection cards as requested. Altogether, data were supplied on 65 isotypes of taxa already in the Smithsonian file and on 125 types of taxa not found in this file. Of these latter, new entries, over half could be entered into the system without further verification; the rest were lacking a literature citation (majority) or needed some type of correction. Among the new entries, there were 14 cases of duplication, i.e. cases where two or more institutions supplied new entries for the same taxon.

Reaction to the concept of a computerized Type Specimen Register has been quite favorable on the whole, as evidenced by specific comments as well as by the high level of initial cooperation. Very few curators take issue with the concept or question the ultimate utility of such a Register, although most are concerned about the lack of trained manpower and of financial resources. By now response would be nearly 100% if it were not for the fact that most herbaria—and especially the largest ones—are seriously understaffed and under-financed. On the whole, however, the larger institutions have responded as quickly as the smaller ones. Few herbaria have maintained card files of their own type collections through the years, as the Smithsonian has done, and most have not segregated their types into separate collections; some herbaria have never attempted to distinguish their type holdings in any manner. Consequently, few herbaria can prepare their type specimen data for automation as quickly and easily as the U.S. National Herbarium. In particular, it usually is necessary to do the bibliographic research at the time of data automation, and this greatly complicates and lengthens the input phase. For this reason, it is critical that ready-made files like the Smithsonian's be incorporated first, so that costly bibliographic research is not repeated unnecessarily.

The Smithsonian is approaching the end of the pilot phase of the Type Specimen Register. The master file now contains about 3000 records, which of course falls far short of the 65 000 records that constitute the complete original file. The pilot phase has proved that curators need and want a Type Specimen Register and are willing to cooperate in producing one, to the extent that their institutional resources permit; it also has proved that the information collection and dissemination TYPES System designed by Creighton works satisfactorily. It has become clear, however, that the Smithsonian will not have sufficient resources to complete its own file in a reasonable period of time, not to mention the lack of adequate resources in the numerous cooperating institutions. If the Type Register Specimen is to become a reality, therefore, it must now enter a second phase of development in which the herbarium community as a whole cooperates to secure the necessary funds and to train the necessary manpower to create and maintain the Register. In the process it will be necessary to establish standards for data collection that are acceptable to curators in general. Minor adjustments may be needed in the data format used by the Smithsonian so far. Steps toward this type of institutional cooperation are already underway, with the FNA Program trying to catalyze and assist.

FNA AUTOMATED BIBLIOGRAPHY

This discussion is intended only to elucidate the general purpose, strategy, and potential of the FNA Automated Bibliography. A detailed documentation of the system will be published later by Shetler and Crockett, who have designed it.

1. Need

(A) PROBLEM OF COMMUNICATION

Today's taxonomist is faced by such a vast body of existing literature and flood of incoming new literature, which appears in many languages, that "keeping abreast" becomes more and more a state of mind, depending on one's satisfaction with one's own bibliographic techniques, and less and less a statement of objective fact with each passing year. Frequently, as we all know, it proves simpler nowadays to go back to nature for the facts than to search the literature, where the needed facts may have been reported not just once but many times.

Depending on how, when, and where they are brought to bear, even trivial facts and ideas can take on great significance. Consequently, all publications are potentially useful, and in developing a literature retrieval system the ideal is to create a system that will always be able to close the communication gap between author and user on demand. If an author fails to communicate with anyone, this may be as much the fault of the communication system as of the form or content of the communication.

Every year many new floristic data are added to the literature in a great miscellany of scattered publications, ranging from half-page notes and local checklists to major Floras and taxonomic monographs. As each author casts his contribution into the gathering sea of knowledge, he is doing so on the assumption that, at least in a small way, he is adding to man's understanding of plants. In practice, however, many contributions get lost in the literature because there is no easy, objective means of measuring each new contribution against the existing body of knowledge. If a flexible, rapid means of literature retrieval were available, then every floristic contribution great or small, could be located when needed and could add, therefore, to the larger picture. In tomorrow's world the very grist for Flora-writing should be stored and manipulated directly by machine, so that publication, if and when it is needed, could come as computer output *after, not before*, the fact of information storage and retrieval.

(B) PROBLEM OF DOCUMENTATION

As stated earlier, when a new Flora appears it represents the static end-point of a long, very dynamic process of scholarship, during which a large amount of evidence has been sifted and evaluated. Space economy usually requires that only a trace of the underlying bibliographic research be left in the published work. Thus the specialized bibliographies that inevitably are created during the preparation of Floras are seldom made available in the end to the general user of the Flora. More often than not, once the taxonomist has completed his Flora and the book or paper has appeared, his bibliographies fall into disuse and sooner or later are discarded, if not by him, by his successors or heirs. Of necessity, the process of scholarship often must be started over again by the next generation if not by a colleague of the same generation, and there is a great need for an information system that permits the capture of such specialized bibliographies and makes them available to the general users along with the published Flora or other work itself.

(c) CURRENT BIBLIOGRAPHIC EFFORTS

Biological Abstracts and similar general bibliographic services have not proved adequate through the years for the specific needs of plant taxonomists. As a result, many specialized bibliographic services have been started and championed, usually by individual botanists whose work sometimes has been taken over eventually by their institutions. Today most major taxonomic institutions have one or more bibliographic efforts going within their walls. These range from the type specimen file of the U.S. National Herbarium (see previous section) to the fantastic, cross-referenced literature index still being compiled at the herbarium in Geneva, Switzerland. Examples of invaluable files that for many years have been published for the benefit of the taxonomic community at large are the Gray Card Index from the Gray Herbarium of Harvard University and the *Index Kewensis* of the Royal Botanic Gardens, Kew, England. Each of these special-purpose bibliographies represents only a partial answer to the taxonomist's information needs, and together they involve extraordinary overlap, hence waste, of effort, because in each case essentially the same body of literature is being combed. A single system involving a single pass through the literature could result in enormous savings.

2. Purpose and strategy

To the FNA Program, with its automation aims, literature access represents an urgent but formidable, indeed frightening, challenge. There are no illusions about developing a panacea for literature retrieval. At the same time, there is concern that today's contributor should have better tools for searching and compiling data from the world's literature, while tomorrow's user should have, in addition, ready access to the full documentational scholarship behind the FNA Data Bank. Automatic literature storage and retrieval obviously cannot be had overnight, and the utility of any system started today will be much greater tomorrow.

The purpose of the pilot "FNA Automated Bibliography" system is to demonstrate the power and usefulness of an integrated literature retrieval system designed specifically to serve the needs of plant taxonomists. A special-purpose system like this one is intended to supplement and complement general-purpose systems like those now being developed by *Biological Abstracts*, not to supplant them. The FNA Automated Bibliography can be used either to develop a central data bank of literature or to automate the personal files of individual taxonomists or institutions; the system merely provides the "box" into which more or less data can be put.

The pilot FNA system employs a formatted file in the unit record. This record is structured for selected, standardized items of taxonomic information abstracted from the document (paper or book) being referenced. At present, the input advantages of using an unformatted file that could accept free text in natural language tend to be offset by the technical problems and high costs of developing an efficient, effective search-and-retrieval strategy. Formatting imposes a degree of selectivity and standardization, and this is essential if the system is not to be rendered ineffective and unpredictable by imprecise terms, synonyms, and meaningless "noise" words. Even if the day had arrived, which in fact seems quite far off at present, when the storage and retrieval of full texts up to book length were economical, the problems of imprecise terms, synonyms, and "noise" words would be no less real.

Retrieval strategy allows for two types of output: (i) book-form bibliographies, indices, and data compilations that can be issued periodically to all members of the user community, perhaps on a subscription basis; (ii) "custom-made" responses that can be issued to individual users who pose direct queries or commands to the master citation file.

3. Data collection

A sample data collection form, as filled out by a bibliographer, is shown in Fig. 3. This sheet represents the unit entry or record in the master citation file. The bibliographer fills out this form from the original literature according to a detailed set of specifications (Shetler, *et al.*, 1970). The bibliographer need not be a professional botanist, although this is a great asset, but at the least he must be well trained and experienced both in the subject matter being abstracted and the requirements of the FNA system. In effect, the bibliographer is a "middle man" between the author and the user, and the day should come when the author is his own bibliographer, providing journal editors and book publishers with appropriate data abstracts of his papers and books for direct automation.

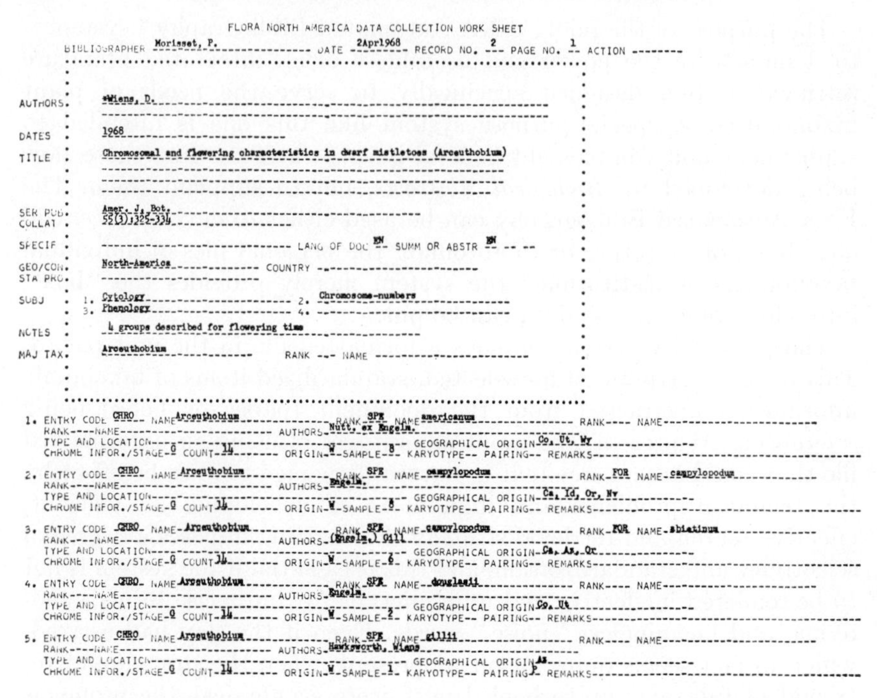

FLORA NORTH AMERICA DATA COLLECTION WORK SHEET

BIBLIOGRAPHER Morisset, P. DATE 2Apr1968 RECORD NO. 2 PAGE NO. 1 ACTION

AUTHORS *Wiens, D.

DATES 1968

TITLE Chromosomal and flowering characteristics in dwarf mistletoes (Arceuthobium)

SER PUB Amer. J. Bot.

COLLAT 55(3):325-334

SPECIF LANG OF DOC EN SUMM OR ABSTR EN

GEO/CON North-America COUNTRY

STA PRO

SUBJ 1. Cytology 2. Chromosome-numbers 3. Phenology 4.

NOTES 4 groups described for flowering time

MAJ TAX Arceuthobium RANK NAME

1. ENTRY CODE CHRO NAME Arceuthobium RANK SPE NAME americanum RANK NAME
RANK NAME AUTHORS Nutt. ex Engelm.
TYPE AND LOCATION GEOGRAPHICAL ORIGIN Co, Ut, Wy
CHROME INFOR./STAGE G COUNT 14 ORIGIN W SAMPLE 8 KARYOTYPE PAIRING REMARKS

2. ENTRY CODE CHRO NAME Arceuthobium RANK SPE NAME campylopodum RANK FOR NAME campylopodum
RANK NAME AUTHORS Engelm.
TYPE AND LOCATION GEOGRAPHICAL ORIGIN Ca, Id, Or, Nv
CHROME INFOR./STAGE G COUNT 14 ORIGIN W SAMPLE 8 KARYOTYPE PAIRING REMARKS

3. ENTRY CODE CHRO NAME Arceuthobium RANK SPE NAME campylopodum RANK FOR NAME abietinum
RANK NAME AUTHORS (Engelm.) Gill
TYPE AND LOCATION GEOGRAPHICAL ORIGIN Ca, Az, Or
CHROME INFOR./STAGE G COUNT 14 ORIGIN W SAMPLE 6 KARYOTYPE PAIRING REMARKS

4. ENTRY CODE CHRO NAME Arceuthobium RANK SPE NAME douglasii RANK NAME
RANK NAME AUTHORS Engelm.
TYPE AND LOCATION GEOGRAPHICAL ORIGIN Co, Ut
CHROME INFOR./STAGE G COUNT 14 ORIGIN W SAMPLE 2 KARYOTYPE PAIRING REMARKS

5. ENTRY CODE CHRO NAME Arceuthobium RANK SPE NAME gillii RANK NAME
RANK NAME AUTHORS Hawksworth, Wiens
TYPE AND LOCATION GEOGRAPHICAL ORIGIN Az
CHROME INFOR./STAGE G COUNT 14 ORIGIN W SAMPLE 1 KARYOTYPE PAIRING P REMARKS

Fig. 3. Bibliographer's data collection sheet for FNA Automated Bibliography.

The collection form has four data sections: (i) control data; (ii) citation; (iii) content classification terms; (iv) data abstract. The control data on the top line are self-explanatory in Fig. 3. The citation and classification sections are included in the 5×8-in rectangular box, while the abstract is given in the numbered entries at the bottom of the form. In addition to author, date, title, serial or publisher (SER PUB), and collation (volumation, pagination, etc.) fields, the citation section includes fields for additional bibliographic specifications and to indicate the languages of the document (DOC = paper or book) and summary or abstract. The classification section provides fields for cross-indexing the document according to the three basic parameters of a taxonomic work: geography, subject, and taxonomic group (i.e. taxon). Three geographic levels can be specified: CON = continent or equivalent region, COUNTRY = country or equivalent region, and STA PRO = state or province or equivalent region. Each of the three geographic fields can accept more than one geographic name; thus by efficient use of standardized abbreviations it is possible to list many geopolitical units (e.g. states of the United States of America). Four fields for four subject terms are provided. The problem of subject "classification" v. "keyword-indexing" has not yet been resolved, and during the pilot phase we have attempted general classification (e.g. cytology, taxonomy, ecology) augmented by specific keyword-indexing; keywords are often taken from the document title. Until now we have worked with open-ended term vocabularies, which have been controlled only to a limited extent. The NOTES field is for brief remarks that further specify the subject content of the paper as a whole. The last three fields inside the box are for designation of the major taxon (MAJ TAX) in the document. Here is entered the name of the taxonomic group, if any, that is dealt with in a substantive way in the document, such that the document is essential reading for anyone doing research on the group.

The data abstract section, below the box, is designed to accommodate annotated lists of taxa, usually species. The form provides only five numbered entries but the checklist can be extended indefinitely by using continuation sheets. The ENTRY CODE specifies the type of annotations to be expected in the several data fields of the entry. For example, CHRO is the code for a newly reported chromosome number; in this case, special fields are provided to indicate: stage of counting (G=gametic, S=somatic); origin of material, whether wild (W), cultivated (C), or unknown (U); number of plants counted (SAMPLE); and whether karyotype or pairing were studied. For most entry codes, the kind of information placed in the REMARKS field is governed by standards; e.g. when the code indicates a new name combination, then the basionym can be expected in this field. Other important entry codes are: NGEN (=new genus), NSPE=(new species), NVAR (=new variety), NFOR (=new form); XSPE, XVAR, etc. (=new name combination at specific, varietal, etc. levels); ILLU (=newly published illustration); DIST (=noteworthy distribution record, with particulars in REMARKS); and DMAP (=newly published distribution map, with type of map in REMARKS).

4. Data automation

The data on the collection form are encoded on paper tape, proof-read, and read onto magnetic tape in a manner quite analagous to that employed in the Type Specimen System (see earlier sections), although the input procedure is considerably more complicated in this case. A set of 5 × 8-in cards are generated from the paper tape for a cross-indexed, "hard-copy", office file before the paper to magnetic tape conversion; these cards, which contain the information in the 5 × 8-in box on the collection form, are invaluable for controlling the input operation and for answering short, local questions.

5. Computer processing and output*

Certain term and authority files are incorporated into validation programs so that the computer can assist in editing and controlling input/output information. The most important of these is the Taxon Catalog (Shetler, *et al.*, 1968), which is kept on a computer disk storage to validate names of taxa during input and output. In this way the master citation file on magnetic tape and the various print-outs can be kept clean of inadvertent typographic errors and orthographic variants.

The Taxon Catalog is an hierarchical file arranged to family level according to the phylogenetic system in the revised *A. Engler's Syllabus der Pflanzenfamilien* (Melchior and Werdermann 1954, Melchior, 1964); at present it includes names only to the rank of genus, and the genera are alphabetical within family. A systematic numericlature paralleling the hierarchical nomenclature assigns a unique number to each name in the file, which numerically identifies the rank of the name to the computer. Names deemed to be taxonomic or nomenclatural synonyms are likewise treated as numericlatural synonyms and assigned the same number. It is possible, therefore, to retrieve references from the file pertaining to specific names, if the search is made by name, or to whole classes or hierarchies of names, if the search is made by number. Thus the search-and-retrieval strategy places extraordinary importance on the Taxon Catalog, in which the hierarchy and synonymy are established.

The present search-and-retrieval capability of the pilot FNA system is already impressive, even though the full potential of the system design has been only partially realized; programming to meet the full design specifications continues. The strategy has been to develop a direct query-response retrieval, so that questions can be answered on demand, and to develop a flexible array of output formats. Answers to questions of general interest (e.g. What recent papers are there on chemotaxonomy?) can be printed in the form of a conventional bibliography (i.e. author, date, title, place of publication, volumation, pagination) and published directly for distribution to the whole user community. A whole series of specialized bibliographies, annotated catalogs, indices, gazetteers, etc. might be issued periodically from the main data base. Special questions, of interest to only one or two users, can be answered directly to them by means of ordinary computer print-out or, eventually, by means of an on-line, time-sharing network in which the individual user can address his own queries to the central file and select the answers he needs. At present, all processing is done in the batch mode.

* The FNA Automated Bibliography system is programmed in COBOL for the Honeywell 1250 computer. See also footnote on p. 287.

To date, a pilot data base of 200 journal references has been entered on the master citation file, and some examples follow of the *kinds* of queries (commands) the system can respond to on the basis of the 200 references:

LIST ALL REFERENCES CONCERNING:

(1) natural hybridization in the family Orchidaceae in the province of Manitoba, Canada;
(2) *Arceuthobium* as the major taxon;
(3) the phenology of salt-marsh plants in the state of Massachusetts, U.S.A.;
(4) flavonoid glycosides in the family Malvaceae;
(5) pollination biology in the family Campanulaceae, published since 1967;
(6) either (or both) apogamy and polyploidy in the order Filicales;
(7) plant-dispersal by birds;
(8) boreal (or arctic-alpine) disjuncts in the province of Nova Scotia, Canada;
(9) floristics in the province of Saskatchewan, Canada;
(10) seed dormancy in the family Scrophulariaceae;
(11) the cytogenetics of the family Poaceae (=Gramineae);
(12) the activities of Thomas Nuttall in the state of Pennsylvania, U.S.A.;
(13) the nomenclature of *Artemisia*;
(14) corrections to *Gray's Manual of Botany*, 8th edition;
(15) the botanist Faustino Miranda (this can be made more specific by asking for only biographies or obituaries).

With a large, comprehensive data base, the query-response possibilities would be unlimited. Documents can be recalled easily, of course, by author and/or date.

The print-out options at present are several. Full entries can be printed out in the format of the original data collection form; selected sections of the entry can be printed; or the references can be reduced and telescoped into orthodox bibliographic citations, as mentioned above.

A different kind of question also can be handled by the pilot system, which involves a request for specific data and citations to document the data:

(1) List all new names or name combinations published during 1966–68, where they were published, and related information as specified. A page from a typical print-out is shown in Fig. 4; it is largely self-explanatory. Here we have, in effect, an *Index Kewensis*, or Gray Card Index as a byproduct of a comprehensive

```
FLORA NORTH AMERICA AUTOMATED BIBLIOGRAPHY     07/03/69  PAGE  7

      SELECTED NEW TAXA AND COMBINATIONS
           PUBLISHED IN 1963-1968

KECKIELLA TERNATA SUBSP. SEPTENTRIONALIS
     (MUNZ, JOHNSON) RICHARD M. STRAW, BRITTONIA 19(3):203-204. 1967.
     P.204.BAS.:PENSTEMON TERNATUS VAR. SEPTENTRIONALIS

LIMONIUM LIMBATUM VAR. GLABRESCENS DONOVAN S. CORRELL,
     RHODORA 68(776):420-428. 1966. TEXAS. HO-LL.  P. 425

LINUM BAHAMENSE F. CORALLICOLA (SMALL) C. M. ROGERS.
     RHODORA 70(783):439-441. 1968.   P. 440. BAS.: CATHARTOLINUM
     CORALLICOLA SMALL

LINUM PERENNE SUBSP. LEWISII VAR. SAXOSUM
     (MAGUIRE, HOLMGREN) JAMES L. REVEAL, RHODORA 70(781):25-54.
     1968.   P. 39. BAS.: L. LEWISII VAR. SAXOSUM

LYCOPODIUM REFLEXUM VAR. RIGIDUM (J. E. GMEL.) GEORGE R. PROCTOR,
     RHODORA 68(776):464-469. 1966. LESSER-ANTILLES.   P.464. BAS.:
     LYCOPODIUM RIGIDUM J. F. GMEL. IN L.

MACHAERANTHERA ARIZONICA R. C. JACKSON, R. R. JOHNSON,
     RHODORA 69(780):476-480. 1967. ARIZONA. HO-KANU, IT-ARIZ.  P.
     476. PIMA CO.

MYRICA PENSYLVANICA F. PARVIFOLIA T. W. WELLS, RHODORA 70(783):453.
     1968. MAINE. HO-NEBC.  P. 453

OENOTHERA MACROCARPA VAR. INCANA (A. GRAY) JAMES L. REVEAL,
     RHODORA 70(781):25-54. 1968.   P. 43. BAS.: OE. MISSOURIENSIS
     VAR. INCANA A. GRAY

OENOTHERA MACROCARPA VAR. OKLAHOMENSIS
     (J.B.S. NORTON) JAMES L. REVEAL, RHODORA 70(781):25-54. 1968.
     P. 43. BAS.: MEGAPTERIUM OKLAHOMENSE J.B.S. NORTON

ORCHIS ROTUNDIFOLIA F. LINEATA (MOUSLEY) EDWARD G. VOSS,
     RHODORA 68(776):435-463. 1966. ONTARIO.   P. 462. BAS.: O.
     ROTUNDIFOLIA PURSH VAR. LINEATA

PARONYCHIA JAMESII VAR. PRAELONGIFOLIA DONOVAN S. CORRELL,
     RHODORA 68(776):420-428. 1966. TEXAS. HO-LL.  P. 423

PERISSOCOELEUM BARCLAYIAE MILDRED E. MATHIAS, LINCOLN CONSTANCE,
     BRITTONIA 19(3):212-226. 1967. COLOMBIA. HO-UC.  P. 223. FIG. 3,
     P. 216

PERISSOCOELEUM CRINOIDEUM (MATH.,CONST.) M.E.MATHIAS, L.CONSTANCE,
     BRITTONIA 19(3):212-226. 1967. COLOMBIA. HO-US.  P.223.BAS.:
     PRIONOSCIADIUM CRINOIDEUM.FIG.4,P.218
```

Fig. 4. Sample page of index to newly published taxa or combinations, retrieved as by-product of FNA Automated Bibliography.

taxonomic literature system, except that we have here more data per entry than either of the mentioned indices. Taking the sixth entry from the top of Fig. 4, for example, we learn that *Machaeranthera arizonica* was described by R. C. Jackson and R. R. Johnson from Pima County, Arizona, and that the holotype is deposited at the herbarium of the University of Kansas (HO-KANU), while an isotype is deposited at the University of Arizona (IT-ARIZ). Not only do we know the page on which his description begins (P.476), but also we have the full pagination of his paper for the event that we wish to order a photocopy of the whole paper from the library.

(2) List by species (or other taxon) a bibliography of recently published (time period as defined by requestor) distribution maps. The format is similar to that in Fig. 4, except that the kind and geographic area of the map replace the information on types.

(3) List by species (or other taxon) a bibliography of recently published distribution records (i.e. range extensions, etc.). This request also is answered in a format similar to that in Fig. 4, and the kind of distribution record, e.g. "1st record for North America, Cape Sabine, Alaska," replaces the information on types.

6. Evaluation

The pilot phase (1967–69) of the FNA Automated Bibliography has demonstrated the technical feasibility of an automated, multi-purpose literature retrieval system designed specifically to serve the needs of plant taxonomists, which currently are met by a patchwork of partial solutions. If this system or some improved version of it is to be implemented, then a brand new phase of activity is required. The first task is to gain the support and cooperation of those persons and institutions who are involved in the current, on-going bibliographic efforts, but much more than a mere pooling of current programs will be required ultimately if we are to succeed with an automated system. It would be foolish to use the computer simply to mimic and perpetuate outdated manual procedures. A whole new scale of bibliographic retrieval is now possible with the technology available to us today, but to achieve this will require an unprecedented degree of standardization and coordination, not to mention a new flexibility concerning content, format, and typography. The costs of an automated program are certain to be large, but no one can calculate the economics until user demand can be tested and evaluated; yet the user can hardly even imagine his demands until he has seen what an automated system can do for him. Obviously, as stated earlier, someone will have to take some calculated risks if we are to push forward in the development of an automated literature retrieval program. Finally, there is the even larger challenge to create a data system that incorporates the very text itself, in its entirety, and thereby erases the distinction between publications and bibliographies.

TAXONOMIC DATA MATRIX SYSTEM

1. Purpose and concept

The purposes of a Flora are to define, describe, and discriminate the kinds of plants in a given geographic region. One of the questions being asked by the FNA Program is this: "Can methods be developed for using the computer to assist the taxonomist in using the data of a Flora more effectively to perform the functions of definition, description, and discrimination (identification)"?

One approach to this problem is to code the diagnostic characteristics of the plants in a matrix of taxa × characters, which then can be operated on with programs for identification, description, key construction, comparison, etc. This kind of "taxonomic data matrix" is central to the system being programmed by Morse, in consultation with Beaman, Shetler, and other members of the FNA Editorial Committee. The initial stages of development have been described briefly already in print (Morse, 1968, 1969; Morse, *et al.*, 1968), and the methods, which are simple but elegant and especially powerful when used on a time-sharing computer system in the conversational mode, have been introduced by Beaman into his undergraduate and graduate courses in plant taxonomy at Michigan State University. The programs are all written in FORTRAN, and most have been implemented on a General Electric (GE) 265 time-sharing computer system; listings, tapes, and sample data and output are available.

The so-called Taxonomic Data Matrix System is intended in the first instance to facilitate rapid, accurate identification of plant specimens, and it utilizes files that differ from those common in numerical taxonomy. Our system presumes that the taxonomy and nomenclature have already been worked out at least in a preliminary way, and the taxonomic characters used in the data matrix are carefully selected to discriminate the species or other taxa. No attempt is made, as in numerical taxonomy, to employ as many characters as possible, regardless of their value in discriminating the OTU's. As in preparing an ordinary dichotomous key, efficient discrimination is the goal.

Other approaches to identification by computer have been made. Bossert (1969) and Peters (*in* Bossert, 1969, pp. 610–613) have shown that conventional keys can easily be stored in the computer, but such systems offer little advantage over the printed page. Polyclave approaches (Leenhouts, 1966, Duke, 1969) are more appropriate for computerization, and the systems of Goodall (1968) and Walker *et al.* (1968) are in many ways equivalent to our own. Unknowns can be identified also by numerical taxonomic methods (Gyllenberg 1965), but these methods appear inefficient. Derivation of minimum-set lists of diagnostic characters from taxonomic matrices (Rypka *et al.* 1967) is an interesting technique linking numerical taxonomy and identification programming. Some experimental techniques in medical diagnosis, especially Bayesian analysis and the likelihood approach (Dybowski and Franklin 1968), are also related to specimen identification. Finally, we should call attention to the paper in this volume by Pankhurst and Walters concerning their method for constructing keys by computers, which is not unlike Morse's method.

A detailed description of the Morse programs and methods is in preparation by him, and in the present section we are confining ourselves to a quick review.

```
LIST

MAPLES     10:08    FRI. 06/06/69

4999 FILE=MAPLES,WIDTH=20,FLAG=MSC,REVISION=05/30/69
5000 MICHIGAN MAPLES MATRIX  --  --  BY LARRY MORSE,  MARCH 13, 1969
5010 USE "LEAF CHARACTERS FOR ACER IN MICHIGAN" LIST OF SAME DATE.
5020 MSC
5100    1 A. CAMPESTRE
5110    2 A. NEGUNDO
5120    3 A. NIGRUM
5130    4 A. PENSYLVANICUM
5140    5 A. PLATANOIDES
5150    6 A. PSEUDO-PLATANUS
5160    7 A. RUBRUM
5170    8 A. SACCHARINUM
5180    9 A. SACCHARUM
5190   10 A. SPICATUM
5200  -99 END TAXON LIST
5400    1 33333323321223321300
5410    2 13232123323233123300
5420    3 33332132212211213300
5430    4 33311133333113333100
5440    5 33133133333131333130
5450    6 32313133332331333300
5460    7 31313133332233323200
5470    8 32323113331223333300
5480    9 32333133231323333300
5490   10 33312133323123323300
5590  -99 99999999999999999999
```

Fig. 5. Taxonomic Data Matrix for species of *Acer* (maples) in the State of Michigan, U.S.A., designed especially for identification.

2. Procedures

A sample matrix for Michigan maples (genus *Acer*), prepared by Morse, is shown in Fig. 5.* The 10 species of *Acer* are listed in lines 5100–5190, and the coded matrix occupies lines 5400–5490. Each row of the matrix represents, as numbered, one of the 10 species, and each column represents one of the 18 couplets of characters in Table I. Each couplet contrasts two character states, and the second state is given the number of the first plus 100 (e.g. 1 and 101; could also be 1A and 1B).

TABLE I. List of leaf characters used in Taxonomic Data Matrix shown in Fig. 5

1. Leaves compound
101. Leaves simple
2. Leaves whitened beneath
102. Leaves light green beneath
3. Leaves shiny-surfaced beneath
103. Leaves dull beneath, whitened or not
4. Leaf margins regularly serrate, with at least 2 teeth per cm.
104. Leaf margins entire or only very coarsely toothed
5. Lower half of blade unlobed
105. Lower half of blade bearing lobes
6. Tips of lobes long and tapering
106. Tips of lobes rounded or blunt
7. Leaves deeply lobed, cleft more than halfway to base
107. Leaves shallowly lobed, cleft less than halfway to base
8. Stipules present
108. Stipules absent
9. Petioles densely pubescent
109. Petioles glabrous or glabrate
10. Leaves noticeably pubescent beneath
110. Leaves glabrous beneath or nearly so
11. Terminal lobe shouldered
111. Terminal lobe not shouldered, tapering uniformly from base to apex
12. Lobes mostly three
112. Lobes mostly five
13. Sinuses between lobes U-shaped at bottom
113. Sinuses between lobes V-shaped at bottom
14. Basal sinus narrow or closed
114. Basal sinus broad and open
15. Bases of petioles enlarged, often enclosing axillary buds
115. Bases of petioles not usually enlarged, buds well-exposed
16. Veins densely pubescent on lower leaf surface
116. Veins glabrous or glabrate
17. Sap milky in freshly broken petiole
117. Sap clear in freshly broken petiole
18. Serrations minute, more than 5 per cm
118. Serrations larger, fewer than 5 per cm

* This is a *sample* matrix; it is not intended in any way to be definitive even for Michigan.

For a given character and species, the character state expressed by that species is coded numerically in the matrix according to the following code:

0 = No information; character state unknown
1 = First character state of couplet applies
2 = Either character state of couplet may apply
3 = Second character state of couplet applies
4 = Neither character state of couplet applies

A sample "conversation" with the computer is recorded in Fig. 6. The program can be written to make the conversation as terse or verbose as the situation demands, and the computer can even be programmed to be polite or humorous, responding with such words as SORRY, YOURE WELCOME, etc. Classroom uses require relatively verbose programs, which can instruct the student step-by-step what to do next. Professional users, by contrast, want to reach an answer as directly and swiftly as possible, therefore, a minimum of conversation or "back talk" is desired.

All the printing shown in Fig. 6 except that which follows question marks (usually numerals only) represents the computer "talking". Only when the computer poses a question, indicated by a question mark, does the user type in a response. This is true of all of the time-sharing programs in the presently developed series. In identifying his last specimen during the run shown in Fig. 6, the user noted that character states 1 and 6 applied, but he mistyped 106 for 6. The computer informed him that no Michigan maple answers to this description, and he resubmitted the correct number and obtained a direct identification.

By contrast with ordinary dichotomous keys, the computerized matrix key permits one to eliminate taxa by using any combination of characters simultaneously, including all characters taken together or as few as one at a time. The typical printed key restricts the user to only two choices at a time and always in a fixed order. Furthermore, most printed keys do not include all of the potentially discriminating features. The value of a numerical coding is that the data are neutral with respect to language (English, German, etc.), and the input/output dialogue can be conducted in any language by translating the appropriate portions of the program into the language desired. Such an approach could have important implications for an international data bank. Furthermore, the program can be written for the layman as well as the specialist.

The Morse system presently includes several other programs than the identification program. A description program translates the numerical

```
  PROGRAM 'IDENT' FOR DIRECT IDENTIFICATION OF SPECIMENS USING
TIME-SHARING COMPUTERS AND TAXONOMIC DATA MATRICES. PREPARED BY
LARRY MORSE, MICHIGAN STATE UNIVERSITY; THIS EDITION REVISED
THRU MAY 29, 1969.

MATRIX CATALOG:
1=MAPLES  2=WINTER TWIGS  3=SANFORD LILIACEAE  4=SPRING FLORA
5=RASPBERRIES  6=MICHIGAN OAKS  7=SANFORD LABIATAE  8=OTHER

WHICH ONE? 1

FILE 'MAPLES', REVISED THRU 05/30/69

    MICHIGAN MAPLES MATRIX  --  --  BY LARRY MORSE,  MARCH 13, 1969
    USE "LEAF CHARACTERS FOR ACER IN MICHIGAN" LIST OF SAME DATE.

NEW SPECIMEN:  HOW MANY CHARACTERS? 3
WHICH ONES? 2,105,7
SUGGESTED IDENTIFICATION OF DESCRIBED SPECIMEN:
    A. SACCHARINUM
    DISTINGUISHING CHARACTERS:    7   11  114  117  101
  NEXT:  3=NEW SPECIMEN  4=NEW MATRIX  5=STOP  --? 3

NEW SPECIMEN:  HOW MANY CHARACTERS? 4
WHICH ONES? 11,104,107,6
 THE FOLLOWING  3 TAXA REMAIN:
    A. NIGRUM
    A. PLATANOIDES
    A. SACCHARUM
  NEXT:  1=CONTINUE  2=LIST  3=NEW SPECIMEN  4=NEW MATRIX  --? 1
HOW MANY CHARACTERS? 0
USEFUL CHARACTERS:  3  10  14  16  17   2   5   8
HOW MANY CHARACTERS? 103
ARE YOU SURE - HOW MANY? 2
WHICH ONES? 103,110
SUGGESTED IDENTIFICATION OF DESCRIBED SPECIMEN:
    A. SACCHARUM
    DISTINGUISHING CHARACTERS:  104   11  112  114  117
  NEXT:  3=NEW SPECIMEN  4=NEW MATRIX  5=STOP  --? 3

NEW SPECIMEN:  HOW MANY CHARACTERS? 2
WHICH ONES? 1,106
NO TAXON IN THE DATA MATRIX HAS THIS COMBINATION OF CHARACTERS.  SORRY.
  NEXT:  3=NEW SPECIMEN  4=NEW MATRIX  5=STOP  --? 3

NEW SPECIMEN:  HOW MANY CHARACTERS? 1
WHICH ONE? 1
SUGGESTED IDENTIFICATION OF DESCRIBED SPECIMEN:
    A. NEGUNDO
    DISTINGUISHING CHARACTERS:    1   15  104  111  113
  NEXT:  3=NEW SPECIMEN  4=NEW MATRIX  5=STOP  --? 5
```

Fig. 6. Sample "conversation" with computer to identify species of *Acer* (maples) using matrix in Fig. 5 and characters in Table I.

```
 VKEY        9:39   FRI. 06/06/69

IN DPRINT


      MICHIGAN MAPLES MATRIX  --  --  BY LARRY MORSE, MARCH 13, 1969


    1. BASAL SINUS NARROW OR CLOSED
      2. LEAF MARGINS REGULARLY SERRATE, WITH AT LEAST
         2 TEETH PER CENTIMETER
              .....A. PSEUDO-PLATANUS
      2. LEAF MARGINS ENTIRE OR ONLY COARSELY TOOTHED
              .....A. NIGRUM
    1. BASAL SINUS BROAD AND OPEN
      3. SAP MILKY IN FRESHLY BROKEN PETIOLE
        4. LEAVES SHINY-SURFACED BENEATH
                .....A. PLATANOIDES
        4. LEAVES DULL BENEATH, WHITENED OR NOT
                .....A. CAMPESTRE
      3. SAP CLEAR IN FRESHLY BROKEN PETIOLE
        5. LEAVES COMPOUND
                .....A. NEGUNDO
        5. LEAVES SIMPLE
          6. LEAVES DEEPLY LOBED, CLEFT MORE THAN HALFWAY
             TO BASE
                  .....A. SACCHARINUM
          6. LEAVES SHALLOWLY LOBED, CLEFT LESS THAN HALFWAY
             TO BASE
            7. LEAF MARGINS REGULARLY SERRATE, WITH AT LEAST
               2 TEETH PER CENTIMETER
              8. LEAVES WHITENED BENEATH
                      .....A. RUBRUM
              8. LEAVES LIGHT GREEN BENEATH
                9. SERRATIONS MINUTE, MORE THAN 5 PER CM
                        .....A. PENSYLVANICUM
                9. SERRATIONS LARGER, FEWER THAN 5 PER CM
                        .....A. SPICATUM
            7. LEAF MARGINS ENTIRE OR ONLY COARSELY TOOTHED
                    .....A. SACCHARUM


NORMAL ENDING OF DICHOTOMOUS KEY
```

Fig. 7. Dichotomous key constructed by computer (on-line) from matrix shown in Fig. 5 and using characters listed in Table I.

codes back into verbal statements of the character states, as given in Table I, and these can be output taxon-by-taxon to yield diagnostic descriptions of the species. Another program causes the computer to construct a dichotomous key of conventional type, as shown in Fig. 7. The taxonomist can decide what algorithm to use in his key-construction program, and the algorithm used in constructing the key in Fig. 7 has been described elsewhere by Morse (1968) (see also Osborne, 1963). Potentially, therefore, the same matrix can yield many different types of keys based on different algorithms. At present, the Morse program for key-construction uses only one character per couplet-lead in the dichotomous key, but this handicap will be overcome shortly in a revised program. The contrast program compares any two taxa and tells their differences; this kind of program obviously approaches numerical taxonomic programs. Finally, the present system includes a program for quizzing students about the most useful (i.e. discriminating) characters in the matrix. The student matches wits with the computer, which scores the student on a percentage basis.

3. Evaluation

The program system described here has proved extremely effective as a means of introducing the concept of man/machine interaction to students and professionals in plant taxonomy. Regardless of its ultimate utility in taxonomy or in the Flora North America Program, Morse's Taxonomic Data Matrix System must be said to represent the kind of data system that, in perfected form, could prove enormously valuable to the researcher who finds himself unable to keep in mind the myriad characters of a large genus or other taxon at the moment of identification or key construction and who at this moment desperately needs and wants the aid of a machine. In Flora-writing, there is a need for a better means of preparing keys and descriptions that are fully consistent with each other and truly diagnostic. Although the FNA Program has not by any means embraced this matrix system, it has embarked on a small-scale trial of the system as a method for generating keys and identifying plants on-line to the computer.

CONCLUSION

The Flora of the future will be a standardized data bank from which all "cutting and pasting" can be done. It will be open-ended, dynamic, and ever growing. If such a bank existed today, then Flora North America could be written overnight by machine, except for the taxonomic judgments required. The specialist of the future doubtless will store his latest revisions and monographs in the computer, not on the printed page. He will deposit his data in a computer that is available nationally if not internationally by a telecommunications network. Such networks already are in operation. Thus the Flora of the future will become a huge memory or series of linked memories available on-line to all users at any place and time.

Computers can and will affect the very process of research and communication in floristics and taxonomy. Today research leads to publication, and publication may lead to information storage and retrieval. Interposed between the researcher-author and the ultimate user is an elaborate editorial-publication process and a hit-and-miss bibliographic indexing process. Tomorrow research will lead directly to information banking, and such banking may lead to publication but only on demand of users. Information systems will eliminate many middle men and processes, because the researcher will be able to add his data directly to the general bank which will be available directly to all users. The users themselves will be able to dictate the parameters and format of retrieval output. Then scientific journals and books can spare their pages of endless descriptive facts and devote them exclusively to scholarly, theoretical discourse.

ACKNOWLEDGMENTS

We appreciate the cooperation of the curators who responded to the trial distribution of *Mimulus* cards for the Type Specimen Register and especially wish to thank Howard S. Irwin of the New York Botanical Garden, Walter H. Lewis of the Missouri Botanical Garden, and James H. Soper of the National Museum of Canada for their highly important contributions to this experiment. Dan H. Nicolson of the Smithsonian Institution has played a very significant role in the development of this pilot project thus far, and thanks are due also to Grace Rickard, Nancy Howard, and Michael J. Harvey for their assistance at the Smithsonian. The programming for the TYPES System has been done largely by Willard Handley at the Smithsonian. To Pierre Morisset, presently of Laval University (Quebec City, Quebec), and Elaine R. Shetler goes major credit for the data specifications and input that have made the FNA Automated Bibliography a reality at the Smithsonian; the programming for this system was done by Shigeko I. Rakosi. We are indebted for general advice and interest to David J. Rogers, University of Colorado, to Theodore J. Crovello, University of Notre Dame, to Gilbert S. Daniels, Hunt Botanical Library, and to Nicholas J. Suszynski, Jr., Smithsonian Institution. Finally, we are deeply indebted to our colleagues on the AIBS-FNA Editorial Committee—Walter H. Lewis, Missouri Botanical Garden; John T. Mickel, New York Botanical

Garden; Peter H. Raven, Stanford University; Roy L. Taylor, University of British Columbia; and John H. Thomas, Stanford University—but especially to Chairman Raven, for their provocative thoughts concerning the development of an FNA Information System which have greatly influenced what we have said here in an attempt to give public voice to the trend of thinking and development now taking place in the Flora North America Program. The pilot systems studies reported in this paper could not have been accomplished without the solid financial assistance from the beginning of the Smithsonian Institution. Directly or indirectly, the Michigan State University also has contributed substantially, especially to the work on the data matrix system, and recently we have been supported in part by the National Science Foundation of the United States of America (grants GN-812, GB-8441, GJ-573).

REFERENCES

Bossert, W. (1969). Computer techniques in systematics. *In* "Systematic Biology", pp. 595–614, publication 1692, National Academy of Sciences, Washington, D.C.

Creighton, R. and King, R. (1969). The Smithsonian Institution Information Retrieval (SIIR) System for Biological and Petrological Data. *Smithson. Instn. Information Systems Innovations* **1**, 1–25.

Crovello, T. J. (1967). Problems in the use of electronic data processing in biological collections. *Taxon* **16**, 481–494.

Duke, J. A. (1969). Family polyclave. Mimeographed, 77 pp. Distributed by author.

Dybowski, W. and Franklin, D. A. 1968. Conditional probability and the identification of bacteria: a pilot study. *J. Gen. Microbiol.* **54**, 215–229.

Goodall, D. W. (1968). Identification by computer. *BioScience* **18**, 485–488.

Gyllenberg, H. G. 1965. A model for computer identification of micro-organisms. *J. Gen. Microbiol.* **39**, 401–405.

Heywood, V. H. (1968). Plant taxonomy today. *In* "Modern Methods in Plant Taxonomy" (V. H. Heywood, ed.) pp. 3–12. Academic Press, New York.

Johnson, M. P. and Holm, R. W. (1968). Numerical taxonomic studies in the genus *Sarcostemma* R. Br. (Asclepiadaceae). *In* "Modern Methods in Plant Taxonomy" (V. H. Heywood, ed.) pp. 199–217. Academic Press, New York.

Lanjouw, J. and Stafleu, F. A. (compilers). (1964). Index herbariorum, Part I: The herbaria of the world. Edit. 5. *Regnum Vegetabile* **31**, 1–251.

Leenhouts, P. W. (1966). Keys in biology. *Proc. Kon. Ned. Ak. Wet.* **69** (ser. C), 571–596.

Manning, R. B. (1969). Automation in museum collections. *Proc. Biol. Soc. Wash.* **82**, 671–686. (In symposium, "Natural History Collections: Past, Present, Future.")

Melchior, H. and Werdermann, E. (eds.) (1954). A. Engler's Syllabus der Pflanzenfamilien. I. Gebrüder Borntraeger, Berlin-Nikolassee.

Melchior, H. (ed.) (1964). A. Engler's Syllabus der Pflanzenfamilien. II. Gebrüder Borntraeger, Berlin-Nikolassee.

Morse, L. E. (1968). Construction of identification keys by computer. *Am. J. Bot.* **55**, 737. (Abstract).

Morse, L. E. (1969). Time-sharing computers as aids to identification of plant specimens. XI International Botanical Congress Abstracts, p. 152.

Morse, L. E., Beaman, J. H., and Shetler, S. G. (1968). A computer system for editing diagnostic keys for Flora North America. *Taxon* **17**, 479–483.

Osborne, D. V. (1963). Some aspects of the theory of dichotomous keys. *New Phytol.* **62**, 144–160.

Rypka, E. W., Clapper, W. E., Bowen, I. G., and Babb, R. (1967). A model for the identification of bacteria. *J. Gen. Microbiol.* **46**, 407–424.

Shetler, S. G. (1966). Meeting of Flora of North America Comittee. *Taxon* **15**, 255–257.

Shetler, S. G. (1968) [1969]. Flora North America Project. *Ann. Mo. bot. Gdn.* **55**, 176–178.

Shetler, S. G. (1969). The herbarium: Past, present, and future. *Proc. Biol. Soc. Wash.* **82**, 687–758. (In symposium, "Natural History Collections: Past, Present, Future.")

Shetler, S. G. (1970). Flora North America as an information system. *Bio-Science*, in press. (In symposium, "North Temperate Floristics and the Flora North America Project.")

Shetler, S. G. (ed.), assisted by Morse, L. E., Crockett, J. J., Rakosi, S. I., and Shetler, E. R. (1968). Preliminary generic taxon catalog of vascular plants for Flora North America. Processed, 69 pp. Distributed by Flora North America Program, Smithsonian Institution, Washington, D.C.

Shetler, S. G., Morisset, P., Shetler, E. R., Crockett, J. J., and Rakosi, S. I. (1970). Automated bibliography for Flora North America: Data collection specifications. Processed, 18 pp. Distributed by Flora North America Program, Smithsonian Institution, Washington, D.C.

Sokal, R. R. and Sneath, P. H. A. (1963). "Principles of Numerical Taxonomy". W. H. Freeman, San Francisco.

Sokal, R. R. and Sneath, P. H. A. (1966). Efficiency in taxonomy. *Taxon* **15**, 1–21.

Soper, J. H. and Perring, F. H. (1967). Data processing in the herbarium and museum. *Taxon* **16**, 13–19.

Squires, D. F. (1966). Data processing and museum collections: A problem for the present. *Curator* **9**, 216–227.

Squires, D. F. (1970). An information storage and retrieval system for biological and geological data: An interim report. *Curator*, in press.

Stearn, W. T. (1968). Observations on a computer-aided survey of the Jamaican species of *Columnea* and *Alloplectus*. *In* "Modern Methods in Plant Taxonomy." (V. H. Heywood, ed.) pp. 219–224. Academic Press, New York.

Taylor, R. L. 1969. Flora North America—Project '67. *In* "Current Topics in Plant Science." (James E. Guncke, ed.) pp. 161–166. Academic Press, New York.

Taylor, R. L. (1970). The Flora North America Project. *BioScience*, in press. (In symposium, "North Temperate Floristics and the Flora North America Project.")

Walker, D., Milne, P., Guppy, J., and Williams, J., (1968). The computer assisted storage and retrieval of pollen morphological data. *Pollen Spores* **10**, 251–262.

19. A Uniform Cataloguing System in the Department of Geology at Cambridge

C. L. Forbes, W. B. Harland and J. L. Cutbill

Department of Geology, Sedgwick Museum, University of Cambridge, England

The collections of the Sedgwick Museum at Cambridge University provide examples of the general diversity of cataloguing methods such as are often encountered among geologists. Systems range from the catalogue and taxonomic index of the palaeontological collection (about 380 000 entries) mostly prepared by A. G. Brighton, through the files of the Spitsbergen research group (about 40 000 specimens, 10 000 localities and 25 000 photographs) organized on a system devised by one of us, to other collections both department and of individual workers, catalogued (or not) according to the ideas and resources available at the time. One obvious drawback of this diversity is that transfer of material from individual collections to the main collection involves much laborious and unproductive reformatting of existing records. Thus a backlog of material accumulates pending recataloguing and the difficulty in finding what material is already present will be increased in future if the collections grow without corresponding improvements in organization.

Automation of clerical work will help but this requires some degree of standardization in temporary cataloguing methods in order to avoid recataloguing. Such procedures must have been applied many times elsewhere, but few published accounts exist so that it may be useful to describe our approach.

This approach starts with the introduction of printed catalogue forms, Figs. 1–5, designed so as to be useful both to the investigator in the course of work and to the curator in the compilation of catalogues and indexes. The forms provide boxes for a very wide range of entries. It is not expected that they will all need to be filled and some initial

Systematics Association Special Volume No. 3. "Data Processing in Biology and Geology," edited by J. L. Cutbill, 1970, pp. 311–320.

DEPARTMENT OF GEOLOGY CAMBRIDGE SPECIMEN CATALOGUE

CATEGORY *Heading* No SET *S M* NUMBER *F20785* TO *F20863*

DERIVED OBJECTS	TYPE	COUNT	SET	NUMBER	TYPE	COUNT	SET	NUMBER

IDENTIFICATION
PRIMARY PURPOSE *Fossil*
GROUP *Ammonoidea* SUBGROUP
NAME
IDENTIFIED BY *Callomon J.H* DATE *1968*

ROCK UNIT GRP FM MB BED

DETAIL

AGE *Jurassic Oxfordian, upper*

ZONE/COMPLEX *Transversarium zone Parandieri subzone*

LOCALITY NUMBER FOUND IN *outcrop*
PLACE *Mandach, Viligen, Brugg* DISTRICT *Aagau Kt.*
COUNTRY *Switzerland* REALM *Europe* HEIGHT

LAT/LONG ° ′ ″ . . . ° ′ ″ . . . LONG ORIGIN ° ′ ″ . . .

GRID REF SYSTEM SQ E N

DETAIL *Cutting on rd twixt Mandach and Viligen*
5 km N. of Brugg

SOURCE COLLECTED BY *Callomon J.H.* DATE *1952*
ACQUIRED FROM *Pres. Callomon J.H.* DATE

UNPUBLISHED DOCUMENTS

TYPE	REFERENCE	DATE	PLACE
Letter	*Callomon J.H.*	*30 Sep 1968*	*Sedgwick Museum file*

PUBLISHED REFERENCES

NOTES *Compare Gygi 1966 Eclog. geol. Helv. 59*
pp 936-8, Section 2

CATALOGUER *C.L. Forbes* DATE *1968* S.M.1 MAY 1969

Fig. 1.

DEPARTMENT OF GEOLOGY CAMBRIDGE SPECIMEN CATALOGUE

CATEGORY No SET NUMBER F20785 TO

DERIVED OBJECTS — TYPE COUNT SET NUMBER TYPE COUNT SET NUMBER

IDENTIFICATION — PRIMARY PURPOSE

GROUP SUBGROUP

NAME *Ochetoceras canaliculatum* von Buch
sexual dimorph of *Glochioceras subclausum* Oppel

IDENTIFIED BY DATE

ROCK UNIT — GRP FM MB BED

DETAIL

AGE

ZONE/COMPLEX

LOCALITY — NUMBER FOUND IN

PLACE DISTRICT

COUNTRY REALM HEIGHT

LAT/LONG — ° ′ ″ . . . | ° ′ ″ . . . | LONG ORIGIN ° ′ ″ . . .

GRID REF — SYSTEM SQ E N

DETAIL

SOURCE — COLLECTED BY DATE

ACQUIRED FROM DATE

UNPUBLISHED DOCUMENTS — TYPE REFERENCE DATE PLACE

PUBLISHED REFERENCES

NOTES

CATALOGUER C.L.Forbes DATE 1968 S.M.1 MAY 1969

Fig. 2.

DEPARTMENT OF GEOLOGY CAMBRIDGE

LOCALITY CATALOGUE SET CSE NUMBER G 882

LOCATION	PLACE Cambellryggen		DISTRICT Bunsow Land	
	COUNTRY Spitsbergen	REALM Arctic		HEIGHT
LAT/LONG	° ′ ″	° ′ ″	LONG ORIGIN	° ′ ″
GRID REF	SYSTEM UTM	SQ 33X/VH	E 140	N 229
DETAIL	Northwest spur above Brucebyen			
TYPES OF DATA COLLECTED	SECTION	BOREHOLE	RADIOMETRIC	PALAEOMAGNETIC
	PLANT	INVERTEBRATE	VERTEBRATE	PALYNOLOGICAL
	MICROFOSSIL	TECTONIC	LITHOLOGICAL	SED STRUCT
	TOPOGRAPHIC	PHOTOGRAPHIC		
ROCK UNITS	Nordenskioldbreen Formation, Minkinfjellet Member, Cadellfjellet Member, Tyrellfjellet Member, Brucebyen Beds.			
ROCK AGE OR COMPLEX	Moscovian, Gshelian, Asselian			
SPECIMEN NUMBERS	G1130 to G1170			
UNPUBLISHED DOCUMENTS	TYPE MS	REFERENCE	DATE 1959	PLACE CSE files
PUBLISHED REFERENCES				
VISITS	BY Gobbett, D.J.		DATE 1959	
NOTES				
CATALOGUER			DATE	S.M.2 MAY 1969

Fig. 3.

DEPARTMENT OF GEOLOGY CAMBRIDGE SPECIMEN CATALOGUE

CATEGORY No SET CSE NUMBER G1137 TO

DERIVED OBJECTS	TYPE	COUNT	SET	NUMBER	TYPE	COUNT	SET	NUMBER
	TS	5						

IDENTIFICATION
PRIMARY PURPOSE
GROUP Foraminifera SUBGROUP Family Fusulinacea
NAME Rugosofusulina arctica (Schellwien)
IDENTIFIED BY Cutbill, J.L. DATE 1962

ROCK UNIT GRP Gipsdalen FM Nordenskioldbr. MB Tyrrellfjellet BED Brucebyen

DETAIL From basal 100 mm.

AGE Permian, Asselian

ZONE/COMPLEX

LOCALITY NUMBER CSE/G 882 FOUND IN situ
PLACE DISTRICT
COUNTRY REALM HEIGHT

LAT/LONG ° ′ ″ . . . | ° ′ ″ . . . | LONG ORIGIN ° ′ ″ . . .

GRID REF SYSTEM SQ E N

DETAIL

SOURCE COLLECTED BY Gobbett, D.J. DATE 1959
ACQUIRED FROM DATE

UNPUBLISHED DOCUMENTS	TYPE	REFERENCE	DATE	PLACE
	Photo	JLC/B17 - A5	1961	J.L.C's files
	Photo	JLC/B17 - A6	1961	"
	Photo	JLC/B11 - C5	1961	"

PUBLISHED REFERENCES

NOTES

CATALOGUER DATE S.M.1 MAY 1969

Fig. 4.

DEPARTMENT OF GEOLOGY CAMBRIDGE PHOTOGRAPHIC CATALOGUE

FILM SET JIC NUMBER B17 FRAME A5 TO

CAMERA	MAKE LENS
FILM	MAKE SPEED
EXPOSURE	TIME STOP FILTER
DEVELOPMENT	TIME DEVELOPER
	PHOTOGRAPHER Cutbill, J.L. DATE 1961
CAMERA LOCATION	NUMBER TAKEN FROM
	PLACE DISTRICT
	COUNTRY REALM HEIGHT
LAT/LONG	° ′ ″ . . . | ° ′ ″ . . . | LONG ORIGIN ° ′ ″ . . .
GRID REF	SYSTEM SQ E N
DETAIL	
SUBJECT	PRIMARY PURPOSE
	Locality
GEOLOGY	ROCK UNIT
	AGE
	OBJECT
PERSON	
EVENT	
OBJECT	
TAXON	Rugosofusulina arctica (Schellwien)
RESTRICTIONS	
NOTES	Axial section
CATALOGUER	DATE

S.M.4 MAY 1969

Fig. 5.

reluctance to use the forms arises from a misunderstanding about this. Only known information can possibly be entered and we have made provision for the avoidance of needless repetition, as will appear below.

However these forms are but one means of entering data to a more complex filing system. Consider the case of specimens, photographs and field notes related together by locality number. The forms may be used in various ways to catalogue this information. It is a matter of convenience whether a locality catalogue is prepared first and details of specimens and photographs extracted later to form specimen and photographic catalogues or the locality catalogue is compiled from data in other files. Provided that the elements of the data form part of a logical system with standard numbering conventions such reorganization is always possible and will be easy if mechanized methods are used. Until mechanization is complete a research worker can sort and arrange the forms himself in the manner of a card index.

The method of transfer of data to a mechanized system is also flexible. Currently we use tape-typewriters but with the use of direct access computer terminals both these and the printed forms will become redundant unless their use happens to be convenient in particular cases. Similarly there is nothing fundamental about the layout of the forms. Provided the underlying logical system is kept in mind other forms could be introduced if this makes recording any easier.

DESIGN AIMS

The system must satisfy at least five requirements:

(1) It must be a single system useful in very diverse fields of research with as few special cases as possible.

(2) There must be no major alterations in existing procedures so that gradual introduction of new methods is possible without re-working existing files.

(3) It must not need much central control, so that curatorial staff may be free for more productive work.

(4) The data must be in a format suitable for automatic processing methods, but must also be capable of manipulation by hand. It must be possible for some parts of the data to be machine processed while other parts are handled manually.

(5) Data must be easy to transfer from one part of the system to another. In particular the catalogues of small research collections should be readily transferable to the permanent catalogues.

INFORMATION RECORDS

Each catalogue entry forms one record and there are several classes of these. Thus the Spitsbergen data files include classes for localities, photographs and objects. We are considering adding a fourth class for unpublished documents. When parts of the system are automated it may be useful to generate other classes of records such as biographies of collectors and publications.

There is considerable latitude in the scope of a record. Thus a measured section may form a single locality record and the positions of individual collecting horizons in the section can be recorded within it. Alternatively each collecting horizon may have a separate record. Likewise an object record may refer to a hand specimen, single fossil, slide, mineral preparation or such like. Specimens may be split so that parts become the subjects of separate records and it is important to have some method for recording such divisions so that parts may be reassembled, slides duplicated and general observations verified. This is accomplished through a system of record numbers which may be used for cross reference.

RECORD NUMBERING

Every record must have a unique identifying mark to avoid confusion. This identifier should also appear on the actual objects and photographs to link them to the record. As our present numbering systems overlap, we have provided each set of records with a prefix such as SM for the main palaeontological collection, CSE for the Spitsbergen records and personal initials for individual collections. This gives us a two part identifier. Provided the second part (the "specimen number" of common usage) gives a unique identifier within its set, it may take any form though we recommend a combination of one letter and a number. We also exclude symbols not on an ordinary typewriter. Records of different types may have the same two part identifier so that the complete unique identifier consists of record type/set code/number, thus:

Locality CSE/C2391
Photograph JLC/C2470·29

If a specimen is split there are two ways of numbering the parts:

(i) Separate series of identifiers may be maintained for particular types of parts, such as thin sections. In this case the data for each derived object forms a separate record. Cross references must be set up using the "Type" "Set" "Number" sections of the "Derived objects" box on the parent form, and a separate form for each derived object with the back reference in the notes section.

(ii) Each part may retain the identifier of the parent and may be distinguished by a suffix "Count" in the "Derived objects" box on the parent form to give the total number of each type of derived object. No cross references are needed.

DATA ASPECTS

Some standardization of the aspects of the data recorded is necessary, though obviously it is neither desirable nor possible to attempt this for all geological data. We have only tried to reach agreement on the commonest kinds recorded about specimens and localities and recognise that at this level the main use of the data recorded is as an index to collections and files rather than as a primary data bank. The main data aspects (Fig. 1) are

identification of material
lithostratigraphical position, geological age, geological association (e.g. fossil zones, structural complexes)
geographical locality
source of material
references (published and unpublished)

The extent to which we subdivide these aspects will be clear from Figs. 1–5.

This system of organization of data is open ended and can be extended to meet particular requirements. Its main drawback is that it makes no provision for recording descriptive data about specimens in a way that allows subsequent manipulation by machine. This is not serious so far as existing collections are concerned since this information is seldom available. However it is rapidly becoming a problem for new material and we are looking at ways of organizing descriptive information in our files so that it can be machine processed without placing unacceptable restrictions on users.

DATA RECORDING

Our standard data sheets (Figs 1–5) are on A4 paper and are large enough to be used as work sheets during research on a collection. When a collection is passed to the curator for archiving, the set of sheets (or copies of them) provides at least the minimum data in a fairly standard format.

Except for boxes on the forms labelled NOTES or DETAIL, all data must be in the form of keywords or key phrases (that is single words or phrases which one might look for in an index and which will therefore be convenient to use in machine searches). The layout is particularly suitable for machine data entry using tape typewriters or teletypes with a tagging system such as that described by Cutbill and Williams (this volume, p. 105). Thus it is relatively cheap to transfer records to machine files at any stage in their existence.

For each sheet to be complete in itself both the specimen and photographic data sheets have sections duplicating the locality data from the locality sheet. It is not necessary to have separate locality records unless a particular user wishes to do so.

To avoid the labour of writing identical entries in series of records, *Heading* may be entered in the "Category" space as in fig. I. All data on the form then apply to the series of records whose identifiers are written in the "Set" ... "Number" ... "To" ... spaces. Entries on the forms for the individuals are then reduced to items in which they differ from each other (Fig. 2).

If only identical data are available for a series of records, only the Heading form is prepared at first and the others are added whenever convenient as data accumulates in the course of research.

If the subject of a single record is divided into parts and data become abundant for these, the parent form may be made a Heading with the part suffixes added to its identifier. Individual forms for the parts have *Part* entered in the "Category" space followed by the suffix in the "No" space and the identifier in "Set" ... "Number" ... Additionally if there is not enough space for a whole record on the face of a form, continuation on the back is acceptable.

EXPERIENCE WITH THE SYSTEM

Full benefits from our system will not be realized for some time and will certainly require a good deal of the data to be on computer files. Only in this way can the benefits of cheap manipulating, sorting and cross referencing be achieved. An immediately obvious result noticed by those now using the recording forms is that owing to the systematic layout more data are recorded than previously and this in itself has already proved useful. It is fair to predict from past experience of the earlier systems that there will be an advantage, on balance, even if some procedures are not as successful as we hope.

20. Remedy for the General Data Handling Failure of Palaeontology

N. F. Hughes

Department of Geology, Sedgwick Museum, University of Cambridge, England

ABSTRACT

For many years now palaeontology has been regarded by geologists as difficult to interpret and relatively ineffective in its main task of refining stratigraphic correlation; this view can be accepted as fair judgement at the present time.

In spite of abundant data, palaeo-palynology has so far provided disappointing correlation results. This leads to the conclusion that normal palaeontologic methods, used hitherto in palynology, are perhaps at fault or outdated. The suggested solution for palynology involves a new way of handling the extensive data, so that it can be retrieved and used independently of Linnéan nomenclature and particularly without any obligation to regard typification, priority or synonymy. This method appears to provide the necessary framework for fine stratigraphic correlation and for further refinement as new study techniques become available for individual fossil groups.

As there is no logical separation of these small abundant fossils of palynology from all other fossils, it is concluded that the method proposed should come to apply with benefit to all stratigraphic palaeontology; some draft rules of procedure are presented to cover this.

Much general stratigraphic work, whether using fossils or not, shows the same kind of fundamental weakness. It is suggested that the proposed method, or one with similar aims, should therefore apply to all correlation means applied in stratigraphy.

So-called biologic objects of palaeontology, for which the Linnéan nomenclature could be retained, would also benefit from the proposed parallel use of the new method in stratigraphic work.

Systematics Association Special Volume No. 3. "Data Processing in Biology and Geology," edited by J. L. Cutbill, 1970, pp. 321–330.

INTRODUCTION

Palaeontology, as a subject for study, has the advantage of the great inherent attractiveness to the human mind of its materials. As a result, however, of teaching and using palaeontology for some years, I have come reluctantly to the conclusion that it has been allowed to sink into a precarious position as a science, at least in Britain.

Up to perhaps sixty years ago palaeontology appears to have held a respected position in both biology and geology. Now with the possible exception of vertebrate studies it appears to have failed biology to the extent of being almost completely ignored. In geology, communication between palaeontologic specialists and with geologists is so poor that very little serious attempt is made to integrate and use its information with general geology; it is often regarded as a separate "mystery" to be taught independently to those who are willing to risk a somewhat hazardous use of their time. This picture unfortunately seems not to be much improved by the efforts of many ingenious hard-working palaeontologists applying new techniques and making clever interpretations of fossils and environments. It is even made worse by the general complacency of, for example, many but not all of the contributors to a recent symposium on the subject (Moore, 1968).

Most distinguishable on the geologic side is the lack of adequate improvement in fine stratigraphic correlation, in spite of the availability of a constantly increasing body of palaeontologic data and techniques. This may involve discipline in that it seems difficult to exchange some of the joys of pure exploration for the close planning that is necessary for genuine exploitation. Some palaeontologists however even express the view that the limit of achievement in this sphere has virtually been reached already, and that there is an unavoidable "fuzziness" in work of this kind due to the nature of the material.

I believe that this relative failure in stratigraphic correlation is mainly due to an unthinking continued use of Linnéan taxonomy and nomenclature for data handling in this field.

APPROACH TO THE PROBLEM THROUGH PALAEO-PALYNOLOGY

The use of fossil miospores in stratigraphy is essentially similar to the use of any other fossils as elements of the evolving biosphere, except perhaps that when they occur fossil spores are easily available in very great numbers and variety. The British Wealden (Early Cretaceous) is up to 2000 feet thick and contains abundant spores, but in several years of study using the entries, peaks of abundance, and ranges etc.

of these fossils for which mostly taxa were already erected, it did not seem possible to achieve a stratigraphic correlation refinement of anything much less than an Age/Stage (say 5–6 m.y.) that would divide the strata into four or five parts.

Somewhat slowly I came to the conclusion that this disappointing result was in the main due to the nature and use of the taxa themselves.

To illustrate the difficulty the distinctive Cretaceous spore *Appendicisporites tricornitatus* Weyland and Greifeld 1953 was taken as an example (Hughes, 1970; paper read at Prague Geological Congress, August 1968) of a "balloon" taxon. From the original holotype, the meaning of the "species" was steadily widened by accretion over the next 15 years by many attributions of other material to it. The resulting time-range of the taxon from Late Jurassic to Santonian (late Cretaceous) was in practice impossible to verify or criticize; it was subject to indefinite expansion by authors of varying competence usually (by custom) without access to the original material. It is not surprising that this and other similar taxa (the selected spore was not an extreme example) fail to provide a satisfactory basis for refinement of stratigraphic correlation.

Although such a fossil taxon is called a "species", and consequently is thought of as comparable with a recent species, there are important differences. With the fossil there is no possibility of limitation by a fertility barrier, and although in Recent species it is by no means always used, this kind of confirmation is potentially available. In the fossil case there is no criterion for any boundary except imitation of the scope of a Recent species which is a purely subjective matter; in truth the time and other limits of fossil taxa must be selected by arbitrary process (for convenience) and this should be openly acknowledged, preferably by refraining from the use of the word "species" for fossils.

The suggested alternative taxonomy (Hughes and Moody-Stuart, 1967 and 1969) is based on the *biorecord* which is in general imitative in scope of a Recent species, but is separately and distinctly defined. The biorecord is based on an agreed (or at least stated) fixed number of specimens (100 for these Wealden miospores) from one sample (i.e. topotype material only), and the variation of the main characters is expressed in terms of mean, standard deviation etc.; this avoids typification. Relevant specimens from all other rock samples are referred to by formal graded comparison with the biorecord; this avoids synonymy and forces the authors to state the degree of comparison observed in each case; nothing can be added to the meaning of the biorecord itself which is thus stable. Details of records of such comparison material used in stratigraphy thus remain available for retrieval instead of being submerged and lost in the "specific name" record of a floral list. The nomenclature includes author initials and the date, and as the biorecord is unique there can be no problems of priority. A biorecord is designed to be neutral in classifications, and may if necessary be superseded by other closely similar or "overlapping" biorecords even based on the same material if the new ones are more convenient. This means that any number of biorecords may be created; they will be used if sound and ignored if not. Each generation of palynologists has equal chance of using new material and techniques to advantage without in any way being tied by older incomplete observations. Present and particularly potential data storage methods make the total number of taxa as irrelevant as the total volume of literature; retrieval for stratigraphic purposes will be by "genus" (but see also below) and by suitable (for the year of study concerned) stratigraphic time divisions.

For stratigraphic correlation purposes biorecords and graded comparisons may be combined as convenient into *events* based on samples (Hughes and Moody-Stuart, 1969); time-correlation is effected by placing the events in sequence with events in a regional or standard time scale.

APPLICATION OF NEW METHOD TO PALAEONTOLOGY AS A WHOLE

The method described above is principally based on the need to divorce all palaeontologic data storage from the personal opinions of a classificatory nature that have to be expressed by any author using nomenclature and taxonomy such as that of the Linnéan system. Such divorce makes possible retrieval of data without having to pass through the single channel of an uncertain and fluctuating concept of a Linnéan taxon and its name.

I believe that the logic of this requirement must be applicable to the whole of palaeontology, and that the method advocated above should apply equally to the whole subject as to palynology alone. Some micropalaeontologists have effected rather finer correlation already, and others who work on larger and far less abundant fossils may not consider any advantages to be worth the trouble of change, but the underlying need is the same throughout if any progress is to be made towards genuine synthesis in the history of the biosphere.

THE DATA HANDLING NEEDS OF PALAEONTOLOGY

Palaeontologic data is virtually unlimited in quantity and very diffuse, comprising the surviving fossil remains (incl. chemical) of all plants and animals of the whole crust over the last 3000 million years. Although it has a quite distinct data structure, palaeontology has never yet had a data handling system designed for it.

The requirement seems to be for a precise and simple system for fossils that takes account of the following factors:

1. There is no possibility of using fertility limits of taxa at "species" level.
2. The biosphere and its environment have evolved jointly and continuously, and any taxonomy on a geologic-time basis must consist of purely arbitrary divisions which in theory must at one point in each case separate parent and progeny into different taxa; there is no other criterion.
3. Such taxonomic divisions on a geologic-time basis are necessary for describing and effecting stratigraphic correlation. Attempts to seek "natural" breaks between taxa in time are misguided in principle in the same sense as is the use of "unit-stratotypes" instead of boundary stratotypes in erecting stratigraphic time-scales.
4. "Geographic" limits of fossil taxa at "species" level are theoretically determinable by tests such as normal distribution of character observations. In practice, however, in palaeontology taxa separated solely on "geography" at one horizon remain subordinate in importance; this is because in most cases time-correlation technique is not sufficiently advanced.
5. For specimens compared with established taxa, a statement of precision of comparison is required in each case.
6. The data handling scheme should be neutral to all classifications.
7. The typification of a genus by a type-species locates only a point within a cluster, the extension of which is certain to be unstable in meaning because any author can add to it. As suggested in Hughes (1969) morphologic knowledge in most groups of fossil organisms is now good enough to state boundaries of genera in terms of limits of characters against the adjacent genus; thus could well lead to reasonable stability of generic concepts.
8. As frustration mounts with the present system, it becomes more difficult to resist the arguments of those who advocate collecting all data afresh from nature to the standard required at the time; such arguments are regrettable but understandable in micropalaeontology.
9. Palaeontology is undermanned and is likely to remain so; this is correlated with greater interest in exploration than in methodology.
10. Knowledge of detailed taxonomy beyond the object of investigation

at any particular time is ultimately an improper and uneconomic employment for a human brain. It is necessary to store information for use by an expert in palaeontologic science who is not an expert in the group concerned. Problems encountered by specialists in many groups are often essentially similar.

11. Later workers with new information need to be preserved from any obligation to regulate their taxonomy on the basis of earlier less well-informed work.
12. The use of evolution for time-stratigraphy and the checking of the validity of the whole concept of doing so, are activities that can and should proceed separately and parallel without normally affecting each other.
13. Although some of the points made above really apply also to the handling of Recent material, the purpose of this discussion is purely concerned on this occasion with palaeontology.
14. Nomenclature should be subordinate to data handling and taxonomy; it should be convenient and neutral, and should allow for data retrieval through several channels; neither great brevity nor latinized names are essential.

PROPOSAL FOR A CODE FOR DATA HANDLING OF PALEONTOLOGIC MATERIAL

The existing International Codes of Botanical (Lanjouw, 1966) and Zoological (Stoll, 1961) Nomenclature, in spite of their titles, provide a far-reaching control of the scope and methods of taxonomy. The proposals made here are only for a Code of Data Handling which, if the factors listed in the last section (above) are accepted as relevant, embraces or replaces what is usually meant by palaeontologic taxonomy. Nomenclature in these circumstances becomes a subsidiary convenience rather than a control; the elaborate case-law support provided in the existing Codes is not required as it is concerned almost entirely with typification, priority and synonymy which are not applicable. Consequently a new code can be expressed very briefly in the following articles:

1. *Paleontologic events* used in time-correlation are based on recorded fossil specimens.
2. *Recorded fossil specimens* are taken from a rock sample with stated geographic limits.
   (A sample may be anything from a hand specimen to a thick rock formation at one locality or even with some stated geographic extent.)

3. A formal *group* of fossil organisms is a file with a name consisting of from four to six letters.
4. A *genusbox* is a formal subfile of a group file that is defined by description of its limits in morphography and in stratigraphic time-scale divisions against the descriptions of other adjacent genus-boxes. The name of a genusbox has seven to ten letters and in citation it is preceded by the relevant group name.
5. The fundamental reference taxon, the *biorecord*, is expressed as a diagnosis and illustrations based on a stated number from one sample, of recorded fossil specimens with normal distribution of variation of the characters studied; the statistical significance of the observations is stated. The treatment (preparation) of the specimens is described. The nomenclature includes formal reference to group file, stratigraphic time-scale division, date of record observation, author initials and serial number, and genusbox subfile.
   (The number of specimens for adequate or acceptable statistical significance is agreed by specialists in each formal group, and this number determines in each instance the size of the sample from which the biorecord topotypes are taken).
6. All other recorded specimens are placed in *comparison records* which are formally compared with one or more reference taxa (biorecords) on a graded scheme that expresses distance.
7. Other terms used are regarded as informal (non-search) until formally included in a paragraph of this code.

The following points are implicit in this code: (a) the principle of binominal nomenclature is retained; (b) the fundamental reference taxon (biorecord) relies only on the original material and description, it is not emended and it bears no priority; (c) a new data-level of description of any biorecord material would simply call for a new bio-record to be erected; (d) new biorecords may be published without reference to earlier ones of similar age and scope, although most authors would probably prefer to comment; (e) specimens used in a biorecord are available for other use such as in a comparison record if desired; (f) it will clearly be better for the science if authors use customary moderation in these matters, but the scheme is in no way damaged and the loss is minimal if they do not.

RECONCILLIATION WITH EXISTING PRACTICES

The proposed code is obviously designed for use with present and particularly with future palaeontologic data. Past data has been observed to be outweighed by the new data gathered in each decade, so that the handling problem ahead is always more serious. Past data can only be stored in a new system by re-working it but when it is required for stratigraphic purposes it is often necessary to do so in any case. Some past data is so incompletely recorded (e.g. as to locality, sample etc.) that its stratigraphic contribution can only be slight.

The changes asked by implication to the average systematic approach of a palaeontologist are minimal in that the erection of a biorecord (e.g. in Hughes and Moody-Stuart, 1969, p. 87) can be almost the same as for a "species". The genusbox is consciously made to be of the same order of scope as the genus which has been found to be convenient. The group is of the scope of a phylum (plant division) or a class, and any other hierarchic grade is informal. The frequent awkwardness of the linnean generic names is due to the perhaps accidental requirement that for search purposes (or is it purely for legal priority purposes) the name be unique in a kingdom (although illogically not in both together).

It is reasonable to ask why the suggested reform could not be accomplished within the existing Rules of Nomenclature by introducing new proposals formally. Such an approach has been attempted both formally (e.g. Hughes, 1963), and informally, but without success. The whole tendency in both codes has been to make the treatment of Recent and palaeontologic material the same. Even when palaeontologists agree to special provisions beneficial for handling fossils, they are outvoted by the natural preponderance of workers on Recent material who doubtless simply opt for apparent simplicity; this was exemplified a few years ago by the failure to introduce the proposed provisions for "parataxa" into the Zoological Code (Stoll, 1961). In the Botanical Code (Lanjouw, 1966), palaeobotanists have retained organ-genera etc. and have just managed to keep themselves (alone) free of the nineteenth century incubus of diagnoses in Latin, but they are not at present moving towards recognition of any further special requirements of palaeobotany. It is necessary to face the fact that palaeontology has very little voice in current decisions on taxonomy as affected by Rules of Nomenclature, and there is no near prospect of a change in this position.

There seems, however, to be every reason to hope that the proposed Data Handling Code for Palaeontology could co-exist with the Linnean codes in the sense of both being used simultaneously (for different purposes) by the same workers. The Linnean codes could continue to be used for expressions of biologic opinion and the taxa used could perfectly well encompass biorecords and comparison records, that had been erected for stratigraphic purposes, as required. Basic data for stratigraphic purposes and perhaps all basic data would be better stored in the first place through the proposed new code. A biorecord appears more likely to approach the position of giving information about past species than does a "fossil species" as at present conceived without any topotype restrictions.

PARALLEL IN STRATIGRAPHY

The recording of stratigraphic data is currently subject to some sporadic but not concerted editorial control but is still largely left to the choice of the individual author. A similar code of practice for handling all stratigraphic data (not merely palaeobiologic data) is required as part of any serious move towards erection and use of a Standard Stratigraphical Scale (George *et al.* 1969), either directly or through regional scales. Some of the difficulties are illustrated in the introductory chapter to the "Fossil Record" published by the Geological Society of London (Harland *et al.*, 1967, pp. 1–11).

In as much as the majority of stratigraphic data are palaeobiologic it should not prove too difficult to devise a code, although the numbers of geologists concerned in arriving at any such decisions are very much greater. A few of the problems have already been surmounted by the Stratigraphic Commission (now of IUGS) in Copenhagen (publ. 1961), but it seems likely that a move in palaeontology could clear the way ahead.

Fundamentally a full acceptance is required of the proposition that evolution of the earth's *crust*, evolution of the biosphere, and most of stratigraphy, are all about the same subject and material to such an extent that eventual restriction to one data handling system is essential.

CONCLUSIONS OF NECESSITY

1. Palaeontology is not currently making any strong contribution to improvement of stratigraphic correlation methods, although it should be in a position to do so from its own advances of techniques and accumulation of data.
2. Palaeontology needs a data-handling system devised solely for its own future requirements. A brief but complete "Code for handling of paleontologic data" is offered here for consideration.
3. Data storage and retrieval prospects are now such that palaeontology could rapidly become much more effective in geology and biology by taking full advantage of this situation.
4. The detailed proposals referred to above were originally made for palaeo-palynology and they should be examined and tested in that context. If they are found to be valid, their extension to all other parts of palaeontology is logical.

5. It is difficulty to predict the level of understanding of the evolutionary history of the biosphere that may be reached in the next two decades but it is clearly necessary to allow in any system for development of much greater refinement. There appears to be no *theoretical* limit to finess of time-correlation short of the size of individual sedimentary rock particles; even the past migrations of organisms are themselves part of the history and knowledge of them could in theory be turned to advantage.
6. The relatively widespread defeatist opinion that there is an unavoidable imprecision or "fuzziness" about palaeontologic work seems to be mainly due to the continued use of Linnean taxonomy for a purpose for which it is no longer suitable.
7. There is no conflict or separation of aim in the allegiance of palaeontology both to biology and to geology; the support to stratigraphy is the primary function of palaeontology, but this support is in itself essential to all progress in palaeobiology.
8. Probably none of the points made above is original; they are merely thought to be appropriate in their association and timing for profitable action in 1970.

REFERENCES

George, T. N. *et al.* (1969). Recommendations on stratigraphical usage. *Proc. geol. Soc. Lond.*, **1656**, 139–166.

Harland, W. B. *et al.* (eds.) (1967). The Fossil Record. London (Geological Society).

Hughes, N. F. (1963). The assignment of species of fossils to genera. *Taxon* **12**, 336–337.

Hughes, N. F. (1969). Suggestions for better handling of the genus in palaeopalynology. *Grana Palynologica*, **9**.

Hughes, N. F. (1970). The need for agreed standards of recording in palaeopalynology and palaeobotany. *Paläont Abh.*,**B3**, (3/4), 357–364.

Hughes, N. F. and Moody-Stuart, J. C. (1967). Proposed method of recording pre-Quaternary palynological data. *Rev. Palaeobotan. Palynol.*, **3**, 347–358.

Hughes, N. F. and Moody-Stuart, J. C. (1969). A method of stratigraphic correlation using early Cretaceous miospores. *Palaeontology*, **12**, 84–111.

Hughes, N. F., Williams, D. B., Cutbill, J. C. and Harland, W. B. (1967). A use of reference points in stratigraphy. *Geol. Mag.* **104**, 634–635.

Hughes, N. F., Williams, D. B., Cutbill, J. C. and Harland, W. B. (1968). Hierarchy in stratigraphical nomenclature. *Geol. Mag.* **105**, 78–79.

Lanjouw, J. (1966). International Code of Botanical Nomenclature. Utrecht.

Moore, R. C. (1968). Developments, Trends and Outlooks in Paleontology. *J. Paleont.* **42**, 1327–1377.

Stoll, N. R. (ed.) (1961). International Code of Zoological Nomenclature. London.

Author Index

Numbers in italic indicate the page on which the reference is listed.

Subject Index

D

E

F

H

T